Effect of Processing on the Structure, Techno-Functional Properties and Nutritional Quality of Animal- and Plant-Based Food Proteins

Effect of Processing on the Structure, Techno-Functional Properties and Nutritional Quality of Animal- and Plant-Based Food Proteins

Guest Editors

Norbert Raak
Laura Roman
Ruifen Li

MDPI

Basel • Beijing • Wuhan • Barcelona • Belgrade • Novi Sad • Cluj • Manchester

Guest Editors
Norbert Raak
Department of Food Science
University of Copenhagen
Frederiksberg C
Denmark

Laura Roman
Department of Agricultural Engineering
University of Valladolid
Palencia
Spain

Ruifen Li
Department of Food Science
Aarhus University
Aarhus N
Denmark

Editorial Office
MDPI AG
Grosspeteranlage 5
4052 Basel, Switzerland

This is a reprint of the Special Issue, published open access by the journal *Foods* (ISSN 2304-8158), freely accessible at: https://www.mdpi.com/journal/foods/special_issues/ADWV4JKF79.

For citation purposes, cite each article independently as indicated on the article page online and as indicated below:

Lastname, A.A.; Lastname, B.B. Article Title. *Journal Name* **Year**, *Volume Number*, Page Range.

ISBN 978-3-7258-3585-0 (Hbk)
ISBN 978-3-7258-3586-7 (PDF)
https://doi.org/10.3390/books978-3-7258-3586-7

Contents

About the Editors

Norbert Raak

Norbert Raak holds a doctoral degree from Technische Universität Dresden, Germany, and has worked as a postdoctoral researcher in food structure design at Aarhus University, Denmark. He is currently holds a position asan assistant professor of dairy chemistry at the University of Copenhagen, Denmark, where he is leading leads a small team investigating processing-induced modifications of proteins and the structure—function -interrelations of food biomacromolecules, in with a particular focus on dairy and plant proteins. Additionally, Norbert Raak is currentlyserves as the secretary of the board of the Nordic Rheology Society and is the Danish representative of for a LINXS theme on Advanced Rheometry for Neutron and X-Ray Sciences (RheoMAXESS).

Laura Roman

Laura Roman holds a doctoral degree in Food Engineering from the University of Valladolid, Spain, where she currently works as a Ramón y Cajal independent researcher (tenure-track). At the University of Valladolid, she leads a research team focused on the structure–function relationships of food polysaccharides, including their digestion and structuring behavior in mixtures with plant proteins, as well as the elucidation of their complex hierarchical structure. Her work aims to understand the structural formation of plant biopolymers, connecting the fields of food biochemistry—especially glycomics—food processing, and materials science, with the goal of designing functional and sustainable ingredients and foods. Laura has previously worked as a postdoctoral researcher at the University of Guelph, Canada, and at Aarhus University, Denmark. In 2021, she was awarded the prestigious L'Oréal-UNESCO For Women in Science research award as one of the top three young women scientists in Denmark.

Ruifen Li

Ruifen Li holds a doctoral degree from the University of Copenhagen, Denmark, and currently works as a postdoctoral researcher at Aarhus University, Denmark, where her main research focus is on food soft matter, including the use of small-angle scattering techniques, tomography, microstructure analysis, and rheology techniques to study food foams and emulsions derived from plant proteins. Her postdoctoral project is funded by the Novo Nordisk Fonden Postdoctoral Fellow program.

foods

MDPI

Editorial

Recent Advances in Processing-Induced Changes in the Structure, Techno-Functional Properties and Nutritional Quality of Animal- and Plant-Based Food Proteins

Ruifen Li [1], Norbert Raak [2,*] and Laura Roman [3]

[1] Department of Food Science and CiFOOD Centre for Innovative Food Research, Aarhus University, Agro Food Park 48, 8200 Aarhus N, Denmark; ruifen@food.au.dk
[2] Department of Food Science, Ingredient and Dairy Technology, University of Copenhagen, Rolighedsvej 26, 1958 Frederiksberg C, Denmark
[3] Food Technology Division, Department of Agricultural Engineering, University of Valladolid, Av. Madrid 50, 34004 Palencia, Spain; lroman@uva.es
* Correspondence: nr@food.ku.dk

1. Introduction

Proteins are essential macronutrients in the human diet and key structural constituents of many food products. In response to global challenges, such as climate change, the lack of sustainability in food systems, and increasing food insecurity, a significant amount of research has focused on food proteins in recent years. Indeed, there is growing interest in finding and utilizing alternative proteins from more sustainable sources, including plant-based proteins from legumes [1,2], oilseeds [3,4], cereals [5], and nuts [6], as well as in further improving the quality of common animal-based proteins (particularly milk proteins). Processing—including protein extraction and isolation techniques [7] and the further conversion of protein ingredients into foods through thermal treatment, high-shear mixing, and homogenization, as well as enzymatic modifications and fermentation—has been shown to profoundly impact protein structure and techno-functional properties, such as solubility, emulsification, foaming, and gelation [8,9], as well as improve their digestibility [10]. These properties are essential for developing food products with desirable textural attributes and flavors and enhancing the nutritional profiles of protein-rich foods. Moreover, the relationship between protein structure and its processing-induced modification is increasingly being elucidated through advanced analytical techniques, such as spectroscopy [11], microscopy [12,13], and computational modeling [14].

The interplay between food processing in both animal- and plant-based foods and its impact on the structural, techno-functional, and nutritional properties of the proteins has attracted considerable attention because of its pivotal role in enhancing global food security and driving food innovation and health. The studies featured in this Special Issue cover a wide range of research advancements, ranging from exploring the heteroprotein complex coacervation mechanism (*contribution 7*); protein extraction and solubilization techniques using high-shear mixing (*contribution 3*), membrane filtration (*contribution 6*), ultrasound (*contribution 9*), and an industrially scalable wet-extraction process (*contribution 10*); the development of more sustainable and/or healthier plant-based foods, like fibrous extruded meat analogs (*contribution 1 & 5*) and plant beverages (*contribution 2*), and low-fat products using protein-based fat replacers (*contribution 11*); and the final step of improving protein digestibility (*contribution 8*) and finding nutritious plant-based foods to promote gut health (*contribution 4*).

Received: 17 February 2025
Accepted: 20 February 2025
Published: 24 February 2025

Citation: Li, R.; Raak, N.; Roman, L. Recent Advances in Processing-Induced Changes in the Structure, Techno-Functional Properties and Nutritional Quality of Animal- and Plant-Based Food Proteins. *Foods* **2025**, *14*, 764. https://doi.org/10.3390/foods14050764

This Special Issue offers critical insights on technological and nutritional applications of animal- and plant-based proteins, linked to the colloidal and structural characteristics of diverse protein ingredients, paramount for developing innovative and nutritious food products that meet both consumer and societal demands, offering more sustainable processing and functional protein ingredients while paving the way for this future research field.

2. Contributions to This Special Issue

This Special Issue features 11 publications including 10 research articles and 1 review, which can be categorized into four main research topics.

2.1. Protein Extraction and Its Effect on the Structure and Technological Properties of the Protein Ingredients

Special attention has been paid to alternative methods for extracting food proteins to enhance their functionality while preserving their native architectures. In this sense, Malterre et al. (*contribution 3*) explored the application of high-shear mixing to lentil protein isolates, reporting significant enhancements in the solubility, surface hydrophobicity, and colloidal stability of lentil protein as well as a reduced particle size compared to isolates not subjected to shear mixing. These authors showcased the potential of more effective and oriented processing techniques to enhance the techno-functional properties of underutilized plant proteins, which could further facilitate their suitable integration into food formulations (i.e., improved dispersion and solubilization within the food matrix). Similarly, Sert et al. (*contribution 9*) applied an innovative technique, namely, ultrasound-assisted alkaline extraction, to pumpkin seed proteins derived from oil press cakes. They found that this technique resulted in higher protein yields and improved the techno-functional properties of the protein isolates, like solubility, foaming capacity, and stability, and that the native state of the isolated protein was better preserved compared to isolates subjected to the normal alkaline extraction procedure. Meanwhile, Hansen et al. (*contribution 10*) developed a scalable wet-extraction process for producing pea protein isolates with better-preserved structural integrity and lower polymerization, bridging laboratory scale extraction with pilot plant processing of the pea protein ingredients for future industrial applications. These authors found that both mild alkaline solubilization combined with isoelectric precipitation and salt solubilization coupled with membrane filtration resulted in superior solubility, gelation, and emulsification properties of the pea protein compared to the commercial pea protein isolate. The authors demonstrated that double solubilization procedures at mild pH conditions can be used to replace single solubilization of pea proteins at high alkalinity.

Apart from plant protein ingredients, emphasis has also been placed on milk protein ingredients. Raak et al. (*contribution 6*) studied the effects of transmembrane pressure on the colloidal structure of casein micelles during microfiltration and diafiltration. Their findings contribute to gaining a better understanding of how processing conditions can modulate the functionality of milk protein concentrates through modifications of the supramolecular structure of proteins.

2.2. Innovations in More Sustainable and Healthier Foods

High-moisture extrusion is one of the most versatile food processing techniques for producing plant-based alternatives whose textural attributes resemble those of animal meat. In this regard, Ribeiro et al. (*contribution 1*) demonstrated how screw speed and barrel temperature during high-moisture extrusion influence the texture, protein structure, and nutritional quality of soy-based extrudates. The extrusion conditions influenced fibrousness, anisotropy, and the reduction of antinutritional factors, with higher processing temperatures resulting in improved tenderness and anisotropic structuring, a result also related to the modification of the protein secondary structure (i.e., increased inter- and intra-molecular aggregation). This work provides valuable insights into tailoring extrusion parameters to achieve meat analogs with optimal textural attributes. Furthermore, Snel et al. (*contribution 5*) examined the feasibility of re-extruding soy and pea protein isolates to reduce waste during production. Their work demonstrates that re-extrusion can maintain fibrous texture formation, indicating that previous processing history does not significantly influence texture and offering a sustainable approach to protein re-processing and reducing food waste. Moreover, Roland et al. (*contribution 2*) investigated the proteolysis and generation of free amino acids in plant-based drinks from several plant sources during storage. Their findings highlight the importance of understanding storage-induced changes to ensure product quality and stability, providing valuable insights for improving shelf-life.

To improve human dietary health and meet consumer demands for more nutritious options, the development of high-protein, low-fat food products with limited deterioration of textural attributes is in demand. Nourmohammadi et al. (*contribution 11*) reviewed the fabrication methods and fat-mimicking mechanisms of protein-based fat replacers. The latest findings reveal several promising fat replacers, which can be classified into four different types. The first one is protein isolates/concentrates, like the commonly used whey protein that benefits from its high compatibility with other dairy ingredients and a matching flavor profile. The second type is microparticles, such as soy protein or bovine serum albumin, which can provide a smooth and creamy mouthfeel. The third type is protein–polysaccharide hydrogels, especially those produced using widely used ionic polysaccharides like gum arabic and pectin. The last type is microgels, which are categorized into fragments of protein hydrogels, protein assembles, emulsified gel droplets, and protein–polysaccharide coacervates. These fat replacers can be fabricated based on their different types, using processing techniques such as thermal–mechanical treatment, anti-solvent precipitation, enzymatic hydrolysis, complexation, and emulsification. This comprehensive review provides a roadmap for developing sustainable fat replacers with minimal caloric impact without compromising texture.

2.3. Protein Digestibility and Nutritional Quality of Plant-Based Ingredients and Foods

Plant-based protein ingredients from walnuts have high nutritional value but are easily oxidized during processing or storage due to their high content of unsaturated fatty acids. Therefore, Zhao et al. (*contribution 8*) explored the impact of oxidative modification on the structure and digestibility of walnut protein isolates. They found that the dual role of oxidation could promote intestinal digestion while it caused gastric digestion resistance. With a focus on nutrition, Calva-Cruz et al. (*contribution 4*) conducted a pilot trial to evaluate the impact of popped amaranth, a protein-rich pseudocereal, on the gut microbiota of stunted children. Their findings suggested that popped amaranth can serve as a nutritious food that supports gut health and combats malnutrition.

2.4. Heteroprotein Complex Coacervation

Soussi Hachfi et al. (*contribution 7*) investigated the influence of ionic strength on the coacervation process between lactoferrin and β-lactoglobulin using direct mixing and desalting protocols. Their study provides new insights into the electrostatically driven mechanisms underlying heteroprotein complex formation. The potential application in food products could be the lactoferrin and β-lactoglobulin complex coacervates work as microcapsules for bioactive and texturing agents.

3. Conclusions

The 11 contributions collectively highlight the significant advancements in the research topics defined in this Special Issue, focusing on understanding the impact of food processing on the structure, functionality, and nutritional quality of animal- and plant-based proteins. The articles in the four groups outlined in this Editorial offer a roadmap for future innovation in the field of food proteins. The advancements in protein extraction and processing techniques open new paths for enhancing the functionality and utilization of plant-based protein ingredients and enable more viable large-scale applications for protein extraction while highly preserving protein nativeness. Furthermore, the optimization of the high-moisture extrusion processing and re-extrusion processes of soy and pea proteins demonstrate the potential to produce high-quality fibrous plant-based meat analogs while reducing waste production, aligning with sustainability goals. Protein-based fat replacers also provide a promising solution for developing high-protein, low-fat products that meet consumer demands without sacrificing texture. At the same time, studies on protein digestion, oxidative stability, and gut health further emphasize the nutritional benefits of plant-based foods. Research on heteroprotein complex coacervation could also offer new strategies for designing innovative food ingredients. Finally, we expect that this Special Issue will inspire further research and innovation, helping the food industry to translate these developments into real-world and scalable technological and nutritional solutions.

Author Contributions: Writing—original draft, R.L.; writing—review and editing, N.R. and L.R. All authors have read and agreed to the published version of the manuscript.

Funding: RL received funding from the Novo Nordisk Foundation (grant number NNF22OC0079708). NR received funding from the Novo Nordisk Foundation (grant numbers NNF21OC0071375 and NNF24OC0087818). LR received funding from the Novo Nordisk Foundation (grant number NNF20OC0064423) and the Ramon y Cajal program (RYC 2021-032926-I) financed by the MCIN/AEI/10.13039/501100011033 and the European Union "NextGenerationEU/PRTR".

Data Availability Statement: No new data were created or analyzed in this study. Data sharing is not applicable to this article.

Acknowledgments: We truly appreciate the opportunity to lead this Special Issue as guest editors and would like to thank all authors for their contributions.

Conflicts of Interest: The authors declare no conflict of interest. The funders had no role in the design of the study; in the collection, analyses, or interpretation of data; in the writing of the manuscript, or in the decision to publish the results.

List of Contributions:

1. Ribeiro, G.; Piñero, M.-Y.; Parle, F.; Blanco, B.; Roman, L. Optimizing Screw Speed and Barrel Temperature for Textural and Nutritional Improvement of Soy-Based High-Moisture Extrudates. *Foods* **2024**, *13*, 1748. https://doi.org/10.3390/foods13111748.
2. Roland, I.S.; Le, T.T.; Chen, T.; Aguilera-Toro, M.; Nielsen, S.D.-H.; Larsen, L.B.; Poulsen, N.A. Storage Stability of Plant-Based Drinks Related to Proteolysis and Generation of Free Amino Acids. *Foods* **2024**, *13*, 367. https://doi.org/10.3390/foods13030367.
3. Malterre, N.; Bot, F.; Lerda, E.; Arendt, E.K.; Zannini, E.; O'Mahony, J.A. Enhancing the Techno-Functional Properties of Lentil Protein Isolate Dispersions Using In-Line High-Shear Rotor-Stator Mixing. *Foods* **2024**, *13*, 283. https://doi.org/10.3390/foods13020283.
4. Calva-Cruz, O.d.J.; Ovando-Vázquez, C.; De León-Rodríguez, A.; Veana, F.; Espitia-Rangel, E.; Treviño, S.; Barba-de la Rosa, A.P. Dietary Supplementation with Popped Amaranth Modulates the Gut Microbiota in Low Height-for-Age Children: A Nonrandomized Pilot Trial. *Foods* **2023**, *12*, 2760. https://doi.org/10.3390/foods12142760.
5. Snel, S.J.; Amroussi, Y.; van der Goot, A.J.; Beyrer, M. Rework potential of soy and pea protein isolates in high-moisture extrusion. *Foods* **2023**, *12*, 2543. https://doi.org/10.3390/foods12132543.

6. Raak, N.; Coşkun, Ö.; Corredig, M. Transmembrane Pressure during Micro-and Diafiltration of Milk Affects the Release of Non-Sedimentable Caseins. *Foods* **2023**, *12*, 2234. https://doi.org/10.3390/foods12112234.
7. Soussi Hachfi, R.; Hamon, P.; Rousseau, F.; Famelart, M.-H.; Bouhallab, S. Ionic Strength Dependence of the Complex Coacervation between Lactoferrin and β-Lactoglobulin. *Foods* **2023**, *12*, 1040. https://doi.org/10.3390/foods12051040.
8. Zhao, J.; Han, M.; Wu, Q.; Mao, X.; Zhang, J.; Lu, Z. Effect of Oxidative Modification by Peroxyl Radical on the Characterization and Identification of Oxidative Aggregates and in vitro Digestion Products of Walnut (*Juglans regia* L.) Protein Isolates. *Foods* **2022**, *11*, 4104. https://doi.org/10.3390/foods11244104.
9. Sert, D.; Rohm, H.; Struck, S. Ultrasound-Assisted Extraction of Protein from Pumpkin Seed Press Cake: Impact on Protein Yield and Techno-Functionality. *Foods* **2022**, *11*, 4029. https://doi.org/10.3390/foods11244029.
10. Hansen, L.; Bu, F.; Ismail, B.P. Structure-Function Guided Extraction and Scale-Up of Pea Protein Isolate Production. *Foods* **2022**, *11*, 3773. https://doi.org/10.3390/foods11233773.
11. Nourmohammadi, N.; Austin, L.; Chen, D. Protein-Based Fat Replacers: A Focus on Fabrication Methods and Fat-Mimic Mechanisms. *Foods* **2023**, *12*, 957. https://doi.org/10.3390/foods12050957.

References

1. Semba, R.D.; Ramsing, R.; Rahman, N.; Kraemer, K.; Bloem, M.W. Legumes as a sustainable source of protein in human diets. *Glob. Food Secur.* **2021**, *28*, 100520. [CrossRef]
2. Zhang, X.; Zhang, Z.; Shen, A.; Zhang, T.; Jiang, L.; El-Seedi, H.; Zhang, G.; Sui, X. Legumes as an alternative protein source in plant-based foods: Applications, challenges, and strategies. *Curr. Res. Food Sci.* **2024**, *9*, 100876. [CrossRef] [PubMed]
3. Kotecka-Majchrzak, K.; Sumara, A.; Fornal, E.; Montowska, M. Oilseed proteins–Properties and application as a food ingredient. *Trends Food Sci. Technol.* **2020**, *106*, 160–170. [CrossRef]
4. Singh, R.; Langyan, S.; Sangwan, S.; Rohtagi, B.; Khandelwal, A.; Shrivastava, M. Protein for human consumption from oilseed cakes: A review. *Front. Sustain. Food Syst.* **2022**, *6*, 856401. [CrossRef]
5. Senarathna, S.; Mel, R.; Malalgoda, M. Utilization of cereal-based protein ingredients in food applications. *J. Cereal Sci.* **2024**, *116*, 103867. [CrossRef]
6. da Silva Matos, J.; Costa, J.E.G.; Krichanã, D.R.G.C.; Azevedo, P.Z.; Nascimento, A.L.A.A.; Stringheta, P.C.; Martins, E.; Campelo, P.H. Nut proteins as plant-based ingredients: Emerging ingredients for the food industry. *Processes* **2024**, *12*, 1742. [CrossRef]
7. Kumar, M.; Tomar, M.; Potkule, J.; Verma, R.; Punia, S.; Mahapatra, A.; Belwal, T.; Dahuja, A.; Joshi, S.; Berwal, M.K. Advances in the plant protein extraction: Mechanism and recommendations. *Food Hydrocoll.* **2021**, *115*, 106595. [CrossRef]
8. Akharume, F.U.; Aluko, R.E.; Adedeji, A.A. Modification of plant proteins for improved functionality: A review. *Compr. Rev. Food Sci. Food Saf.* **2021**, *20*, 198–224. [CrossRef] [PubMed]
9. Nasrabadi, M.N.; Doost, A.S.; Mezzenga, R. Modification approaches of plant-based proteins to improve their techno-functionality and use in food products. *Food Hydrocoll.* **2021**, *118*, 106789. [CrossRef]
10. Sá, A.G.A.; Moreno, Y.M.F.; Carciofi, B.A.M. Food processing for the improvement of plant proteins digestibility. *Crit. Rev. Food Sci. Nutr.* **2020**, *60*, 3367–3386. [CrossRef] [PubMed]
11. Wang, K.; Sun, D.-W.; Pu, H.; Wei, Q. Principles and applications of spectroscopic techniques for evaluating food protein conformational changes: A review. *Trends Food Sci. Technol.* **2017**, *67*, 207–219. [CrossRef]
12. Jonic, S.; Vénien-Bryan, C. Protein structure determination by electron cryo-microscopy. *Curr Opin. Pharmaco.* **2009**, *9*, 636–642. [CrossRef]
13. Scott, G.; Awika, J.M. Effect of protein–starch interactions on starch retrogradation and implications for food product quality. *Compr. Rev. Food Sci. Food Safety* **2023**, *22*, 2081–2111. [CrossRef] [PubMed]
14. Barroso da Silva, F.L.; Carloni, P.; Cheung, D.; Cottone, G.; Donnini, S.; Foegeding, E.A.; Gulzar, M.; Jacquier, J.C.; Lobaskin, V.; MacKernan, D. Understanding and controlling food protein structure and function in foods: Perspectives from experiments and computer simulations. *Annu. Rev. Food Sci. Technol.* **2020**, *11*, 365–387. [CrossRef] [PubMed]

foods

MDPI

Article

Optimizing Screw Speed and Barrel Temperature for Textural and Nutritional Improvement of Soy-Based High-Moisture Extrudates

Gabriela Ribeiro [1,2], María-Ysabel Piñero [1], Florencia Parle [2], Belén Blanco [1] and Laura Roman [2,*]

1 CARTIF Technology Centre, Boecillo, 47151 Valladolid, Spain; gabrieladaconceica.ribeiro@estudiantes.uva.es (G.R.); marpin@cartif.es (M.-Y.P.); belbla@cartif.es (B.B.)
2 Food Technology Area, Department of Agricultural and Forestry Engineering, University of Valladolid, Av. Madrid, 50, 34004 Palencia, Spain; florenciaagustina.parle22@estudiantes.uva.es
* Correspondence: laura.roman.rivas@uva.es

Abstract: Soy remains the legume protein of excellence for plant-based meat alternatives due to its fiber-forming potential. In this study, protein-rich powders from soy protein isolate (SPI), concentrate (SPC), and their mixture (SPM) were thoroughly characterized for their proximate composition, nutritional quality, and physicochemical properties to understand their structuring behavior during high-moisture extrusion. SPI presented higher degrees of protein denaturation and aggregation, least gelation concentration and lower essential amino acid contents. Thus, an SPI:SPC combination (1:9 ratio, 70% protein) was extruded at three different screw speeds (300, 350, and 400 rpm) and two temperature profiles (120 and 140 °C maximum temperature). The effects of the processing parameters on the extrudates were evaluated for their appearance (fibrousness), texture (TPA, cutting force, and anisotropy), color, protein structure (FTIR), and trypsin inhibitors. Higher temperatures resulted in softer and darker extrudates, with increased visual and instrumental anisotropy. Increasing screw speeds led to softer and lighter extrudates, without a clear fibrousness effect. β-sheet structures decreased and intermolecular aggregates (A1) increased after extrusion, especially at 140 °C, together with the formation of intramolecular aggregates (A2). Extrusion also significantly decreased the amount of trypsin inhibitors (>90%). This study demonstrates that extrusion parameters need to be carefully selected to achieve meat analogs with optimal textural and nutritional characteristics.

Keywords: high-moisture extrusion; fibrous structuring; meat analogue; protein texturization; plant-based

Citation: Ribeiro, G.; Piñero, M.-Y.; Parle, F.; Blanco, B.; Roman, L. Optimizing Screw Speed and Barrel Temperature for Textural and Nutritional Improvement of Soy-Based High-Moisture Extrudates. *Foods* **2024**, *13*, 1748. https://doi.org/10.3390/foods13111748

Academic Editor: Carlos W. P. Carvalho

Received: 8 May 2024
Revised: 24 May 2024
Accepted: 28 May 2024
Published: 2 June 2024

1. Introduction

In recent years, there has been a notable increase in plant-based meat innovation, driven by the convergence of consumer demands, market growth, and sustainable future food supply [1]. The use of plant proteins in food formulations is emerging as a popular option to sustain the food system against environmental limitations [2]. However, consumers do not want to compromise on the typical texture and flavor of meat. In an attempt to meet consumer expectations, meat analogs are produced using plant proteins through sustainable technologies such as extrusion (low- and high-moisture processes), aiming to mimic the characteristic fibrous anisotropic structure of meat [3]. High-moisture extrusion is currently the most predominant industrial technology for producing whole-cut meat analogs from plant proteins, involving extrusion processing at moisture levels of 40–80% coupled with a prolonged cooling die for fibrous texture development [3–5]. Despite the high initial investment cost of the equipment needed for high-moisture extrusion processing, this technology offers significant advantages, including scalability and versatility, reduced energy consumption and cost-effectiveness, minimized waste production, and the superior quality of the texturized products [5,6]. The resulting extruded products aim to achieve a meat-like texture by using plant-based material with a fibrous and dense structure, strong elasticity, and a high moisture content. In fact, these extruded products have been shown to possess sensory descriptors similar to those found in different animal whole-cuts, including beef, pork and chicken, although some other descriptors still represent a challenge (i.e., drier and less fibrous sensory perceptions) [7].

Extrusion is a high-temperature, short-time processing technique, inducing various changes to the shape/texture, structure, and composition of the raw materials. During extrusion, the raw materials are exposed to the effects of temperature, shear force, and pressure, leading to changes such as denaturation, oxidation, degradation, reassociation, and aggregation of proteins, carbohydrates, and/or lipids, and to the interaction between molecules that alter their molecular conformation, forming new structural assemblies and textures [8]. During extrusion, proteins undergo denaturation, unfolding, and molecular realignment to form a fibrous anisotropic structure upon cooling due to the directional shear force in the cooling die [4]. However, the molecular and colloidal mechanism for these fibers' formation is still unclear due to the great differences in the purity and characteristics of the raw materials and the complexity of the extrusion process itself [9]. Therefore, evaluating both raw material characteristics and extrusion conditions is crucial to understand product quality [10]. Extrusion parameters such as moisture, barrel temperature, and screw speed, as well as indirect response parameters like the specific mechanical energy (SME), torque, and melt temperature and pressure, are key points for controlling product quality [11]. In addition, the properties of legume proteins used as raw materials, influenced by factors such as protein composition, purity, conformation, and extraction methods, are essential parameters to consider [9,12]. These determine physicochemical properties like solubility, water affinity, viscosity and gel formation, greatly influencing the extrusion process responses and impacting the overall quality of the texturized proteins.

In recent years, high-moisture cooking has been used to produce meat analogs with a variety of different sources of plant proteins [9,12–15]. Nonetheless, soy protein is by far the most studied plant protein in extrusion due to its gelation and fiber-forming properties, more balanced amino acid profiles, and more neutral flavor and color [16]. There are several soy protein ingredients commercially available, including flours, protein concentrates, isolates, as well as textured and hydrolyzed proteins. Soy protein isolate (SPI), the most refined form of soy with a protein content exceeding 90%, is commercially produced via protein solubilization at high temperatures, salts, or pHs, followed by its isoelectric precipitation and drying to remove unwanted components and improve protein yield. However, this process can lead to extensive protein denaturation, resulting in significant changes in its physicochemical properties. As a cost-effective alternative with reduced processing requirements to SPI, soy protein concentrate (SPC) has been used as a main raw material in the production of meat analogs [17] as well as soy flour combinations with SPI at differing ratios [5]. Although SPI has proved to result in fibrous anisotropic textures [4], comparatively, SPC-based meat analogs are considered to result in enhanced textural attributes with improved fibrousness under similar extrusion or shear cell processing conditions compared to SPI-based formulations [18–20]. In this sense, Grabowska et al. [20] reported that anisotropy during shear cell processing is fostered by the presence of two separate phases, either proteins-proteins or proteins-macromolecule (i.e., polysaccharides), which may be thermodynamically incompatible, preventing transverse protein aggregation upon longitudinal arrangement in the cooling die and forming elongated fibers. Similarly, Sandoval Murillo et al. [21] also reported that anisotropy during the extrusion of peas forms as a result of the presence of protein fractions with different hydration properties, leading to a multiphasic system, even when using a simple single-protein concentrate. However, although less-refined protein ingredients are generally preferred from the sustainability and processing standpoint, they usually contain a higher content of antinutritional factors, such as trypsin inhibitors [22,23], that should not be disregarded when formulating animal-meat alternatives, especially if these possess digestive enzyme-inhibitory activities. Although the effect of extrusion on reducing trypsin inhibitors (TIs) has been addressed at low-moisture extrusion conditions [24–26], there are very limited number of studies determining TI reductions under high-moisture extrusion conditions.

Several studies have focused on evaluating the effect of high-moisture extrusion process parameters on either the textural attributes or protein structural characteristics of soy-based extrudates. Moisture's influence has been widely studied and recognized as one of the most important factors affecting the textural properties and/or fibrousness of the final product [5,27,28]. Others like Fang et al. [19] have examined the effect of SME on the extrudates' quality, indicating that an increased SME resulted in darker products with a higher tensile strength and hardness, but a lower texturization degree. Regarding the effect of extrusion temperature, diverse results have been reported. In this regard, Chen et al. [27] indicated that cooking temperature only had a significant effect on the tensile strength of SPI extrudates, while hardness and chewiness were not affected, and no clear effect was seen for the degree of texturization. Conversely, Lin et al. [28] reported that higher temperatures produced softer and less chewy products, only slightly changing protein solubility and with no results reported for the extrudates' fibrousness. In a more recent study, Wittek et al. [4] indicated that increasing the extrusion temperature enhanced the alignment of the proteins, resulting in a more pronounced visual fibrousness of the SPI. However, Dekkers et al. [29] reported that temperature effect was hardly visible when processing a pure SPI, while it did influence the shear-induced fibrous formation of pectin-SPI blends at both macroscopic and microscopic levels. Regarding the effect of screw speed, fewer studies have addressed its effect on the textural and protein structural characteristics of soy-based high-moisture extrudates. Pietsch et al. [17] evaluated both the effect of screw speed and barrel temperature on an SPC, reporting that, at the studied conditions, anisotropy occurred with higher temperatures, and no changes in protein–protein interactions were observed between the tested conditions. What is more, these authors reported that at screw speeds higher than 180 rpm and temperatures above 100 °C, the visual anisotropic fibrous structure was disrupted by the appearance of porous structures. Despite all the above studies denoting the influence of extrusion parameters on the quality of high-moisture extrudates, there is a pressing need for holistic studies that provide a more comprehensive understanding of these parameters, not only assessing macrostructural and textural changes but also integrating insights from the protein structural level and their aggregation state in conjunction with their nutritional quality.

Therefore, to better understand the effect of extrusion processing parameters on the textural and nutritional quality of high-moisture extruded meat analogs, this study evaluated the macro- and microstructure (i.e., visual fibrousness), color, textural attributes (TPA, cutting force and anisotropy), and trypsin inhibitor contents of soy protein-based extrudates. Moreover, the changes in the secondary structure of the proteins (identified via FTIR) occurring upon the extrusion of the meat analogs were also evaluated. Specifically, the high-moisture extrudates were produced from a well-characterized mixture of a soy protein isolate and concentrate, which was extruded at two different barrel temperature profiles (a 120 and 140 °C maximum temperature) and three screw speeds (300, 400, and 450 rpm). Furthermore, to understand soy protein structuring during high-moisture extrusion conditions, the protein-rich powders from a soy protein isolate, concentrate, and their mixture were previously characterized for their proximate composition, nutritional quality and physicochemical properties.

2. Materials and Methods

2.1. Materials

A commercial soy protein isolate (SPI) (ProFam 974) and concentrate (SPC) (Arcon SM) were purchased from Archer Daniels Midland Company (ADM, Chicago, IL, USA). A mixture (SPM) of both SPI and SPC protein powders was prepared by mixing SPI and SPC at a 1:9 ratio, to obtain ~77% (*w*/*w*, dry basis) of protein content.

2.2. Methods

2.2.1. Proximate Composition of the Protein Ingredients

The proximate composition of SPI, SPC, and SPM was determined following AOAC standard methods for moisture, ash, total dietary fiber (TDF), and fat measurements. Moisture content was determined according to the AOAC 934.01 method [30] by drying the samples in an oven set at 105 °C for 16 h. Ash content was determined following the gravimetric AOAC 923.03 method [31]. TDF was determined according to the AOAC 991.43 method [32]. Total fat content was measured using a Soxhlet extraction in accordance with the AOAC 920.39 method [33]. Protein was determined following the combustion method according to AACC 46-30.01 [34] with an automated DUMAS protein analyzer (Leco CHN-828, St. Joseph MI, USA), using a conversion factor of 6.25 to calculate protein content from nitrogen. Total carbohydrates were calculated via weight difference. All determinations were conducted in duplicate.

2.2.2. Protein Molecular Profile (SDS-PAGE)

Protein molecular profile was obtained via sodium dodecyl sulphate gel electrophoresis (SDS-PAGE) of the samples under reducing conditions. SDS-PAGE analysis was conducted using a vertical gel electrophoresis device (Mini-PROTEAN® TGX™ Precast Protein Gels Bio-Rad®) following the method of Laemmli [35]. The molecular weight of the samples was estimated with a standard marker (Dual Color Standard, Bio-Rad®, Hercules, CA, USA), ranging from 10–250 kDa.

Briefly, 5 μL of each protein ingredient (20 mg protein/mL of ultrapure water) was collected in an Eppendorf tube and mixed with 5 μL of Sample Buffer Solution (consisting of 190 μL of Laemmli Sample Buffer and 10 μL of Native Sample Buffer (Bio-Rad)). Then, the mix was heated to 100 °C for 10 min. After that, 1 μL was collected and loaded into the wells. The loading volume of the marker was 5 μL. The buffer solution was loaded into the electrophoretic chamber, and consisted of a premixed electrophoresis buffer solution (25 mM Tris, 192 mM glycine, 0.1% SDS, pH 8.3, Bio-Rad®) diluted with distilled water in a 1:10 ratio. The electrophoretic separation was conducted by applying 150 V for 50 min. The gels were stained by adding 50 mL of Bio-Safe Coomassie Brilliant Blue dye (Bio-Rad®) for 1 h under continuous mixing. Finally, the gels were rinsed several times with distilled water and were allowed to discolor for 30 min before capturing the gels. The gels were prepared in duplicate.

2.2.3. Amino Acid Composition of the Protein Ingredients

To determine the amino acid content of the SPI, SPC and SPM powders, 50 mg (±0.2 mg) of sample was weighed in quartz vials for microwave hydrolysis. In the hydrolysis rotor vessel of the microwave, 30 mL of 6 N HCl containing 400 μL of 90% phenol were added. The phenol was added to avoid the degradation of the tryptophan under acidic conditions. Following this, 200 μL of the hydrolysis solution (6 N HCl containing 1.2% phenol) was added to each quartz vial containing the samples to be hydrolyzed, then the vials were sealed and inserted into the rotor vessel. An inert atmosphere was created inside the rotor vessel using nitrogen and the microwave digestion protocol was performed. The microwave digestion was programmed at 650 W and the cycle consisted of a 3 min ramp to 160 °C, a hold period of 10 min and a final cooling down period of 20 min. After microwave digestion, the sample was completely dried using a dry heater set at 60 °C using nitrogen as a carrier gas, dissolved in 1 mL of 0.1 N HCl, filtered through a 0.22 μm nylon membrane, and transferred to HPLC vials for analysis.

Amino acid composition was analyzed according to Guideline 98/64/EG (1998) of the European Union [36]. The amino acid content was analyzed using high-performance liquid chromatography (HPLC) equipment (Agilent 1200, Agilent Technologies, Santa Clara, CA, USA) equipped with a UV detector (G1314B) and a Zorbax Eclipse Plus AAA column (4.6 × 150 mm × 3.5 μm) (Agilent Technologies, CA, USA) placed in a thermostatted column compartment kept at 40 °C. The separation and quantification of amino acids was conducted as precolumn derivatization with an OPA (o-phthalaldehyde) and FMOC reagent (2.5 mg/mL of 9-fluorenylmethylchloroformate in acetonitrile). Separation was obtained at a flow rate of 2 mL/min with a flow gradient consisting of 1.9 min at 0% B followed by a 18.1 min time to raise eluent B to 53%. Then, the column was washed at 100% B, followed by equilibration at 0%B in a total analysis time of 30 min. Mobile phase A consisted of 40 mM of NaH_2PO_4 (pH 7.8), while mobile phase B was a mix of acetonitrile: methanol: water (45:45:10, *v*/*v*). Calibration curves were obtained for each amino acid of interest by using different concentrations of a standard amino acid mix (Amino Acid Standard, AAS18, Sigma Aldrich, San Luis, MO, USA). Protein hydrolysis and HPLC analyses were conducted in duplicates. The amount of each individual amino acid was expressed in g/100 g of protein and the content of the essential amino acids was compared to the FAO/WHO (2013) reference pattern of essential amino acids requirements for adults. The ratio of essential to total amino acids was reported as E/T (%).

2.2.4. Thermal Properties

A differential scanning calorimeter Q-20 (TA instruments, Crawley, UK) equipped with a refrigerated cooling system (RCS 40) was used to evaluate the native state of the protein ingredients. The calibration for enthalpy and temperature was completed using indium. Tzero aluminum hermetic DSC pans (TA-Instruments, New Castle, DE, USA) were employed. An empty pan was used as a reference and dry nitrogen at a flow rate of 50 mL/min was used as a purge gas. For DSC analysis, protein powders and ultra-pure water were mixed in a 1:3 ratio (*w*/*w*) and hermetically sealed. Samples were kept at 25 °C for 24 h and heated from 30 to 120 °C at 5 °C/min. The enthalpy of protein denaturation (ΔH_d in J/g protein) and the temperature of denaturation at the peak maximum (T_d) were reported. Three replicates per sample were conducted.

2.2.5. Physicochemical Properties

Water-Binding and Oil-Binding Capacity

The water-binding capacity (WBC) measurement was carried out following the method AACC 56.30.01 [37] with small modifications. An amount of 1 g of protein powder was transferred into a centrifuge tube and 5 g of distilled water was added. After vortex mixing for 1 min, the sample was centrifuged for 10 min at 2000× *g* and unbound water was discarded. Results were reported as the weight of the water retained per 1 g of sample (g/g).

The oil binding capacity (OBC) measurement was carried out following the procedure used by Lin et al. [38], with slight modifications; 0.5 g of protein powder was transferred into a centrifuge tube and mixed with 3 mL of sunflower oil. Samples were vortex-mixed for 1 min, followed by centrifugation for 25 min at 1610× *g*. The unbound oil after centrifugation was discarded with the help of a pipette. The density of the sunflower oil was used to express the result in mL of the oil bound per g of sample. Duplicate measurements were performed for WBC and OBC of each sample type.

Least Gelation Concentration

The least gelation concentration (LGC) was determined as reported by Adebiyi and Aluko [39]. Briefly, 5 mL dispersions of SPI, SPC, and SPM were prepared at concentrations ranging from 6 to 20% (*w*/*v*) of protein powder in distilled water. Subsequently, samples were vortex-mixed and the mixtures were heated for 1 h at 95 °C in a water bath (BOE-4 series, Raypa, Barcelona, Spain), cooled down rapidly under tap water, and kept at 4 °C for 2 h. After cooling, the tubes were inverted and gel formation was visually determined at each concentration. The LGC was expressed as the lowest concentration at which the sample did not slip or fall down the side of the test tubes. LGC analyses were performed in duplicates.

Foaming and Emulsifying Properties

Foaming capacity (FC) and foaming stability (FS) were determined following the procedure described by Lin et al. [38] with some modifications. Briefly, 3 g of protein powder was suspended in 100 mL of distilled water. The mixture was homogenized with a T18 homogenizer (IKA, Staufen, Germany) at a speed of 10.000 rpm for 2 min and subsequently transferred to a graduated cylinder. Then, the total volume and the foam volume (1 min after homogenization) were recorded. Measurements of the total and foam volumes were taken at 1 min (the initial foam volume), 30 min and 120 min intervals. Foaming capacity was expressed as the relative percentage of the initial foam volume (after 1 min) to the total volume of the sample. Foam stability was defined as the relative percentage of the foam volume remaining after 120 min to the initial foam volume.

The emulsifying activity index (EAI) and emulsion stability index (ESI) were determined following the method described by Pearce and Kinsella [40] with small modifications. An amount of 6 mL of a 0.5% (*w*/*w*) protein solution and 2 mL of sunflower oil were homogenized for 2 min at 10.000 rpm using a T18 homogenizer (IKA, Staufen, Germany). A 0.1% solution of SDS was prepared and used as a blank. Subsequently, 50 µL of the homogenized emulsion sample was diluted in 5 mL of SDS solution (0.1% *m*/*v*) and mixed using a vortex mixer. The absorbance of the mixture was measured at 500 nm using a UV–visible spectrophotometer (UV-1603, Shimadzu, Kyoto, Japan), using plastic cuvettes (1 cm path length). Measurements were taken immediately after sample preparation and after 10 min. The EAI and ESI were calculated according to the equations reported in Karaca et al. [41]:

$$\text{EAI } (\text{m}^2/\text{g}) = (2 \times 2.303 \times \text{A}_0 \times \text{N})/(\text{c} \times ø \times 10{,}000)$$

$$\text{ESI (min)} = \text{A}_0/(\text{A}_0 - \text{A}_{10}) \times 10$$

where, 2 and 2.303 are constant values [40], A_0 is the absorbance of the diluted emulsion immediately after homogenization, N is the dilution factor (×100), c is the weight of the protein per volume (g/mL), ø is the oil volume fraction of the emulsion, A_{10} is the absorbance of the diluted emulsion after 10 min. Duplicate measurements were performed for the FC/FS and the EAI/ESI for each sample.

Pasting Properties

The viscosity properties of SPI, SPC and SPM protein powders under a heating–holding–cooling cycle were analyzed following the standard 95 °C method (RVA Method 50.01 reported by Perkin Elmer (Waltham, MA, USA) [42] using a Rapid Visco Analyzer (RVA-4800, Perten Instruments, Sydney, Australia). For the three protein powders subjected to the standard 95 °C method, samples were first maintained at 50 °C under a shear rate at 960 rpm for 10 s for adequate mixing, and after that the speed was maintained at 160 rpm for the rest of the test. Subsequently, the slurry was held at 50 °C for 60 s and then heated to 95 °C with a temperature increase of 6 °C/min, held at 95 °C for 5 min, and finally cooled to 50 °C at 6 °C/min, and kept at 50 °C for 2 min. Furthermore, the viscosity profile of the mixed protein powder (SPM), used for further extrusion experiments, was obtained at three different maximum hold temperatures, following the standard pasting cycle at 95 °C, as described above, as well as the high-temperature pasting cycle at 120 °C and 140 °C, described in the RVA Method 50.01-Pulse Flour and Pulse Starch Method [42]. The parameters used in the three pasting cycles at a maximum heating temperature of 95, 120, and 140 °C are reported in Table 1. For all the experiments, samples were prepared by mixing the protein powders (4.5 g) with distilled water (24 g) at a 14% moisture basis. From the viscosity profiles, the maximum peak viscosity (PV) and final peak viscosity were recorded (FV) and included in the Supplementary Materials (Table S1). The viscosity tests were performed in duplicates.

Table 1. Pasting profiles obtained at 95, 120, and 140 °C maximum temperatures (adapted from [42]).

Parameter	Value	Time (min:s)		
		95 °C	120 °C	140 °C
Initial Temperature (°C)	50	00:00	00:00	00:00
Speed (rpm)	960	00:00	00:00	00:00
Speed (rpm)	160	00:10	00:10	00:10
Temperature (°C)	50	01:00	01:00	01:00
Temperature (°C)	Hold *	08:30	12:40	16:00
Temperature (°C)	Hold *	13:30	17:40	21:00
Temperature (°C)	50	21:00	29:10	36:00
End Temperature (°C)	50	23:00	31:10	38:00

* Hold period for 5 min at maximum temperature of 95, 120, or 140 °C, respectively, for each pasting cycle.

2.2.6. Extrusion Processing

Extrusion experiments were conducted on a semi-industrial Evolum 25 corotating twin-screw extruder (CLEXTRAL, Firminy, France), able to work at a maximum solid feeding of 25 kg/h and a maximum liquid feeding of 40 l/h via volumetric dosing. The screw diameter (D) was 25 mm, and the screw length (L) was 600 mm, resulting in a length/diameter (L/D) ratio of 24:1. The barrel consisted of six independent temperature-controlled zones (Z). A LWFD5-20 volumetric feeder (K-Tron Corp., Pitman, NJ, USA) was used to feed the dry protein powder into the barrel. The feeder was calibrated and adjusted beforehand. Water was injected directly into the feeding zone, in the second barrel section, using a Super K pump (DKM-CLEXTRAL, Firminy, France) to maintain the target moisture content. The process conditions were monitored from a control panel (see Table 2). A thermoregulated cooling die (450 × 30 × 4, L × W × H) with an internal rectangular section of 30 mm × 4 mm was attached to the end of the extruder. The temperature of the cooling die was kept at 80 °C.

Table 2. The operational extrusion conditions for screw speed and temperature profiles, specific mechanical energy (SME) and product melt temperature and pressure resulting from the extrusion experiments.

Barrel Segment Temperatures (°C)						Screw Speed (rpm)	Motor Torque (%)	Melt Temperature (°C)	Melt Pressure (bar)	SME (Wh/Kg)	Extrusion Conditions
Z1	Z2	Z3	Z4	Z5	Z6						
30	50	75	105	115	120	300	11.2	115	18.0	41.1	LT-LS
						350	12.3	117	20.4	52.6	LT-MS
						400	9.2	115	14.3	44.9	LT-HS
30	70	95	115	125	140	300	10.4	131	16.5	38.1	HT-LS
						350	9.8	130	15.5	41.9	HT-MS
						400	7.9	130	11.6	38.7	HT-HS

SME: specific mechanical energy; LT: low temperature; HT: high temperature; LS: low speed; MS: medium speed; HS: high speed.

An experimental design was used to investigate the effect of screw speed and extrusion temperature on the quality of the high-moisture extruded meat analogs. The extrusion conditions are summarized in Table 2. A moisture content of 69% was selected and kept constant throughout all the experiments as this moisture content resulted in the most fibrous appearance in previous trials. The solid feed rate and water flow rate were 2.8 kg/h and 5.1 l/h, respectively. The screw speed was set to 300, 350, and 400 rpm, resulting in low (LS)-, medium (MS)-, and high (HS)-speed conditions, respectively. Furthermore, two different temperatures profiles were used. The extruder temperatures in the six-barrel zones (from the first barrel to the end plate) were set at 30, 50, 75, 105, 115, 120 °C (low temperature processing, LT) and 30, 70, 95, 115, 125, and 140 °C (high temperature processing, HT) to achieve two different thermal energies during extrusion.

Furthermore, the extruder responses including the motor torque, the melt pressure and melt temperature at the extruder exit, and Specific Mechanical energy (SME) were recorded online during the trials. The specific mechanical energy (SME; kJ/kg), defined as the amount of work supplied from the driving motor into the material being extruded, was used to quantify the severity of the shear forces during the extrusion process. The SME was calculated according to the following equation:

$$\text{SME (kJ/kg)} = 2\pi \times \text{n} \times \text{T/MFR}$$

where n is the screw speed (rpm), T is the torque (Nm), and MFR is the mass flow rate (kg/h).

The samples were collected after the stabilization of the operating conditions for each set of parameters. The protein extrudates were cut into rectangular strips as they exited the cooling die. The samples were allowed to cool down, stored in sealed plastic bags and kept at −18 °C until further analysis was conducted. The moisture of the extruded products was determined after thawing, before textural and protein structural analyses, according to the AOAC 934.01 method [30].

2.2.7. Evaluation of the Visual, Textural, Nutritional, and Protein Structural Changes upon Extrusion

Macro- and Microscopical Evaluation

The macrostructure of the samples was evaluated using a phone camera. A piece of each extruded sample was manually broken into halves parallel to the flow direction to observe the fibrous anisotropic formation. The microstructure analysis of the extrudates was conducted using a scanning electron microscope (SEM) (Quanta 200, FEI, Eindhoven, The Netherlands). Frozen, extruded samples were thawed 4 h before analysis at room temperature, cut transversal to the flow direction, and placed on double-sided adhesive. Images of the samples were taken with the SEM at different magnifications, in low vacuum conditions, and with an accelerating voltage of 12.50 kV.

Color Evaluation

The color attributes of the protein powders and the produced extrudates were measured using a spectrophotometer (CM-3500d, Konica Minolta, Osaka, Japan). Calibration was performed with a zero calibration box and a white calibration plate (CM-A124; CM-A120, Konica Minolta, Osaka, Japan) before the measurements. Samples were measured using a Petri dish CM-A128, and results were obtained using a D65 standard illuminator and 2° observer. The color of the extrudates and protein powders was expressed in the CIELab color space, as parameters of L^* (lightness, 0 = black and 100 = white), a^* (−a/+a, greenness or redness), and b^* (−b/+b, blueness or yellowness). Four measurements were taken at random locations on the surface of each sample. The total color differences (ΔE) of the extrudates were calculated using the following equation:

$$\Delta E = \sqrt{(\Delta L^*)^2 + (\Delta a^*)^2 + (\Delta b^*)^2}$$

where, ΔL^*, Δa^*, and Δb^* are the differences between the extrudates and the SPM, the protein powder used for the extrusion experiments.

Protein Secondary Structure

The secondary structure of the native and extruded proteins was studied using Fourier-Transform Infrared Spectroscopy (FTIR) by analyzing the amide I band (the stretching vibrations of C=O in the peptide bond) using a Nicolet IS50 FTIR spectrometer (Thermo Scientific, Waltham, MA, USA) coupled to an attenuated total reflectance (ATR) cell. The spectrum was taken by averaging 64 scans at a 4 cm^{-1} resolution in the wavenumber range of 600 to 4000 cm^{-1}. A blank was performed before each spectrum collection. Extruded proteins were freeze-dried and milled before FTIR analysis to remove water interference from the amide I region. A Gaussian/Lorentzian deconvolution procedure of the Amide I band (1600–1700 cm^{-1}) was conducted using OMNIC FTIR software (version 9.2.86, ThermoFisher Scientific, USA), enabling the identification of various secondary structures, including intermolecular aggregated structures (A1~range 1612–1620 cm^{-1}), parallel β-sheets (range 1630–1638 cm^{-1}), random coils and α-helices (range 1640–1660 cm^{-1}), β-turns (range 1660–1680 cm^{-1}), antiparallel β-sheets (range 1680–1688 cm^{-1}) and intramolecular aggregated structures (A2~range 1690–1695 cm^{-1}) [43–45]. As reported by Fevzioglu et al. [46], an overlapping of random coil and α-helix peak structures occurs in the range of 1640–1660 cm^{-1}. Given that correctly assigning these peaks has been described as challenging [47], in the present study, peaks in that range were considered as α-helices. The relative proportions of the identified structures were calculated by integrating their respective deconvoluted bands and dividing them by the sum of the areas of all identified structures. Examples of the deconvoluted amide I band are reported in the Supplementary Materials (Figure S1). Baseline correction and normalization was performed before deconvolution. FTIR analyses were performed at least in triplicate.

Textural Properties

The textural properties of the extruded samples were evaluated by texture profile analysis (TPA) and a cutting test using a TA.XTPlusC 650 texture analyzer (Stable Micro-Systems, Godalming, UK). All samples were thawed at 25 °C and cut into a square shape of 20 mm × 20 mm before analysis. For TPA analyses, samples were subjected to a double compression test using a P/36 probe to compress 50% of its original thickness at a speed of 1 mm/s. The textural properties for hardness (N), springiness, resilience and cohesiveness were then recorded [27].

For cutting test evaluation, the sample was cut using a Meullenet Owens Razor Shear Blade (A/MORS) (Stable Micro-Systems, UK) to 75% of its original thickness at a speed of 2 mm/s [27,48]. Samples were evaluated parallel (with longitudinal cutting force, FL) and perpendicular (with transversal cutting force, FT) to the flow direction of the extrudates from the cooling die. The maximum cutting strength was calculated using the maximum force (N) obtained from Exponent connect software version 8.7.07 (Stable Micro-Systems, Godalming, UK) at the maximum fracture point per the flow direction. The maximum cutting strength at the FL and FT and the anisotropic index, defined as the ratio of the maximum cutting force between the FL and FT, was used to evaluate the fibrous formation. Equal values (FL = FT) indicate uniform texture with low material anisotropy. All textural determinations were performed with at least five replicates.

Antinutritional Factors (Trypsin Inhibitor Activity)

The quantitative evaluation of trypsin inhibitor activity in fresh native and extruded samples was performed in duplicates via a spectrophotometric method according to the standardized ISO method 14902 [49]. Briefly, 1 g of sample was suspended in 50 mL of 0.01 M NaOH, and the pH was adjusted to 9.4–9.6 with HCl, stored and further diluted with distilled water. A 1 mL sample was withdrawn and properly diluted to achieve 40–60% inhibition of the trypsin enzymatic activity. For the enzymatic reaction, a standard solution of trypsin was added to a mixture containing the properly diluted sample solution and a solution of α-N-benzoyl-l-arginine-4-nitroanilide hydrochloride (BAPA) and incubated at 37 °C for 10 min. After incubation, the reaction was quenched with the addition of 30% acetic acid. Blank reactions were simultaneously carried out. The reaction mixtures were then centrifuged, and their absorbance was measured at 410 nm, calculating the trypsin inhibitor activity (TIA) as the mg of the inhibited pure trypsin per g of sample.

2.3. Statistical Analysis

Statistical analysis from at least two independent measurements was conducted using Statgraphics Centurion 19-X64 software (Statgraphics Technologies, Inc., Warrenton, VA, USA). Data were expressed as mean ± standard deviation (SD). Significant differences between the parameters for the different samples were studied via a one-way analysis of variance (ANOVA). Fisher's least significant difference (LSD) was used to describe means with 95% confidence intervals.

3. Results and Discussion

3.1. Proximate Composition of Protein Powders

The proximate composition, amino acid and protein composition, as well as the physicochemical properties of the protein powders are key parameters in the selection of raw materials for high-moisture extrusion processing, as these will determine the behavior of the protein-rich matrix during thermo-mechanical processing and, in turn, the characteristics of the final product.

The proximate compositions of the soy protein isolate (SPI) and concentrate (SPC), used to prepare the protein mixture (SPM) for extrusion experiments, are listed in Table 3. The moisture content ranged between 6 and 8% for all protein powders. The protein content was significantly higher for SPI (95% protein), compared to SPC (74% protein). The soy protein mix, SPM (SPI:SPC 1:9, *w*/*w*), had a similar protein content to SPC as SPC was the major component in the mixture. Regarding the fiber contents, while it was negligible in SPI (<1%), up to 4% of the fiber was found in SPC. This fiber fraction would mainly consist of cellulose, hemicellulose and pectin polysaccharides. Furthermore, in SPC and, hence, SPM, an important amount of other carbohydrates was present, which may correspond to small sugars, starch or oligosaccharides not accounted for in the crude fiber fraction [50]. The oil content was very low (<1%) in all samples denoting a high efficiency in fat removal from the oilseed during oil extraction and further protein extraction and purification. Similar to fat content, no significant differences were found for ash content (<5%).

Table 3. The proximate composition and physicochemical properties of soy protein ingredients.

	SPI	SPC	SPM
	Proximate composition		
Moisture (g/100 g, w.b.)	*7.6 ± 0.2 b*	*6.1 ± 0.1 a*	6.3 ± 0.2 a
Protein (g/100 g d.b.)	*95.4 ± 1.1 c*	*73.9 ± 0.1 a*	76.5 ± 0.1 b
Crude Fiber (g/100 g d.b.)	*0.8 ± 0.0 a*	*3.5 ± 0.1 c*	3.1 ± 0.1 b
Carbohydrates (g/100 g d.b.)	*0.2 ± 0.0 a*	*11.7 ± 0.1 c*	10.3 ± 0.1 b
Lipids (g/100 g d.b.)	*0.4 ± 0.1 a*	*0.6 ± 0.1 a*	0.6 ± 0.1 a
Ash (g/100 g, d.b.)	*4.0 ± 0.1 a*	*4.4 ± 0.2 a*	4.3 ± 0.2 a
	Physicochemical properties		
T_{d1} (°C)	nd	86.4 ± 0.7 a	86.1 ± 1.9 a
T_{d2} (°C)	nd	116.6 ± 0.1 a	117.4 ± 0.6 a
ΔH_1 (J/g $_{protein}$)	nd	1.8 ± 0.6 a	1.4 ± 0.1 a
ΔH_2 (J/g $_{protein}$)	nd	3.4 ± 0.1 b	3.2 ± 0.1 a
WBC (g/g)	4.7 ± 0.1 a	5.0 ± 0.1 b	5.0 ± 0.0 b
OBC (mL/g)	3.4 ± 0.1 a	3.0 ± 0.2 a	3.0 ± 0.1 a
LGC (%)	14 ± 0 c	12 ± 0 a	13 ± 0 b
FC (%)	43.1 ± 2.0 a	50.0 ± 0.0 b	49.2 ± 0.2 b
FS (%)	79.2 ± 5.9 b	62.7 ± 0.2 a	64.7 ± 0.9 a
EAI (m^2/g)	99.5 ± 0.0 b	94.5 ± 0.3 a	95.1 ± 0.2 a
ESI (min)	1350 ± 0.0 c	80.2 ± 0.2 a	555.0 ± 0.2 b

Mean ± standard deviation followed by different letters within each row indicate significant differences ($p < 0.05$). SPI: soy protein isolate; SPC: soy protein concentrate; SPM: soy protein mix; T_d: denaturation temperature at peak maximums of peaks 1 and 2; ΔH: the enthalpy of protein denaturation of peaks 1 and 2; nd: not detected; WBC: water-binding capacity; OBC: oil-binding capacity; LGC: least gelation concentration; FC: foaming capacity; FS: foaming stability, EAI: emulsion activity index, ESI: emulsion stability index.

3.2. Protein Molecular Properties

To identify differences in the protein compositions between the SPI and SPC, the gel electrophoresis (SDS-PAGE) under reducing conditions (Figure 1) and the amino acid composition (Table 4) of the powders was evaluated. Electrophoretic profiles denoted that the soybean protein powders were mainly composed of subunits from their glycinin (11S) and β-conglycinin (7S) protein fractions. The 11S/7S ratio has been found to affect the degree of texturization of the soy protein [51]. In this sense, the three protein powders exhibited similar protein profiles with bands of different intensities between ~15 and 75 KDa. The three intense bands between 75 and 50 KDa were associated with the subunits of the trimeric β-conglycinin, α′ (~75 KDa), α (~70 KDa) and β (~50 KDa) and the other two bands corresponded to the acidic (~40–32 KDa) and basic (~20 KDa) subunits of glycinin. Similar bands were reported by Liu et al. [52], Flory and Alavi [53], and Xia et al. [54] in soy protein concentrates and isolates. In addition to the bands from the protein fractions mentioned above, bands of lower intensities were observed, including faint low-molecular-weight bands between 10 and 15 kDa. These bands could correspond to albumins and/or other low-molecular-weight proteins [9].

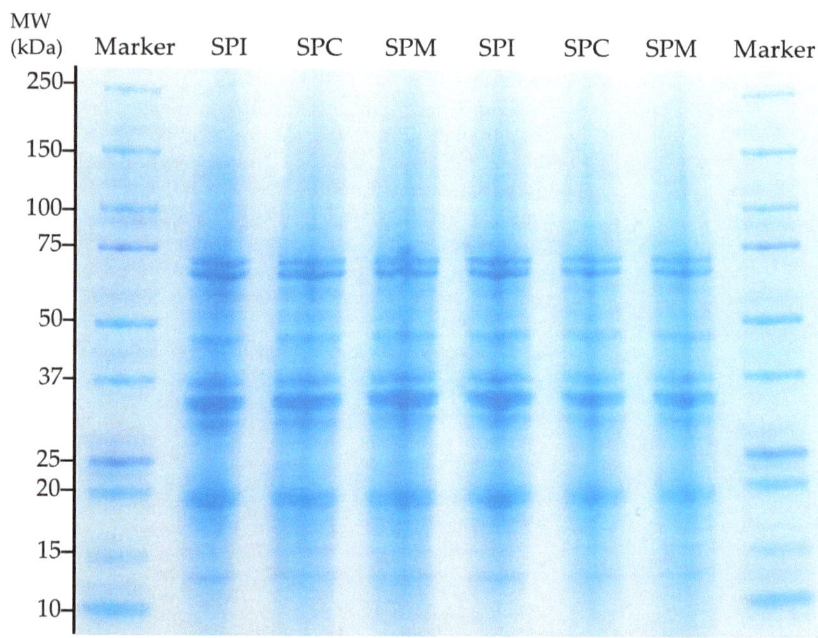

Figure 1. Gel electrophoresis profile at disassociating (SDS-PAGE) and reducing conditions for the soy protein isolate (SPI), concentrate (SPC) and their mixture (SPM) performed in duplicate.

The amino acid compositions of the protein powders are shown in Table 4. The amino acid compositions were similar among the protein powders, denoting no significant differences between the SPI, SPC and SPM for histidine, valine, methionine, phenylalanine, isoleucine, leucine, tryptophan, glycine, arginine, tyrosine, cysteine, aspartic acid, glutamic acid, serine, and alanine. All samples showed high aspartic acid and glutamic acid contents, and a low content of methionine, tryptophan, and cysteine. Similar results were reported for the SPI by Fernández-Quintela et al. [55] and Mateen and Singh [56]. Regarding the proportion of essential to non-essential amino acids (E/NE), a decrease in the E/NE ratio was observed in SPI as compared to SPC, due to the lower presence of the essential amino acid threonine and lysine in the SPI, which could be attributed to factors such as the extraction process and the initial raw material composition [57]. As for the daily adult intake requirements recommended by the FAO, soy protein powders met or exceeded the patterns for amino acid requirements for all the essential amino acids except for methionine [56,58].

Table 4. Amino acid profile of protein powders expressed in g of amino acid per 100 g of protein and the daily requirement of essential amino acids.

Amino Acid	SPI	SPC	SPM	Daily Requirement **
Histidine *	2.83 ± 0.28 a	3.02 ± 0.25 a	2.91 ± 0.25 a	1.50
Threonine *	2.97 ± 0.23 a	3.24 ± 0.25 b	3.11 ± 0.25 a	2.30
Valine *	3.12 ± 0.42 a	3.26 ± 0.46 a	3.19 ± 0.47 a	3.90
Methionine *	1.33 ± 0.21 a	1.36 ± 0.22 a	1.38 ± 0.22 a	2.20 ***
Phenylalanine *	4.18 ± 0.24 a	4.14 ± 0.25 a	4.05 ± 0.25 a	4.00
Isoleucine *	2.94 ± 0.30 a	2.97 ± 0.33 a	2.89 ± 0.33 a	3.00
Leucine *	6.60 ± 1.02 a	6.75 ± 1.75 a	6.57± 1.26 a	5.90
Lysine *	6.22 ± 0.14 a	6.51 ± 0.11 b	6.37 ± 0.11 ab	4.50
Tryptophan *	0.15 ±0.70 a	0.12 ±0.86 a	0.11 ±0.94 a	0.60
Glycine	3.69 ± 0.35 a	3.89 ± 0.38 a	3.77 ± 0.38 a	
Arginine	5.87 ±1.27 a	5.49 ± 0.92 a	5.49 ± 0.95 a	
Tyrosine	3.15 ± 0.8 a	2.99 ± 0.7 a	2.91 ± 0.8 a	
Cysteine	0.88 ± 0.73 a	1.00 ± 0.67 a	0.98 ± 0.70 a	
Proline	4.87 ± 0.56 a	3.56 ± 0.51 b	4.00 ± 0.25 ab	
Aspartic acid	10.57 ± 1.69 a	10.59 ± 1.81 a	10.31 ± 1.76 a	
Glutamic acid	16.93 ± 1.81 a	16.46 ± 1.92 a	16.16 ± 1.94 a	
Serine	4.56 ± 0.43 a	4.69 ± 0.75 a	4.56 ± 0.55 a	
Alanine	3.83 ± 0.40 a	4.22 ± 0.42 a	4.04 ± 0.43 a	
E/T (%)	30.32	31.35	30.58	
E/NE (%)	57.44	59.93	58.54	

Mean ± standard deviation followed by different letters within each row indicate significant differences ($p < 0.05$). SPI: soy protein isolate, SPC: soy protein concentrate, SPM: soy protein mix. E/T = the proportion of essential amino acids to total amino acids; E/NE = the proportion of essential amino acids to non-essential amino acids. * Essential amino acid. ** Values of the daily requirement corresponding to the adult (>18 y) maintenance amino acid pattern (g/100 g protein) recommended in the dietary protein quality evaluation in the human nutrition report (FAO, 2013) [59]. *** Daily requirement for sulfur-containing amino acids (Methionine + Cysteine = 2.2 g/100 g).

Differential scanning calorimetry (DSC) was used to study the native state of the protein ingredients (Table 3). Calorimetric results indicated extensive protein denaturation in the SPI as no enthalpic transitions were observed in this sample, suggesting important bonds in the creation and maintenance of the protein structure were disrupted during its purification. Nonetheless, the SPC and SPM still exhibited two endothermic peaks in the DSC thermogram. This would align with the results observed for the secondary structure components of the different unextruded soy proteins (see Section 3.4.2), with higher intermolecular protein aggregated structures (A1~1617–1624 cm^{-1}) observed for the SPI as compared with its SPC counterpart. The two endothermic peaks found in the SPC and SPM were attributed to the main protein components of the soybean protein, β-conglycinin (7S) and glycinin (11S), the main protein fractions identified in the SDS-PAGE profiles. The transition temperature of the peak corresponding to 7S was observed at ~86 °C for the SPC and the SPM. This peak exhibited significantly lower enthalpy (1.4–1.8 J/g protein) than the peak attributed to 11S, in agreement with previous observations [44,60]. The denaturation temperature of 11S was much higher than that of 7S, with a T_d at ~116 °C and an enthalpy of 3.2–3.4 J/g, which was similar between the SPC and the SPM. Temperatures of denaturation were in the range of those reported in the literature for 7S (78–90 °C) and 11S (95–109 °C) [44,60–62] with variations attributed to the soybean characteristics, such as processing conditions or salt contents [60,63], or the moisture contents during DSC analyses [64], among others.

3.3. Physicochemical Properties of Protein Powders

The physicochemical properties of protein powders were evaluated in terms of their water- and oil-binding capacity, gel-forming ability and viscosity, and foaming and emulsifying properties (Table 3). Regarding the WBC, no significant differences were observed between SPC and SPM. Both protein powders showed a higher WBC than SPI, which can be attributed to the higher nativity state of the SPC protein (fewer exposed hydrophobic patches) [65] and/or the higher amount of non-protein components, especially high-molecular fibers, with high water-binding affinities [66]. The OBC of proteins is partly related to their intrinsic hydrophobicity, as non-polar amino acids can interact with aliphatic chains of lipids, with extended surface hydrophobicity also fostering protein aggregation rather than protein–lipid interaction [65]. However, no significant differences ($p < 0.05$) were observed between the protein powders.

The foaming and emulsifying properties were evaluated as an indicator of the surfactant properties of the protein ingredients. These properties depend on intrinsic factors such as protein structure and composition, their colloidal properties [67], as well as the presence of some other components, such as carbohydrates [68]. In this regard, conglycinin has better surfactant properties than glycinin due to its smaller molecular mass, higher structural flexibility (being easier to unfold), the presence of glycosylated groups and its higher hydrophobicity [69]. SPC exhibited a higher initial foaming capacity than SPI, while the latter was characterized by a greater foam stability, which could be related to their different nativity state, as native plant proteins have limited foaming properties due to their compact structure [67]. In addition, SPI presented both a higher emulsion activity and stability. Meanwhile, no significant differences ($p < 0.05$) were observed between SPC and SPM. The better emulsifying properties of SPI could be related to its higher effective protein concentration or extended denaturation rather than differences in the 7S/11S ratio between SPC and SPI. In this sense, pulse protein concentrates have been reported to possess lower emulsifying activity than soy protein isolates [70,71].

Globular proteins are able to form gels when they are heat-denatured in aqueous solutions. The least gelation concentration (LGC) indicates the protein ability to create a protein network that can entrap water upon heating and subsequent cooling, forming a heat-set gel [12]. Therefore, a lower LGC indicates that the protein has a better capacity to form gels. Despite its lower protein content, SPI showed a higher LGC after heating than the SPM and especially SPC, with LGC values ranging from 12–14% solids, similar to those found by Flory and Alavi [53] for a commercial SPI and SPC (14%). The lower LGC of the SPI could be related to the lower native and greater aggregation states of the denatured SPI, whose extensive aggregation would result in a higher protein concentration needed to form the network [12]. Therefore, limited thermal denaturation during processing may have a beneficial effect on the functionality of the protein ingredients, especially if further heat processing is not performed at temperatures high enough to disrupt the aggregates.

To further evaluate the gelation potential of the protein ingredients, their viscosity profiles under a heating–holding–cooling cycle were evaluated under low shear stirring conditions. The viscosity profile was evaluated at a 16% solids concentration, and, hence, was above the LGC of all protein ingredients. When protein powders were heated up to 95 °C (Figure 2), all protein powders exhibited cold-swelling properties, which further solubilized upon heating, decreasing their viscosity, which was later increased upon the final cooling. Flory and Alavi [53] obtained similar results for the viscosity profile of an SPI and SPC. These authors classified protein ingredients based on the peak viscosity development during the heating cycle, those proteins with peak viscosity before 4 min were categorized as having cold-swelling properties while peaks later than 4 min were indicative of heat-swelling properties. In the heating–cooling cycle, the SPI was characterized by a high initial viscosity that decreased with the temperature [72]. SPI showed a higher initial swelling, a faster solubilization and a higher final viscosity than SPC or SPM. SPI was characterized as having cold-swelling proteins, what allows it to easily form gels [53]. When soy proteins unfold due to shearing or heat and are exposed to water, their non-polar and sulfhydryl groups readily form hydrophobic and disulfide bonds that cause the proteins to aggregate and produce an increase in viscosity [73]. In the case of SPC, the formed network was weaker (i.e., at a lower final viscosity) which could be attributed to the presence of other components such as carbohydrates that interfere with protein aggregation and the formation of the protein network as well as with its lower protein content (see Table 3). Mixing the two protein powders (the SPI and the SPC) had a synergetic effect on the viscosity profile of the SPM, showing a viscosity profile similar to that of the SPC, but with enhanced viscosity.

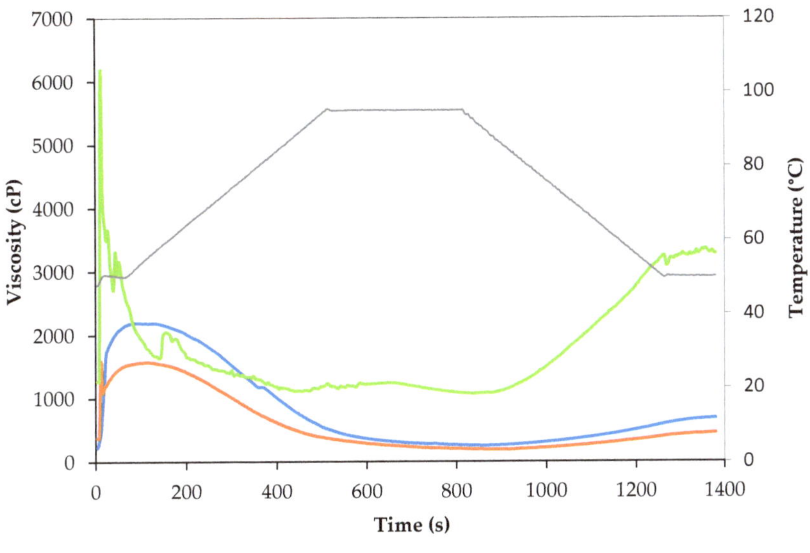

Figure 2. Viscosity profile of the protein powders. The green line is pasting profile of the soy protein isolate; the red line is the pasting profile of the soy protein concentrate; the blue line is pasting profile of the soy protein mix; the grey line represents the temperature profile during the viscosity test.

Viscosity plays a crucial role in altering the flow behavior and the mechanical energy input in extrusion cooking. For this reason, the viscosity profiles of the SPM powder used for the extrusion process were further determined at three different maximum temperatures; in addition to 95 °C, profiles were also determined at 120 °C and 140 °C, which were the maximum barrel temperatures reached during the extrusion process at low (LT)- and high (HT)-temperature conditions (120 and 140 °C), respectively. Figure 3 shows the three viscosity profiles at 95, 120 and 140 °C maximum heating temperatures. No significant differences ($p < 0.05$) were observed in the initial cold-peak viscosity as cooking was performed at the same heating rate. Nonetheless, after the cold-peak viscosity, a higher drop in viscosity (i.e., the minimum viscosity) was obtained with the temperature profile of 120 and 140 °C, during the maximum heating temperature compared with the profile at 95 °C, with even the appearance of a second peak at 140 °C conditions. The sudden increase in viscosity at a 140 °C holding temperature could be explained partly by undenatured protein fractions or the disruption of protein aggregates that may not be disrupted at milder heating conditions, based on the higher thermal stability of glycinin observed in DSC results. Interestingly, a second peak during heating has also been observed in a hemp protein concentrate processed at 130 °C [72] which possesses a predominance of the 11S edestin fraction, similar to soy glycinin. Upon cooling, viscosity results showed that as the maximum cooking temperature increased, SPM exhibited a greater capacity to form a structured gel, increasing its final consistency, which may be the result of an increased extent of protein denaturation at elevated temperatures (glycinin) and/or a structure breakdown with the rupture of aggregates. Thus, as the temperature increased, the proteins denatured and interacted with water; and the unfolded proteins led to stronger interactions, resulting in greater water retention and a higher capacity to form a structured gel.

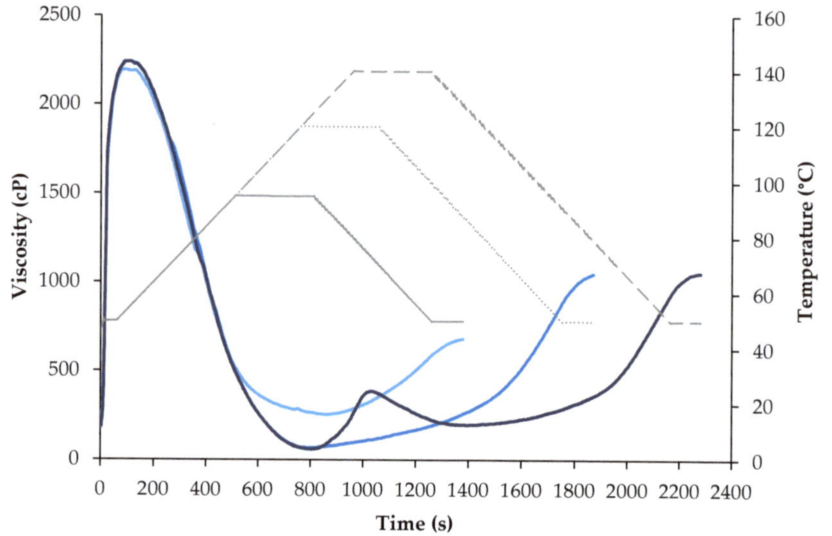

Figure 3. Viscosity profile of the soy protein powder mix, SPM, at three different maximum temperature profiles (95, 120, and 140 °C). The light blue line indicates the pasting profile with a maximum holding temperature at 95 °C; the medium blue line is the pasting profile with a maximum holding temperature of 120 °C; the dark blue line is the pasting profile with a maximum holding temperature of 140 °C, and the continuous, dotted, and line-broken grey lines represent the temperature profiles for the 95, 120, and 140 °C viscosity cycles, respectively.

3.4. Extrusion Processing of Soy Protein Powders

Extrusion experiments to produce high-moisture extrudates were carried out using the SPM as the raw material, which was produced by mixing the SPC and the SPI. The use of less refined ingredients has been reported to be beneficial for allowing phase separation and a better fibrousness of high-moisture extruded meat analogs [20]. During extrusion processing, the effect of two different processing parameters, barrel temperatures and screw speed, were analyzed (see Table 2). Extrusion tests were carried out using two different barrel temperature profiles, low (LT) at 120 °C, and high (HT) at a 140 °C maximum temperature, that resulted in product temperatures of 115 and 130 °C, respectively. Furthermore, three different speeds were used, low, medium, and high, at 300 (LS), 350 (MS), and 400 (HS) rpm, respectively. This resulted in six different extrudates, LT-LS, LT-MS, LT-HS, HT-LS, HT-MS, and HT-HS. The influence of these process parameters on the visual anisotropy, textural and color attributes as well as the antinutritional factors of the extruded meat analogs were evaluated, together with the changes in the protein secondary structure occurring upon extrusion.

3.4.1. Macro- and Microscopic Evaluation of Extruded Meat Analogs

Figure 4 shows the macrostructural appearance of the extrudates obtained at different process conditions. All extruded meat analogs had visible anisotropy in the form of fibrous or layered structures, with predominantly lengthwise-oriented fibers upon tearing [48]. It is worth noting that SPM contained significant amounts of carbohydrates (>10%), especially fibers (see Table 1), which would have facilitated the phase separation of protein domains and the formation of layered structures upon cooling [9,74]. As two or more thermodynamically incompatible phases can prevent a transverse aggregation of the proteins and benefit the longitudinal arrangement forming the fibers [20]. As for the processing parameters, more lengthwise-oriented fibers were formed after extrusion at a higher barrel temperature, in agreement with Osen et al. [48]. Higher temperatures have been reported to promote more disassembled protein subunits, which could realign more efficiently via the directional shear force inside the cooling die creating more elongated fibers [75], something that was also suggested when evaluating the viscosity profile of SPM at 120 and 140 °C (see Figure 3). Furthermore, at the same temperature conditions, smaller differences in fibrousness could be observed according to the screw speed.

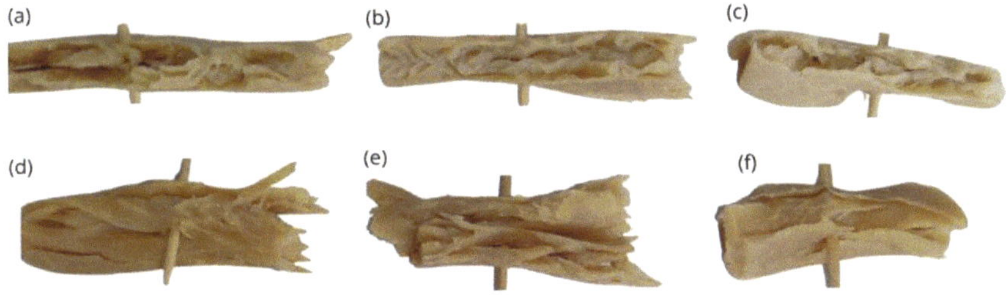

Figure 4. Visual appearance of the high-moisture extruded meat analogs cut longitudinally to the flow direction under different temperatures and screw speed conditions. (**a**) LT-LS; (**b**) LT-MS; (**c**) LT-HS; (**d**) HT-LS; (**e**) HT-MS; (**f**) HT-HS. LT: low temperature; HT: high temperature; LS: low speed; MS: medium speed; HS: high speed.

Scanning microscopy provided more detailed information on the protein network at the microscopic level (Figure 5). At the microscopic level, it was also observed that soy extrudates displayed anisotropy in the form of fibrous/layered structures, without clear differences between the different processing conditions.

From the visual appearance (Figure 4), darker meat analogs were also noted at high-temperature extrusion conditions, in agreement with colorimetric results. Thus, the extrusion parameters also influenced the color attributes. The color parameters of the protein powders and the extrudates are reported in Table 5. Both barrel temperature and screw speed showed a significant effect ($p < 0.05$), especially the former, on the lightness L^* and the chromatic parameter b^* (yellowness). However, only temperature had a significant effect on the a^* values (redness). Thus, the total color change (ΔE) of the extrudates compared to that of the raw material (SPM) was especially influenced by the thermal energy applied. All the extrudates showed a decrease in L^* and increase in b^* and a^* values compared to the SPM powder, resulting in darker and more reddish/yellowish products. Among the extrudates, L^* values ranged between 60.84 and 64.25, a^* values ranged between 2.87 and 4.69, and b^* values ranged between 17.42 and 20.15. At the same barrel temperature, lowering the screw speed generally resulted in an increased chromatic value (b^*) and decreased lightness (L^*). At the same screw speed conditions, a higher barrel temperature resulted in increased redness/yellowness (a^*, b^*), but decreased lightness (L^*). All in all, the total color difference, ΔE, increased with higher temperatures (120 to 140 °C) and with lower screw speeds (400 to 300 rpm) at LT conditions, probably associated with increased Maillard/caramelization reactions with higher product temperatures and longer resident times, respectively. Wang et al. [76] reported similar results for the effect of barrel temperature on the color parameters, showing a significant drop in lightness as the barrel temperature rose from 120 to 160 °C.

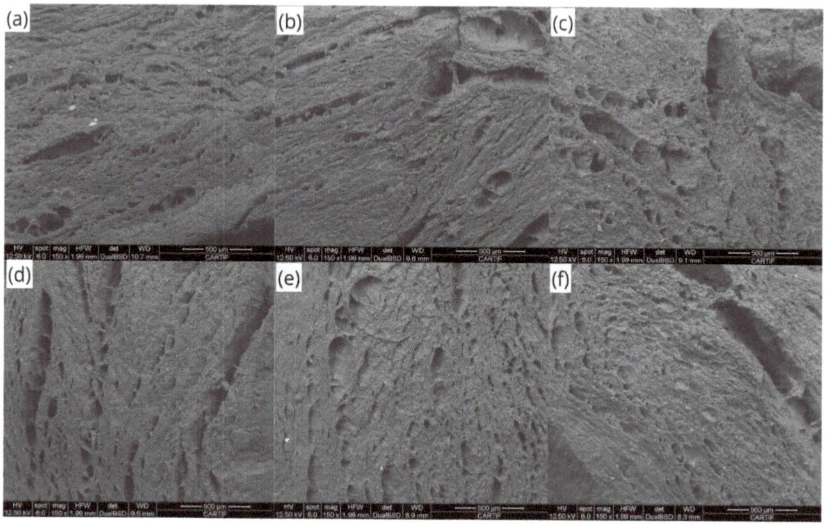

Figure 5. Microstructure of the high-moisture extrudates cut transversal to the flow direction under different temperatures (T) and screw speed (S) conditions. (**a**) LT-LS; (**b**) LT-MS; (**c**) LT-HS; (**d**) HT-LS; (**e**) HT-MS; (**f**) HT-HS. The scale bar in each image indicates 500 μm. LT, low temperature; HT, high temperature; LS, low speed; MS, medium speed; HS, high speed.

Table 5. Results of the CIE of the L^*, a^*, and b^* color parameters for protein powders and high-moisture extruded meat analogs and the ΔE of extrudates as compared to the soy protein mixture.

	L^*	a^*	b^*	ΔE
SPI	86.07 ± 0.06 e	−0.27 ± 0.01 a	16.33 ± 0.09 b	-
SPC	88.05 ± 0.00 g	−0.55 ± 0.01 a	14.58 ± 0.02 a	-
SPM	87.38 ± 0.05 f	0.10 ± 0.01 a	14.68 ± 0.08 a	-
LT-LS	64.25 ± 0.23 c	3.22 ± 0.03 b	18.85 ± 0.19 de	23.71 ± 0.25 b
LT-MS	64.55 ± 0.22 c	3.29 ± 0.07 b	18.59 ± 0.05 d	23.38 ± 0.28 b
LT-HS	66.08 ± 0.17 d	2.87 ± 0.01 b	17.42 ± 0.25 c	21.65 ± 0.19 a
HT-LS	60.84 ± 0.25 a	4.31 ± 0.25 c	19.66 ± 0.24 ef	27.33 ± 0.22 d
HT-MS	61.83 ± 0.32 b	4.69 ± 0.47 c	20.15 ± 0.93 f	26.53 ± 0.51 cd
HT-HS	61.74 ± 0.4 b	4.24 ± 0.88 c	18.95 ± 0.79 de	26.32 ± 0.72 c

Mean ± standard deviation followed by different letters within each column indicate significant differences ($p < 0.05$). SPI: soy protein isolate; SPC: soy protein concentrate; SPM: soy protein mix. ΔE: total color difference of the high-moisture extrudates compared to their unextruded protein powder (SPM). LT: low temperature; HT: high temperature; LS: low speed; MS: medium speed; HS: high speed.

3.4.2. Changes in Protein Secondary Structures upon Extrusion

The changes in the secondary structures of the proteins after extrusion were studied using FTIR. The amide I region, C=O-stretching vibration band at 1600–1700 cm^{-1} was resolved to reveal the secondary structure components present in the protein powders and high-moisture extrudates. Examples of the deconvoluted FTIR spectra are included in the Supplementary Materials (Figure S1). The relative proportions of the secondary structures are shown in Figure 6. Before extrusion treatment, the SPM, the SPC:SPI mixture used for extrusion, denoted a secondary structure more similar to the SPC. the SPM was composed of bands associated with intermolecular protein-aggregated structures and the absorption of amino acid side chains (A1), β-sheets (βS), α-helices (α-H), β-turns (βT) and antiparallel β-sheets (A-β) as the main secondary structure components. These components were in agreement with other results in soy proteins [43,44]. In these protein powders, intermolecular aggregates A1 predominated, followed by β-sheets, while lower proportions of α-helices were visible, in agreement with Carbonaro et al. [43].

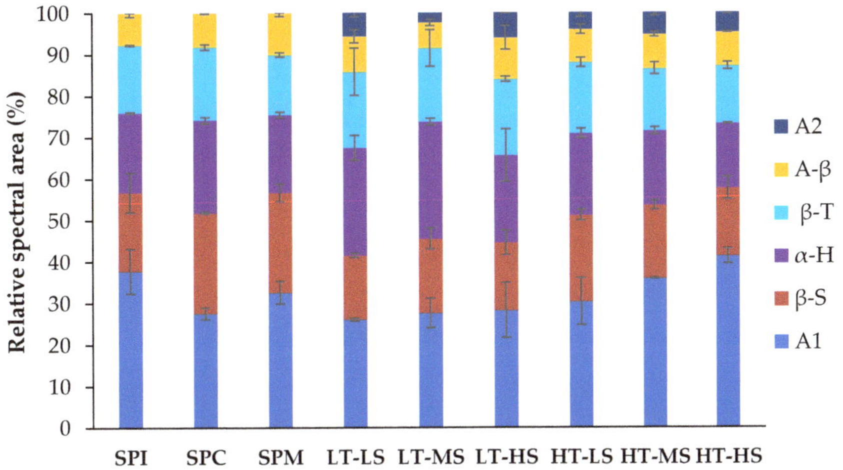

Figure 6. Relative FTIR spectral areas, expressed as percents of the total area, of the secondary structure bands of protein powders and high-moisture extruded meat analogs obtained via the Gaussian/Lorentzian deconvolution of the amide I band (between 1600 and 1700 cm^{-1}). A1: aggregated structures; β-S: β-sheets; α-H: α-helix; β-T: β-turns; A-β: antiparallel β-sheets; A2: aggregated structures; SPI: soy protein isolate; SPC: soy protein concentrate; SPM: soy protein mix; LT: low temperature; HT: high temperature; LS: low speed; MS: medium speed; HS: high speed.

Upon extrusion, a significant reduction in the proportion of the band attributed to β-sheets structures, together with an increase of the A1 band, corresponding to intermolecular aggregates, was generally observed in the extruded meat analogs. Furthermore, a small proportion of a new band (~1690–1695 cm^{-1}), related to intramolecular aggregates (A2), was also visible in all extrudates. Similar results were reported in soy concentrates after low-moisture extrusion [44]. The high temperature would have caused protein molecular unfolding and the formation of new protein structures or aggregates, including those of an intramolecular nature. Still, intermolecular aggregates were more abundant than intramolecular ones after extrusion treatment [44]. Protein molecular weight aggregation in soy protein isolates has been reported as a consequence of extrusion [77]. Regarding processing parameters, results denoted a higher presence of A1 aggregates at HT and HS extrusion conditions, with no clear differences for β-sheets or A2 aggregates among the different extruded meat analogs. Nonetheless, the sum of aggregated structures (A1 + A2) was higher at HT and HS conditions. Beck et al. [78] also reported that intermolecular aggregates in heat-treated pea protein isolates, which could be made of β-turn structures or antiparallel β-sheets, increased during heat treatment when shear force was applied. However, they also reported no clear trend in the proportion of intra-molecular aggregates with processing parameters. Meanwhile, Carbonaro et al. [43] also reported that legume proteins mainly possessed β-sheet and β-turn structures and lower proportions of α-helices before processing, with the loss of the β-sheet contribution in parallel with an increased formation of heat-induced A1 and/or A2 stable aggregates after autoclaving.

3.4.3. Textural Properties of Extruded Meat Analogs

The textural attributes of the high-moisture extrudates obtained in the TPA analysis and cutting test are listed in Table 6. Regarding TPA analyses, hardness values ranged from 86 to 128 N, while smaller differences were observed for resilience, cohesiveness, and springiness, the latest not being significantly different among the extruded meat analogs. The process parameters, screw speed, and barrel temperature had an effect on the textural attributes, with higher hardness values found at lower processing temperatures (LT > HT) and screw speeds (LS > HS). The hardest extrudate was obtained at 120 °C (LT) and 300 rpm (LS) combined conditions. These results are consistent with those obtained by Lin et al. [28], showing that at a fixed moisture content, a higher cooking temperature resulted in softer and less chewy soy protein extrudates, partially related to an increase in protein solubilization. Higher barrel temperatures (140 °C) may result in lower melt viscosities due to the higher molecular mobility of the SPM components [17]. Thus, higher extrusion temperatures and resident times at these high temperatures could have fostered polysaccharides degradation, protein denaturation, and/or the rupture and reformation of new aggregates, in line with the different viscosity results observed when heat-processing the SPM at different maximum temperatures (Figure 3) and with the changes in the protein secondary structure of the extrudates (Figure 6). In fact, the motor toque and melt pressure at the extruder exit (Table 2) were lower at HT (140 °C) and HS (400 rpm) conditions, also supporting the lower hardness of these extruded products, as less viscous melted extruded materials would result in lower melt pressure conditions [17]. For resilience and cohesiveness, small changes were noticed, and only low screw speed conditions seemed to result in more resilient and cohesive extrudates, which may be related to the longer resident times of the melted extruded material inside the heated barrels. Zahari et al. [72] also found that the increase in hardness was generally reflected with an increase in the resilience of the high-moisture extrudates.

Table 6. Textural properties of the high-moisture extruded meat analogs obtained at different screw speeds and barrel temperatures.

Conditions	Hardness	Resilience	Cohesiveness	Springiness	Cutting Force Longitudinal (FL)	Cutting ForceTrans-verse (FT)	FL/FT
	N	%	%	%	N	N	
LT-LS	127.62 ± 8.37 c	0.41 ± 0.03 bc	0.81 ± 0.02 ab	0.78 ± 0.02 a	3.18 ± 0.15 b	3.73 ± 0.25 b	0.85 ± 0.05 c
LT-MS	113.60 ± 3.52 b	0.40 ± 0.03 ab	0.80 ± 0.02 ab	0.78 ± 0.04 a	2.95 ± 0.31 b	3.65 ± 0.22 b	0.81 ± 0.07 c
LT-HS	111.48 ± 1.35 b	0.35 ± 0.04 a	0.78 ± 0.04 a	0.74 ± 0.04 a	2.02 ± 0.08 a	2.46 ± 0.09 a	0.82 ± 0.05 c
HT-LS	109.06 ± 6.97 b	0.47 ± 0.07 c	0.83 ± 0.04 b	0.78 ± 0.04 a	5.23 ± 0.27 d	7.23 ± 0.44 e	0.73 ± 0.05 b
HT-MS	89.25 ± 4.41 a	0.39 ± 0.06 ab	0.76 ± 0.05 a	0.78 ± 0.07 a	4.05 ± 0.16 c	6.31 ± 0.27 d	0.64 ± 0.04 a
HT-HS	86.16 ± 4.90 a	0.37 ± 0.06 ab	0.76 ± 0.05 a	0.78 ± 0.09 a	4.01 ± 0.15 c	5.62 ± 0.24 c	0.72 ± 0.03 b

Mean ± standard deviation followed by different letters within each column indicate significant differences ($p < 0.05$). FL: longitudinal; FT: transverse; LT: low temperature; HT: high temperature; LS: low speed; MS: medium speed; HS: high speed.

The cutting force or cutting strength, defined as the energy required to cut the sample, can be used to analyze the extent of fiber formation in the extrudates if cutting is performed parallel (L) and perpendicular (T) to the flow direction of the extrudates in the cooling die. The fibrous degree is also referred to as the anisotropic index (FL/FT) [27]. Equal values of FL = FT indicate a uniform isotropic texture, while FT > FL or FL > FT values indicate the predominance of longitudinal or transverse fibers, respectively [79]. The effects of screw speed and temperature on the cutting strength and degree of texturization are reported in Table 6 In all cases, transversal cutting force was higher than the longitudinal one, denoting the presence of anisotropic structures (FT > FL), with fibers oriented longitudinal to the flow direction. This would agree with the macroscopic observations of fiber directions in

Figure 4. Values obtained for the cutting force lengthwise were between 2.02 and 5.23 N, and for the transverse force were between 2.46 and 7.23 N. The degree of texturization (FL/FT) was lower for low-temperature conditions (120 °C), as values closer to 1 were obtained, and hence, were closer to more isotropic structures (FL = FT). However, no clear effect for screw speed conditions was observed, which was also in agreement with visual observations. According to Bouvier and Campanella [80], an increase in melt viscosity results in a flatter velocity profile in the cooling die channel, while lowering the viscosity (i.e., by increasing cooking temperature) has the opposite effect, resulting in an extended velocity profile, facilitating the development of flow lines within the melt that, upon cooling, results in elongated fiber formation. This would be in line with the lower pressure (indicative of a lower melt viscosity) observed at HT conditions, when more fibrousness was observed.

3.4.4. Reduction of Trypsin Inhibitors upon Extrusion

Figure 7 shows the trypsin inhibitor (TI) content of soy protein powders and high-moisture extruded meat analogs obtained at different screw speed and temperature conditions. Regarding protein powders, SPC showed significantly higher trypsin inhibitor contents than SPI, with values of 7.4 and 4.6 mg/g (14.1 and 8.7 U/mg), respectively. Likewise, previous research reported values in the range of 7.3–5.4 mg/g for an SPC and 3.6–7.1 mg/g for an SPI [81]. Similarly, other studies have reported a 70–90% reduction in trypsin inhibitor concentrations or activity during the concentration and purification of protein from soybean meals [22,23].

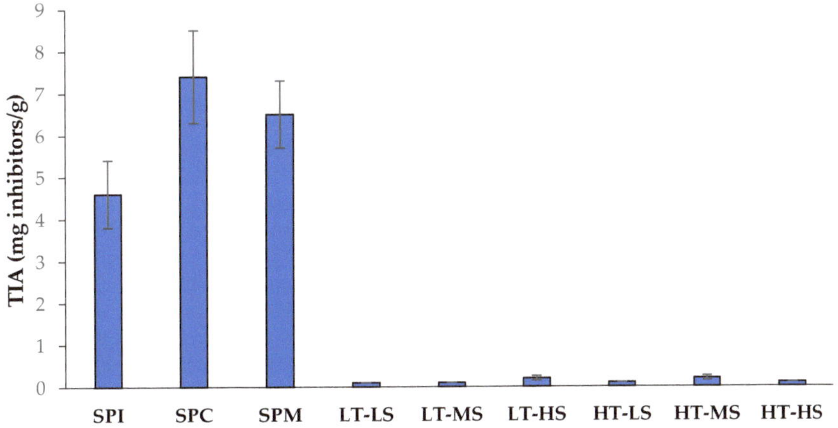

Figure 7. Trypsin inhibitors (TIA, mg inhibitors/g solids of product, in a wet basis) for protein powders and extruded meat analogs produced at different temperatures and screw speed conditions. SPI: soy protein isolate; SPC: soy protein concentrate; SPM: soy protein mix; LT: low temperature; HT: high temperature; LS: low speed; MS: medium speed; HS: high speed.

All high-moisture extruded products showed a decrease in the amount of trypsin inhibitors as compared to the SPM (6.5 to 0.1–0.2 mg/g, w.b.). This effect was not only the result of the dilution of the SPM in the extruded product as a consequence of the high-water content used for extrusion (~70%), but was also the result of the high thermal instability of trypsin inhibitors [82,83]. Thus, considering SPM concentration in the extruded meat analogs (i.e., correcting to their solid contents), trypsin inhibitors were reduced by more than 90% after extrusion treatments. The combination of high temperatures, pressure, and mechanical forces during the extrusion process leads to certain non-reversible structural changes in trypsin inhibitor conformations, leading to their denaturation, and the loss of inhibitory enzymatic activity, consequently improving the nutritional quality of the final products [82,84]. Likewise, other authors have reported ~90% reductions in trypsin inhibitor activities after the low-moisture extrusion of breadfruit–corn–soy composites, with no significant effect of screw speed on the TIA [25]. Conversely, in another study at low-moisture conditions, Petres and Czukor [26] reported the destruction of trypsin inhibitors in soybean flour as a function of temperature, screw speed (residence time), and moisture level. However, in our study, performed in high-moisture conditions, it seems that the processing variables selected were severe enough to achieve a high degree of removal of the trypsin inhibitors, not finding significant differences among the studied parameters. In line with these results, other authors have reported that higher temperatures are needed to inactivate TIs under low-moisture conditions, being that higher moisture conditions are recommended to effectively remove TIs, especially if low extrusion temperatures are applied (<130 °C) [24,85].

4. Conclusions

To better understand the effects of extrusion processing parameters on the textural and nutritional quality of high-moisture extruded meat analogs, this study examined the micro- and macrostructure, textural attributes, color, protein secondary structure, and trypsin inhibitors of soy protein-based extrudates. The high-moisture extrudates were produced at two different barrel temperature profiles and three screw speeds from a well-characterized mixture (SPM) of a soy protein isolate (SPI) and concentrate (SPC). SPC presented a lower extent of protein denaturation and aggregation (observed in DSC and FTIR), a lower least gelation concentration, and higher total essential amino acid contents compared to the SPI. Nonetheless, the SPI exhibited a higher gel strength after gelation in a heating–cooling cycle, and enhanced foaming and emulsifying properties, which may be related to its higher protein purity (>90%). Therefore, the combined SPM used for extrusion exhibited enhanced functional properties compared to those of SPC and SPI, respectively.

During high-moisture extrusion of soy protein, a high level of both visual and instrumental anisotropy was observed in all extruded meat analogs, especially when higher extrusion temperatures were selected. The higher level of anisotropy at higher extrusion temperatures could be related to a greater level of denaturation of the 11S or the molecular disassembling of protein aggregates in the raw material, which could then realign more efficiently via the directional shear force in the cooling die, creating more fibrous structures. These fibers would presumably be stabilized by increased protein aggregation, as suggested by the more effective reassociation visible in the FTIR, by increasing the level of inter- and intramolecular aggregates (A1 and A2 structures) reformed in these extrudates. It is noteworthy that both process parameters, screw speed and temperature, greatly influenced both the textural attributes, with softness increasing at higher temperatures and higher speeds, and the color attributes, with darker extrudates at higher temperatures and lower screw speeds (i.e., longer resident times). Regarding the antinutritional factors, trypsin inhibitors were significantly reduced after extrusion (>90% reduction), without any clear effect of the temperature or speed parameters, which may be related to the thermolabile nature of this antinutrient and the high temperatures used during extrusion (>120 °C). Findings indicated that the quality of the extrudates was heavily influenced by the selected extrusion parameters, highlighting the importance of carefully selecting extrusion conditions to achieve meat analogs with optimal textural and nutritional characteristics. Thus, higher temperatures are recommended to obtain a greater fibrousness for soy protein extruded products, while the effect of screw speed on fibrousness was less noticeable, although differences in resident times (screw speeds) influenced the hardness and color attributes of the product.

Supplementary Materials: The following supporting information can be downloaded at: https://www.mdpi.com/article/10.3390/foods13111748/s1, Table S1: Peak maximum viscosity and final viscosity obtained from the pasting profiles of SPI, SPC and SPM at 95 °C, and for SPM at 120 °C and 140 °C; Figure S1: FTIR-ATR spectra of the soy protein mix and extruded soy.

Author Contributions: Conceptualization, G.R. and L.R.; methodology, G.R., M.-Y.P., F.P. and L.R.; formal analysis, G.R., M.-Y.P. and F.P.; investigation, G.R., M.-Y.P., F.P. and L.R.; data curation, G.R.,

F.P. and L.R.; writing—original draft, G.R. and L.R.; writing—review and editing, G.R., M.-Y.P., F.P., B.B. and L.R.; visualization, L.R. and B.B.; supervision, L.R. and B.B.; project administration, M.-Y.P. and L.R.; funding acquisition, M.-Y.P., B.B. and L.R. All authors have read and agreed to the published version of the manuscript.

Funding: This research was funded by the European Regional Development Fund (FEDER) of the European Union and the Board of Castile and León, through the Instituto para la Competitividad Empresarial de Castilla y León (ICE), "Proyecto MEATING PLANTS" (Grant number CCTT4/20/0003) and the "Proyecto Generación de Conocimiento" (Grant number PID2022-139425OA-I00) financed by MICIU/AEI/10.13039/501100011033 and FEDER, UE. L.R. would like to acknowledge the financial support from the Ramon y Cajal funding (RYC2021-032926-I) granted by the MICIU/AEI/10.13039/501100011033 and the European Union "NextGenerationEU/PRTR".

Data Availability Statement: Data are contained within the article or in the Supplementary Materials. Further inquiries can be directed to the corresponding author.

Acknowledgments: Some of the results from this research were obtained through accessing the research infrastructure at the Laboratory of Instrumental Techniques at the University of Valladolid.

Conflicts of Interest: The authors declare no conflicts of interest. All authors have read and agreed to the published version of the manuscript. The funders had no role in the design of the study; in the collection, analyses, or interpretation of data; in the writing of the manuscript; or in the decision to publish the results.

References

1. Sha, L.; Xiong, Y.L. Plant Protein-Based Alternatives of Reconstructed Meat: Science, Technology, and Challenges. *Trends Food Sci. Technol.* **2020**, *102*, 51–61. [CrossRef]
2. Springmann, M.; Clark, M.; Mason-D'Croz, D.; Wiebe, K.; Bodirsky, B.L.; Lassaletta, L.; de Vries, W.; Vermeulen, S.J.; Herrero, M.; Carlson, K.M.; et al. Options for Keeping the Food System within Environmental Limits. *Nature* **2018**, *562*, 519–525. [CrossRef] [PubMed]
3. Fu, L.; Wang, Z.; Adhikari, B.; He, Z.; Zeng, M.; Qin, F.; Chen, J. Soluble and Insoluble Fractions of Soy Protein Isolate Affect the Properties of Its High-Moisture Extrudates. *Food Biosci.* **2024**, *59*, 103850. [CrossRef]
4. Wittek, P.; Zeiler, N.; Karbstein, H.P.; Emin, M.A. High Moisture Extrusion of Soy Protein: Investigations on the Formation of Anisotropic Product Structure. *Foods* **2021**, *10*, 102. [CrossRef]
5. Mateen, A.; Mathpati, M.; Singh, G. A Study on High Moisture Extrusion for Making Whole Cut Meat Analogue: Characterization of System, Process and Product Parameters. *Innov. Food Sci. Emerg. Technol.* **2023**, *85*, 103315. [CrossRef]
6. Ryu, G.-H. Extrusion cooking of high-moisture meat analogues. In *Extrusion Cooking*; Woodhead Publishing: Cambridge, UK, 2020; pp. 205–224. [CrossRef]
7. Nasrollahzadeh, F.; Alexi, N.; Skov, K.B.; Roman, L.; Sfyra, K.; Martinez, M.M. Texture Profiling of Muscle Meat Benchmarks and Plant-Based Analogues: An Instrumental and Sensory Design Approach with Focus on Correlations. *Food Hydrocoll.* **2024**, *151*, 109829. [CrossRef]
8. Day, L.; Swanson, B.G. Functionality of Protein-Fortified Extrudates. *Compr. Rev. Food Sci. Food Saf.* **2013**, *12*, 546–564. [CrossRef] [PubMed]
9. Nasrollahzadeh, F.; Roman, L.; Skov, K.; Jakobsen, L.M.A.; Trinh, B.M.; Tsochatzis, E.D.; Mekonnen, T.; Corredig, M.; Dutcher, J.R.; Martinez, M.M. A Comparative Investigation of Seed Storage Protein Fractions: The Synergistic Impact of Molecular Properties and Composition on Anisotropic Structuring. *Food Hydrocoll.* **2023**, *137*, 108400. [CrossRef]
10. Guerrero, P.; Beatty, E.; Kerry, J.P.; De La Caba, K. Extrusion of Soy Protein with Gelatin and Sugars at Low Moisture Content. *J. Food Eng.* **2012**, *110*, 53–59. [CrossRef]
11. Wei, Y.; Kang, L.; Zhang, C. *Food Extrusion Theory and Technology*, 2nd ed.; China Light Industry Press: Beijing, China, 2009.
12. Nasrollahzadeh, F.; Roman, L.; Swaraj, V.J.S.; Ragavan, K.V.; Vidal, N.P.; Dutcher, J.R.; Martinez, M.M. Hemp (*Cannabis sativa* L.) Protein Concentrates from Wet and Dry Industrial Fractionation: Molecular Properties, Nutritional Composition, and Anisotropic Structuring. *Food Hydrocoll.* **2022**, *131*, 107755. [CrossRef]
13. Kyriakopoulou, K.; Dekkers, B.; van der Goot, A.J. *Plant-Based Meat Analogues*; Elsevier Inc.: Amsterdam, The Netherlands, 2019. [CrossRef]
14. McClements, I.F.; McClements, D.J. Designing Healthier Plant-Based Foods: Fortification, Digestion, and Bioavailability. *Food Res. Int.* **2023**, *169*, 112853. [CrossRef]
15. Webb, D.; Dogan, H.; Li, Y.; Alavi, S. Physico-Chemical Properties and Texturization of Pea, Wheat and Soy Proteins Using Extrusion and Their Application in Plant-Based Meat. *Foods* **2023**, *12*, 1586. [CrossRef] [PubMed]
16. Bakhsh, A.; Lee, E.Y.; Ncho, C.M.; Kim, C.J.; Son, Y.M.; Hwang, Y.H.; Joo, S.T. Quality Characteristics of Meat Analogs through the Incorporation of Textured Vegetable Protein: A Systematic Review. *Foods* **2022**, *11*, 1242. [CrossRef]

17. Pietsch, V.L.; Bühler, J.M.; Karbstein, H.P.; Emin, M.A. High Moisture Extrusion of Soy Protein Concentrate: Influence of Thermomechanical Treatment on Protein-Protein Interactions and Rheological Properties. *J. Food Eng.* **2019**, *251*, 11–18. [CrossRef]
18. Cheftel, J.C.; Kitagawa, M.; Queguiner, C. New Protein Texturization Processes by Extrusion Cooking at High Moisture Levels. *Food Rev. Int.* **1992**, *8*, 235–275. [CrossRef]
19. Fang, Y.; Zhang, B.; Wei, Y. Effects of the Specific Mechanical Energy on the Physicochemical Properties of Texturized Soy Protein during High-Moisture Extrusion Cooking. *J. Food Eng.* **2014**, *121*, 32–38. [CrossRef]
20. Grabowska, K.J.; Zhu, S.; Dekkers, B.L.; De Ruijter, N.C.A.; Gieteling, J.; Van Der Goot, A.J. Shear-Induced Structuring as a Tool to Make Anisotropic Materials Using Soy Protein Concentrate. *J. Food Eng.* **2016**, *188*, 77–86. [CrossRef]
21. Sandoval Murillo, J.L.; Osen, R.; Hiermaier, S.; Ganzenmüller, G. Towards Understanding the Mechanism of Fibrous Texture Formation during High-Moisture Extrusion of Meat Substitutes. *J. Food Eng.* **2019**, *242*, 8–20. [CrossRef]
22. Coscueta, E.R.; Pellegrini Malpiedi, L.; Pintado, M.M.; Nerli, B.B. Production of Soy Protein Concentrate with the Recovery of Bioactive Compounds: From Destruction to Valorization. *Food Hydrocoll.* **2023**, *137*, 108314. [CrossRef]
23. Waggle, D.H.; Steinke, F.H.; Shen, J.L. Isolated soy proteins. Ch. 4. In *Legumes*; Matthews, R.H., Ed.; Marcel Dekker: New York, NY, USA, 1989; pp. 99–138.
24. Clarke, E.; Wiseman, J. Effects of Extrusion Conditions on Trypsin Inhibitor Activity of Full Fat Soybeans and Subsequent Effects on Their Nutritional Value for Young Broilers. *Br. Poult. Sci.* **2007**, *48*, 703–712. [CrossRef]
25. Nwabueze, T.U. Effect of Process Variables on Trypsin Inhibitor Activity (TIA), Phytic Acid and Tannin Content of Extruded African Breadfruit-Corn-Soy Mixtures: A Response Surface Analysis. *LWT* **2007**, *40*, 21–29. [CrossRef]
26. Petres, J.; Czukor, B. Investigation of the Effects of Extrusion Cooking on Antinutritional Factors in Soybeans Employing Response Surface Analysis Part 1. Effect of Extrusion Cooking on Trypsin-inhibitor Activity. *Food/Nahrung.* **1989**, *33*, 275–281. [CrossRef]
27. Chen, F.L.; Wei, Y.M.; Zhang, B.; Ojokoh, A.O. System Parameters and Product Properties Response of Soybean Protein Extruded at Wide Moisture Range. *J. Food Eng.* **2010**, *96*, 208–213. [CrossRef]
28. Lin, S.; Huff, H.E.; Hsieh, F. Texture and Chemical Characteristics of Soy Protein Meat Analog Extruded at High Moisture. *J. Food Sci.* **2000**, *65*, 264–269. [CrossRef]
29. Dekkers, B.L.; Nikiforidis, C.V.; van der Goot, A.J. Shear-induced fibrous structure formation from a pectin/SPI blend. *Innov. Food Sci. Emerg. Technol.* **2016**, *36*, 193–200. [CrossRef]
30. AOAC. AOAC Official Method 934.01: Proximate analysis and calculations moisture. In *Official Methods of Analysis of AOAC International*; AOAC International: Washington, DC, USA, 2002.
31. AOAC. AOAC Official Method 923.03: Ash of Flour. Gravimetric. In *Official Methods of Analysis of AOAC International*; AOAC International: Washington, DC, USA, 2002.
32. AOAC. AOAC Official Method 991.43: Total, soluble and insoluble dietary fiber in foods. Gravimetric. In *Official Methods of Analysis of AOAC International*; AOAC International: Washington, DC, USA, 2002.
33. AOAC. AOAC Official Method 920.39: Fat (crude) or ether extract in animal feed. Gravimetric. In *Official Methods of Analysis of AOAC International*; AOAC International: Washington, DC, USA, 2002.
34. AACC International. *Approved Methods of the American Association of Cereal Chemists, Methods 46–30.01 (Protein)*, 11th ed.; American Association of Cereal Chemists: St. Paul, MN, USA, 2015.
35. Laemmli. *Cleavage of Structural Protein during the Assembly of the Head of Bacteriophage T4*; Nature Publishing Group: London, UK, 1970; Volume 227, pp. 680–685. [CrossRef]
36. Guideline 98/64/EG (1998) of the European Union: Commission Directive 98/64/EC of 3 September 1998 Establishing Community Methods of Analysis for the Determination of Amino-Acids, Crude Oils and Fats, and Olaquindox in Feedingstuffs and Amending Directive 71/393/EEC (Text with EEA Relevance) (OJ L 257 19.09.1998, p. 14, ELI). Available online: http://data.europa.eu/eli/dir/1998/64/oj (accessed on 5 April 2024).
37. AACC International. *Approved Methods of the American Association of Cereal Chemists International*, 11th ed.; Methods: 56-30.01 (WBC); American Association of Cereal Chemists: St. Paul, MN, USA, 2012.
38. Lin, M.J.Y.; Humbert, E.S.; Sosulski, F.W. Functional Properties Sunflower Meal Products of. *J. Food Sci.* **1974**, *39*, 5–7. [CrossRef]
39. Adebiyi, A.P.; Aluko, R.E. Functional Properties of Protein Fractions Obtained from Commercial Yellow Field Pea (*Pisum sativum* L.) Seed Protein Isolate. *Food Chem.* **2011**, *128*, 902–908. [CrossRef]
40. Pearce, K.N.; Kinsella, J.E. Emulsifying Properties of Proteins: Evaluation of a Turbidimetric Technique. *J. Agric. Food Chem.* **1978**, *26*, 716–723. [CrossRef]
41. Karaca, A.C.; Low, N.; Nickerson, M. Emulsifying Properties of Chickpea, Faba Bean, Lentil and Pea Proteins Produced by Isoelectric Precipitation and Salt Extraction. *Food Res. Int.* **2011**, *44*, 2742–2750. [CrossRef]
42. Moore, E.; Jennifer, M.C.; Hallin, M. Physical Testing—Food RVA Method 50.01 Pulse Flour and Pulse Starch Method. Perkin Elmer, Inc.: Waltham, MA, USA, 2022. Available online: www.perkinelmer.com (accessed on 30 May 2024).
43. Carbonaro, M.; Maselli, P.; Nucara, A. Relationship between Digestibility and Secondary Structure of Raw and Thermally Treated Legume Proteins: A Fourier Transform Infrared (FT-IR) Spectroscopic Study. *Amino Acids* **2012**, *43*, 911–921. [CrossRef]
44. Vidal, N.P.; Roman, L.; Swaraj, V.J.S.; Ragavan, K.V.; Simsek, S.; Rahimi, J.; Kroetsch, B.; Martinez, M.M. Enhancing the Nutritional Value of Cold-Pressed Oilseed Cakes through Extrusion Cooking. *Innov. Food Sci. Emerg. Technol.* **2022**, *77*, 102956. [CrossRef]
45. Wang, C.; Jiang, L.; Wei, D.; Li, Y.; Sui, X.; Wang, Z.; Li, D. Effect of Secondary Structure Determined by FTIR Spectra on Surface Hydrophobicity of Soybean Protein Isolate. *Procedia Eng.* **2011**, *15*, 4819–4827. [CrossRef]

46. Fevzioglu, M.; Ozturk, O.K.; Hamaker, B.R.; Campanella, O.H. Quantitative Approach to Study Secondary Structure of Proteins by FT-IR Spectroscopy, Using a Model Wheat Gluten System. *Int. J. Biol. Macromol.* **2020**, *164*, 2753–2760. [CrossRef] [PubMed]
47. Singh, B.R. Basic Aspects of the Technique and Applications of Infrared Spectroscopy of Peptides and Proteins. *ACS Symp. Ser.* **2000**, *750*, 2–37. [CrossRef]
48. Osen, R.; Toelstede, S.; Wild, F.; Eisner, P.; Schweiggert-Weisz, U. High Moisture Extrusion Cooking of Pea Protein Isolates: Raw Material Characteristics, Extruder Responses, and Texture Properties. *J. Food Eng.* **2014**, *127*, 67–74. [CrossRef]
49. *ISO 14902:2001*; ISO Method 14902—Animal Feeding Stuffs—Determination of Trypsin Inhibitor Activity of Soya Products, n.d. ISO: Geneva, Switzerland, 2001.
50. Bainy, E.M.; Tosh, S.M.; Corredig, M.; Poysa, V.; Woodrow, L. Varietal Differences of Carbohydrates in Defatted Soybean Flour and Soy Protein Isolate By-Products. *Carbohydr. Polym.* **2008**, *72*, 664–672. [CrossRef]
51. Ning, L.; Villota, R. Influence of 7S and 11S globulins on the extrusion performance of soy protein concentrates. *J. Food Process. Preserv.* **1994**, *18*, 421–436. [CrossRef]
52. Liu, Q.; Geng, R.; Zhao, J.; Chen, Q.; Kong, B. Structural and Gel Textural Properties of Soy Protein Isolate When Subjected to Extreme Acid PH-Shifting and Mild Heating Processes. *J. Agric. Food Chem.* **2015**, *63*, 4853–4861. [CrossRef] [PubMed]
53. Flory, J.; Alavi, S. Use of Hydration Properties of Proteins to Understand Their Functionality and Tailor Texture of Extruded Plant-Based Meat Analogues. *J. Food Sci.* **2024**, *89*, 245–258. [CrossRef]
54. Xia, W.; Pan, S.; Cheng, Z.; Tian, Y.; Huang, X. High-Intensity Ultrasound Treatment on Soy Protein after Selectively Proteolyzing Glycinin Component: Physical, Structural, and Aggregation Properties. *Foods* **2020**, *9*, 839. [CrossRef]
55. Fernández-Quintela, A.; Macarulla, M.T.; Del Barrio, A.S.; Martínez, J.A. Composition and Functional Properties of Protein Isolates Obtained from Commercial Legumes Grown in Northern Spain. *Plant Foods Hum. Nutr.* **1997**, *51*, 331–341. [CrossRef] [PubMed]
56. Mateen, A.; Singh, G. Evaluating the Potential of Millets as Blend Components with Soy Protein Isolate in a High Moisture Extrusion System for Improved Texture, Structure, and Colour Properties of Meat Analogues. *Food Res. Int.* **2023**, *173*, 113395. [CrossRef] [PubMed]
57. van den Berg, L.A.; Mes, J.J.; Mensink, M.; Wanders, A.J. Protein Quality of Soy and the Effect of Processing: A Quantitative Review. *Front. Nutr.* **2022**, *9*, 1004754. [CrossRef] [PubMed]
58. Ma, C.Y. Soybean: Soy Concentrates and Isolates. In *Encyclopedia of Food Grains*, 2nd ed.; Elsevier: Amsterdam, The Netherlands, 2015; Volumes 3–4, pp. 482–488. [CrossRef]
59. FAO. *Dietary Protein Quality Evaluation in Human Nutrition. Report of an FAO Expert Consultation*; FAO: Rome, Italy, 2013; Volume 92.
60. Tang, C.H.; Choi, S.M.; Ma, C.Y. Study of Thermal Properties and Heat-Induced Denaturation and Aggregation of Soy Proteins by Modulated Differential Scanning Calorimetry. *Int. J. Biol. Macromol.* **2007**, *40*, 96–104. [CrossRef] [PubMed]
61. Lee, K.H.; Ryu, H.S.; Rhee, K.C. Protein Solubility Characteristics of Commercial Soy Protein Products. *JAOCS J. Am. Oil Chem. Soc.* **2003**, *80*, 85–90. [CrossRef]
62. Li, S.; Wei, Y.; Fang, Y.; Zhang, W.; Zhang, B. DSC Study on the Thermal Properties of Soybean Protein Isolates/Corn Starch Mixture. *J. Therm. Anal. Calorim.* **2014**, *115*, 1633–1638. [CrossRef]
63. Damodaran, S. Refolding of Thermally Unfolded Soy Proteins during the Cooling Regime of the Gelation Process: Effect on Gelation. *J. Agric. Food Chem.* **1988**, *36*, 262–269. [CrossRef]
64. Tang, X.; Shen, Y.; Zhang, Y.; Schilling, M.W.; Li, Y. Parallel Comparison of Functional and Physicochemical Properties of Common Pulse Proteins. *LWT* **2021**, *146*, 111594. [CrossRef]
65. Zhang, L.; Li, Q.; Zhang, W.; Bakalis, S.; Luo, Y.; Lametsch, R. Different Source of Commercial Soy Protein Isolates: Structural, Compositional, and Physicochemical Characteristics in Relation to Protein Functionalities. *Food Chem.* **2024**, *433*, 137315. [CrossRef]
66. Farooq, Z.; Boye, J.I. Novel Food and Industrial Applications of Pulse Flours and Fractions. In *Pulse Foods*; Academic Press: Cambridge, MA, USA, 2011; pp. 283–323. [CrossRef]
67. Day, L. Proteins from Land Plants—Potential Resources for Human Nutrition and Food Security. *Trends Food Sci. Technol.* **2013**, *32*, 25–42. [CrossRef]
68. Du, S.-K.; Jiang, H.; Yu, X.; Jane, J.-L. Physicochemical and Functional Properties of Whole Legume Flour. *LWT* **2014**, *55*, 308–313. [CrossRef]
69. Keerati-U-Rai, M.; Corredig, M. Heat-Induced Changes Occurring in Oil/Water Emulsions Stabilized by Soy Glycinin and β-Conglycinin. *J. Agric. Food Chem.* **2010**, *58*, 9171–9180. [CrossRef]
70. Kinsella, J.E. Functional Properties of Soy Proteins. *J. Am. Oil Chem. Soc.* **1979**, *56*, 242–258. [CrossRef]
71. Wang, C.; Johnson, L.A. Functional Properties of Hydrothermally Cooked Soy Protein Products. *JAOCS J. Am. Oil Chem. Soc.* **2001**, *78*, 189–195. [CrossRef]
72. Zahari, I.; Ferawati, F.; Helstad, A.; Ahlström, C.; Östbring, K.; Rayner, M.; Purhagen, J.K. Development of High-Moisture Meat Analogues with Hemp and Soy Protein Using Extrusion Cooking. *Foods* **2020**, *9*, 772. [CrossRef]
73. McClements, D.J.; Weiss, J.; Kinchla, A.J.; Nolden, A.A.; Grossmann, L. Methods for Testing the Quality Attributes of Plant-Based Foods: Meat-and Processed-Meat Analogs. *Foods* **2021**, *10*, 260. [CrossRef]

74. Dekkers, B.L.; Boom, R.M.; van der Goot, A.J. Structuring Processes for Meat Analogues. *Trends Food Sci. Technol.* **2018**, *81*, 25–36. [CrossRef]
75. Chiang, A. Protein-Protein Interaction of Soybean Protein from Extrusion Processing. Master's Thesis, University of Missouri-Columbia, Columbia, MO, USA, 2007.
76. Wang, H.; van den Berg, F.W.J.; Zhang, W.; Czaja, T.P.; Zhang, L.; Jespersen, B.M.; Lametsch, R. Differences in Physicochemical Properties of High-Moisture Extrudates Prepared from Soy and Pea Protein Isolates. *Food Hydrocoll.* **2022**, *128*, 107540. [CrossRef]
77. Fang, Y.; Zhang, B.; Wei, Y.; Li, S. Effects of Specific Mechanical Energy on Soy Protein Aggregation during Extrusion Process Studied by Size Exclusion Chromatography Coupled with Multi-Angle Laser Light Scattering. *J. Food Eng.* **2013**, *115*, 220–225. [CrossRef]
78. Beck, S.M.; Knoerzer, K.; Sellahewa, J.; Emin, M.A.; Arcot, J. Effect of Different Heat-Treatment Times and Applied Shear on Secondary Structure, Molecular Weight Distribution, Solubility and Rheological Properties of Pea Protein Isolate as Investigated by Capillary Rheometry. *J. Food Eng.* **2017**, *208*, 66–76. [CrossRef]
79. Osen, R.J.J. Texturization of Pea Protein Isolates Using High Moisture Extrusion Cooking. Ph.D. Thesis, Technische Universität München, Munchen, Germany, 2017.
80. Bouvier, J.; Campanella, O.H. *Extrusion Processing Technology*; John Wiley & Sons, Ltd.: Hoboken, NJ, USA, 2014. [CrossRef]
81. Peace, R.W.; Sarwar, G.; Touchburn, S.P. Trypsin Inhibitor Levels in Soy-Based Infant Formulas and Commercial Soy Protein Isolates and Concentrates. *Food Res. Int.* **1992**, *25*, 137–141. [CrossRef]
82. Avilés-Gaxiola, S.; Chuck-Hernández, C.; Serna Saldívar, S.O. Inactivation Methods of Trypsin Inhibitor in Legumes: A Review. *J. Food Sci.* **2018**, *83*, 17–29. [CrossRef] [PubMed]
83. Vanga, S.K.; Wang, J.; Raghavan, V. Effect of Ultrasound and Microwave Processing on the Structure, in-Vitro Digestibility and Trypsin Inhibitor Activity of Soymilk Proteins. *LWT* **2020**, *131*, 109708. [CrossRef]
84. Nikbakht Nasrabadi, M.; Sedaghat Doost, A.; Mezzenga, R. Modification Approaches of Plant-Based Proteins to Improve Their Techno-Functionality and Use in Food Products. *Food Hydrocoll.* **2021**, *118*, 106789. [CrossRef]
85. Rackis, J.; Honig, D.H.; Sessa, D.J.; Moser, H.A. Lipoxygenase and Peroxidase Activities of Soybeans as Related to the Flavor Profile During Maturation. *Cereal Chem.* **1972**, *49*, 586–597.

foods

MDPI

Article

Storage Stability of Plant-Based Drinks Related to Proteolysis and Generation of Free Amino Acids

Ida Schwartz Roland [1], Thao T. Le [2], Tony Chen [2], Miguel Aguilera-Toro [1], Søren Drud-Heydary Nielsen [1], Lotte Bach Larsen [1] and Nina Aagaard Poulsen [1,*]

[1] Department of Food Science, Aarhus University, Agro Food Park 48, DK-8200 Aarhus, Denmark
[2] Department of Food Science and Microbiology, Auckland University of Technology, Auckland 1010, New Zealand
* Correspondence: nina.poulsen@food.au.dk

Abstract: The market for plant-based drinks (PBDs) is experiencing a surge in consumer demand, especially in Western societies. PBDs are a highly processed food product, and little is known about this relatively new food product category when compared to bovine milk. In the present study, the storage stability, proteolysis and generation of free amino acids were investigated in commercially available PBDs over the course of a one-year storage period. Generally, pH, color and protein solubility were found to be stable in the PBDs during storage, except for the pea-based product, which showed less protein solubility after storage. The pea-based drinks also had higher initial levels of free N-terminals prior to storage compared with levels for the other plant-based drinks, as well as significantly increasing levels of total free, and especially bitter free, amino acids. The development of free amino acids in the oat-based drink indicated that the released amino acids could be involved in various reactions such as the Maillard reaction during the storage period.

Keywords: plant-based drinks; storage; free amino acids; proteolysis; color; shelf-life

Citation: Roland, I.S.; Le, T.T.; Chen, T.; Aguilera-Toro, M.; Nielsen, S.D.-H.; Larsen, L.B.; Poulsen, N.A. Storage Stability of Plant-Based Drinks Related to Proteolysis and Generation of Free Amino Acids. *Foods* **2024**, *13*, 367. https://doi.org/10.3390/foods13030367

Academic Editor: Aimin Shi

Received: 8 December 2023
Revised: 10 January 2024
Accepted: 22 January 2024
Published: 23 January 2024

1. Introduction

Over the past decade, there has been a growing interest in and consumption of plant-based drinks (PBDs), driven by factors, such as lactose intolerance and allergy towards bovine milk, ethical reasons including animal welfare as well as environmental reasons such as lower carbon footprint connected to PBDs compared to bovine milk consumption [1–4]. PBDs refer to beverages made from various plant sources milled with water that resemble a milk-like texture and appearance, and they are therefore often marketed as milk replacers [5]. From a nutritional point of view, PBDs lack quality compared to bovine milk in terms of protein content and levels of essential amino acids (AAs) as well as vitamins and minerals [5,6]. However, some PBDs hold other health-promoting properties such as containing dietary fibers and being low in cholesterol [2,7] and, therefore, replacing one with another can be difficult [6].

The most predominant plant sources for PBDs are oat, almond, soy and rice, with oat-based products being the most predominant variant in the Danish market based on the number of commercial products available [8]. PBDs are highly processed food products and the processing steps usually include milling, filtration, homogenization and heat treatment [9]. Ultra-high-temperature (UHT) treatments are usually a part of the processing of PBDs to ensure food safety, but this severe treatment can produce unwanted changes in the food matrix [10]. These changes include molecular processing modifications on the present proteins and carbohydrates [8] and could be expected to develop further during storage periods, as seen in bovine UHT milk [11]. The processing-induced changes in UHT milk are known to influence nutritional quality, lead to the development of off-flavors and color changes, e.g., from the Maillard reaction, as well as potentially leading to sedimentation or age gelation [12,13]. Age gelation happens under the presence of heat-stable proteases that hydrolyze the milk proteins during storage, ultimately leading to the gelation of milk [13]. This mechanism is related to the unique properties of milk proteins, but proteolysis in PBDs might also occur during long-term storage, especially when products are stored outside the cold chain [12]. Proteolysis and the generation of free AAs could potentially also occur in PBDs and thereby influence the taste and odor of the products. This could happen both via their properties as free AAs i.e., sweet or bitter taste [14], and as part of Maillard reactions involving free AAs as observed in milk systems [11]. Free AAs as well as intact proteins can react with reducing sugars during the initial stage of the Maillard reaction, and they can play a role in the advanced stage forming Strecker aldehydes, leading to unwanted properties in the product [11]. These parameters, which can be particularly affected by the UHT treatment, i.e., gelation, proteolysis, color and off-flavor development, have not been studied as extensively in plant-based systems. Therefore, the aim of this paper was to investigate the storage stability in commercial PBDs in terms of pH and particle size, as well as color development, protein solubility, proteolysis and free AA content over a storage period of one year. We expected to observe an increase in the level of free AAs as well as in browning through color changes in PBDs that could be correlated with proteolysis and the Maillard reaction. For this study, commercially available PBDs on the Danish market were used.

2. Materials and Methods

2.1. Samples of Plant-Based Drinks and Experimental Set-Up

Seven different commercially available PBDs were selected for this study. The plant sources of the drinks comprised one oat, one oat/hemp in combination, two almond drinks, one pea, one soy and one soy/rice combination (Table 1). The drinks were selected to represent the Danish market of PBDs, as well as a variety of plant sources for PBDs. Declared nutritional contents and shelf-life of the studied drinks are provided in Table 1, and product names and manufacturers are listed in Supplementary Materials Table S1. All drinks were produced within 1–3 weeks prior to the start of the experiment.

Table 1. The nutritional composition of the studied PBDs, as stated on their packaging.

	Oat	Oat/Hemp	Almond	Roasted Almond	Pea	Soy	Soy/Rice
Energy (kJ/kcal)	219/52	205/48	87/21	89/21	142/34	147/35	231/55
Protein (g/100 g)	0.4	1.1	0.5	0.4	2	3.7	3.5
Fat (g/100 g)	1.9	1.0	1.2	0.9	2.1	2.1	2.0
Saturated fat (g/100 g)	0.3	0	0.1	0.1	0.2	0.4	0.3
Carbohydrates (g/100 g)	8	8.7	2	2.7	2	0.6	5.6
Sugars (g/100 g)	3.2	3.3	1.9	2.4	2	0.1	4.0
Fibre (g/100 g)	1	N/A	0.1	0.3	N/A	0.6	0.5
Salt (g/100 g)	0.08	0.1	0.1	0.1	0.2	0.04	0.19
Calcium (mg/100 g)	N/A	N/A	120	120	120	N/A	120
Shelf life [1]	9 mo., RT	1–2 mo., 5 °C	9 mo., RT	9 mo., RT	12 mo., RT	9 mo., RT	9 mo., RT
Heat treatment	UHT	ESL	UHT	UHT	UHT	UHT	UHT

Room temperature (RT). Ultra-high temperature (UHT). Extended shelf life (ESL). [1] From manufacturers.

For investigation of storage stability and potential storage-induced changes, the PBDs were stored at 21 °C for 364 days, comprising 11 sampling time points on days 0, 14, 28, 49, 91, 133, 175, 217, 259, 301 and 364. Fresh sample analyses conducted on each day of sampling comprised determination of color and pH. Samples were then aliquoted, frozen at −80 °C and then transferred to −20 °C and stored until further analysis for other parameters.

2.2. Color and pH Measurement

Color development of stored samples of PBDs was monitored using a Chroma-meter (CR-400, Konica Minolta Inc., Tokyo, Japan) on 10 mL samples of PBDs held in a ceramic crucible. Before the measurements, calibration of the chroma-meter was conducted with a standard plate, and the measurements were taken in technical triplicates. The CIELAB system was used as the color space application [15], where L* is the black–white scale, a* is the red–green scale and b* is the yellow–blue scale. Hence, L* = 0 is black and L* = 100 is white; negative a* is green and positive a* is red; and negative b* is blue and positive b* is yellow. Data for days 0, 175 and 364 are shown in the results section, and the remaining data points are available in Supplementary Materials Table S2. The pH was measured using a pH meter (PHM 92, Radiometer, Copenhagen, Denmark) calibrated with pH standards (Hanna Instruments, Woonsocket, RI, USA). Data for days 0, 175 and 364 are shown in the results section, and the rest of the data points are available in Supplementary Materials Table S2.

2.3. Protein Content and Protein Stability during Storage

Protein contents were stated on the package, Table 1. Furthermore, total nitrogen content was assessed through Dumas analysis, employing Dumatherm (Gerhardt, Königswinter, Germany). Values of total N were measured in the PBDs on day 0, as well as days 0 and 364 of storage for PBD supernatants after centrifugation for 10 min at 10,000× *g* at 4 °C to investigate protein solubility before and after storage. Via comparison of protein content on day 0 before and after centrifugation and after centrifugation at day 364, protein solubility was investigated. The samples were measured in technical duplicates using 0.5% tris(hydroxymethyl)aminomethane as a standard. For each plant source, the applied nitrogen to protein conversion factor was oat = 5.83, almond = 5.18, pea = 5.3, soy = 5.71 and hemp seeds = 5.3 [16].

2.4. Particle Size Distribution

Development in particle size distribution in the PBDs was determined using Zetasizer Lab (Malvern Panalytical, Malvern, UK) [8]. The PBD samples were diluted at a ratio of 1:50 in Milli-Q water and assessed using a 1 × 1 cm cuvette. Milli-Q water (refractive index of 1.33) was in the continuous phase while PBD constituted the scattered phase. Measurement and analysis were conducted using the Zetasizer Lab instrument and processed through the ZS Xplorer software (version 2.0.0.98, Malvern Panalytical, Malvern, UK).

2.5. Free N-Terminals via o-Phthalaldehyde Assay

Development of free N-terminals was measured with the spectrophotometric method o-phthalaldehyde (OPA) assay. Samples of the stored PBDs were precipitated on ice for 30 min with 5% trichloroacetic acid and centrifuged at 13,000× *g* for 20 min at 4 °C. A 10 µL supernatant was mixed with 200 µL OPA reagent (0.1 M $Na_2B_4O_7$, 0.1% SDS, 5.7 mM DTE, 1% OPA) in triplicate in a 96-well microtiter plate and incubated for 15 min at 21 °C. The absorbance was measured at 340 nm in a plate reader (BioTek Synergy 2, Holm & Halby, Brøndby, Denmark) and the standard curve was glutamine (1.25–40 mM).

2.6. Quantification of Free Amino Acids on Days 0, 175 and 364 of Storage Period Using Triple-Quadrupole Mass Spectrometry

Ten mg freeze-dried PBD was weighed off and precipitated on ice with 250 µL of 50% *v*/*v* methanol and centrifuged at 10,000× *g* for 10 min at 4 °C to extract free AAs. The supernatant was collected, and the pellet was resuspended in 250 µL 80% *v*/*v* methanol and precipitated on ice, then being centrifuged a second time at 10,000× *g* for 10 min at 4 °C. The two supernatants were combined and 40 µL was mixed with an equal volume of methanol containing 10 mg/L d4-alanine internal standard. The mixture was vortexed and centrifuged at 10,000× *g* for 5 min at 4 °C. The supernatant was derivatised based on the AccuTag method as previously described [17,18]. The AccuTag reagent was

6-aminoquinolyl-N-hydroxysuccinimidyl carbamate, 2.8 mg in 1 mL dry acetonitrile, and the derivatization reaction occurred in 70 µL borate buffer with 10 µL sample, standard or blank and 10 µL Accutag reagent. The reaction was incubated for 15 min at 55 °C followed by addition of 400 µL 90% formic acid, and the samples were ready for analysis. The standard solution of AA was A9906 (Sigma-Aldrich, Darmstadt, Germany).

The liquid chromatography–mass spectrometry (LC-MS) analysis method was conducted on an Agilent 1260 Infinity Quaternary LC system (Santa Clara, CA, USA) coupled to a 6420 Triple-Quadrupole Mass Spectrometer (triple Q-MS) with an electrospray ionization source. The LC-MS was equipped with a Kinetex Evo C18 2.1 × 150 mm, 1.7 µm (Phenomenex, Torrance, CA, USA) and ran at 25 °C with a flow rate of 0.25 mL/min. The mobile phases were 0.1% formic acid in Milli-Q water (A) and 0.1% formic acid in acetonitrile (B), and the gradient program was set as follows: 0–8 min: 95% A, 8–13 min: 90% A, 13–16.5 min: 85% A, 16.5–18.5: 20% A, 18.5–29 min: 95% A. The MS ionisation source conditions were as follows: capillary voltage of 4 kV, drying gas temperature of 300 °C, drying gas flow of 10 L/min, nebulizer pressure of 30 psi. Positive ion mode was utilized with multiple reaction monitoring (MRM) for quantitative analysis. Data analysis was conducted in MassHunter Qualitative Analysis version 10.0 and MassHunter Quantitative Analysis version 10.1 for QQQ (Agilent Technologies, Santa Clara, CA, USA). Results for proteinogenic AAs are shown in the results section whereas results for non-proteinogenic AAs are available in Supplementary Materials Table S3.

2.7. Statistical Analysis

Statistical analysis was conducted using the RStudio software (version 2023.6.1., Boston, MA, USA) and Microsoft Excel (2016). Significant differences were tested for Z-average particle sizes over time with one-way analysis of variance (ANOVA) in Excel as well as one-tail *t*-test for analysing the protein solubility. Variances in the free amino acid content were evaluated through mean comparisons utilizing Tukey's HSD post hoc test subsequent to conducting a one-way ANOVA test the data. Statistical significance was established at $p \leq 0.05$.

3. Results

3.1. Color, pH and Protein Solubility

pH and color were measured at each sampling time and are summarized in Table 2 for sampling days 0, 175 and 364. Data for all 11 sampling days can be seen in the Supplementary Materials Table S2. The off-set pH for the samples ranged from 6.77 (oat) to 8.08 (almond), while the end pH after 364 days of storage ranged from 6.39 (oat/hemp) to 7.61 (pea). The stored samples slightly decreased in pH over the storage period except for pea and soy PBDs, where the pH did not change during storage. For the color measurements, the oat/hemp showed the lowest values for L* (black and white scale), meaning that it was the darkest PBD at the end of the storage period. The rest of the samples had L*-values between 52 and 64, and all samples decreased in whiteness over time. The soy sample was the most green and yellow according to the a* and b* values, whereas the almond samples were the least yellow. All samples showed decreasing values of b* (yellowness) over the storage period, whereas parameter a* (red-green scale) changed for oat/hemp and the two soy drinks over time.

Protein content was measured based on total N in the original PBDs, and the contents ranged from 0.44% (roasted almond) to 4.22% (soy) (Table 2). Protein solubility and thereby indications of sedimentation involving protein were measured as an expression of the protein remaining in the supernatant after centrifugation, both on day 0 and day 364. Apart from the soy/rice, the PBDs had significantly higher protein content measured as total N in the original PBD, as compared with the supernatant after centrifugation on day 0 prior to storage (Table 2), indicating the presence of proteinaceous sediment. The soy/rice PBD was the only sample containing the same levels of protein in the intact sample, as in the centrifuged samples on day 0, thereby indicating no sedimentation involving protein.

Comparing total protein contents in the supernatants of the PBDs, day 0 and day 364 after storage showed a statistically different increase for pea, while increases for roasted almond and soy were not statistically different. This showed that there was no decrease in protein solubility over time but rather, that in some PBDs it seemed that the protein solubility increased with storage, especially for pea. For the soy/rice, there was no difference in total protein contents in supernatants from day 0 to day 364 of storage, indicating a very stable protein matrix in relation to solubility.

Table 2. pH, color parameters (L*, a*, b*) and protein solubility measurement of stored PBDs on sampling days 0, 175 and 364. For pH and color parameters, the standard deviations were all under 3% and not included for better readability. Different subscript letters for protein solubility for each PBD indicate statistical significance, $p \leq 0.05$).

Sample	Day	pH	L*	a*	b*	Protein PBD (%)	Protein Supernatant (%)
Oat	0	6.77	58.89	−0.33	10.02	0.456 ± 0.007 [b]	0.164 ± 0.018 [a]
	175	6.45	54.41	0.20	9.93		
	364	6.54	53.97	0.44	9.92		0.226 ± 0.001 [a]
Oat/hemp	0	6.80	54.42	−1.51	8.46	0.930 ± 0.003 [b]	0.328 ± 0.000 [a]
	175	6.32	49.55	−0.67	7.60		
	364	6.39	48.55	−0.40	7.47		0.334 ± 0.002 [a]
Roasted Almond	0	7.60	55.84	0.48	7.87	0.441 ± 0.005 [b]	0.265 ± 0.000 [a]
	175	7.08	52.32	0.65	7.18		
	364	7.37	52.43	0.44	6.93		0.368 ± 0.002 [ab]
Almond	0	8.08	59.32	−0.19	2.84	0.449 ± 0.019 [b]	0.310 ± 0.003 [a]
	175	7.42	55.61	−0.01	2.26		
	364	7.50	55.60	−0.19	2.25		0.312 ± 0.125 [a]
Pea	0	7.62	63.75	−1.40	8.41	1.702 ± 0.035 [c]	1.144 ± 0.003 [a]
	175	7.45	59.74	−1.02	8.28		
	364	7.61	59.57	−1.03	8.01		1.173 ± 0.000 [b]
Soy	0	7.18	62.48	−3.00	11.57	4.223 ± 0.182 [b]	3.201 ± 0.033 [a]
	175	7.05	58.71	−2.49	10.27		
	364	7.24	58.40	−2.16	9.42		3.222 ± 0.001 [ab]
Soy/rice	0	7.94	58.70	−1.10	10.73	2.617 ± 0.584 [a]	2.884 ± 0.065 [a]
	175	7.56	53.37	0.16	9.80		
	364	7.54	52.85	0.52	9.68		2.984 ± 0.000 [a]

3.2. Particle Size

The development in particle size for each PBD during the storage period was measured and expressed as a Z-average (Figure 1). The Z-average particle size of the samples ranged from 1633 nm (almond) to 393 nm (oat), exhibiting large heterogeneity between the samples. The final Z-average particle sizes showed less heterogeneity, with the largest size being 897 nm (oat hemp) and the smallest particle size being 399 nm (soy). All PBDs changed significantly over time ($p \leq 0.05$), with oat and pea increasing in Z-average over time, especially after 259 days of storage, whereas the other five PBDs decreased during the storage period. Almond rapidly decreased within the first 28 days, whereas oat/hemp increased in the same period and generally displayed a large fluctuation in particle size. All Z-average particle sizes changed significantly over time ($p \leq 0.05$); oat and pea PBDs displayed an increase in particle size during storage, whereas oat/hemp, roasted almond, almond, soy and soy/rice more or less clearly decreased in particle size.

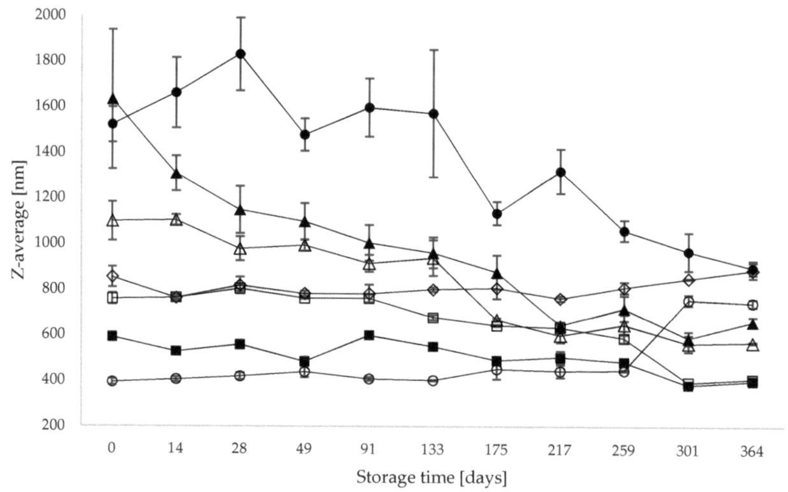

Figure 1. Particle size measured as Z-average (nm) on stored PBD. Oat (○), oat/hemp (•), roasted almond (∆), almond (▲), soy (□), soy/rice (■) and pea (◊). All samples were measured in technical triplicates (n = 3).

3.3. Proteolysis and Generation of Free Amino Acids during Storage

Protein degradation during storage as a result of potential proteolysis was determined as the level of free N-terminals using the OPA assay (Figure 2) and free AA content (Table 3). Almond and roasted almond had the lowest concentrations of free N-terminals, with average values < 1 µM on day 0 and for the whole storage period. The pea sample had the highest levels of free AAs, with an initial concentration of 18.1 µM and a final concentration of 23.8 µM (cf., right *y*-axis). The oat PBD and soy PBD were similar in free N-terminal content and remained constant over the storage period, except for the oat PBD, which increased after 259 days of storage.

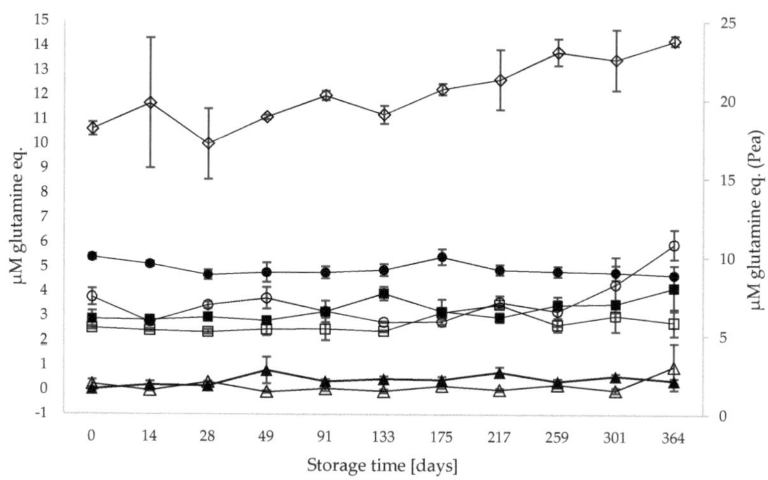

Figure 2. Level of free N-terminals in stored PBD expressed as glutamine equivalents (µM). Right *y*-axis refers to PBD pea (◊), whereas left *y*-axis refers to all other PBDs, oat (○), oat/hemp (•), roasted almond (∆), almond (▲), soy (□), and soy/rice (■). Results represent technical triplicates (n = 3).

Table 3. Contents of free AAs in PBD at days 0, 175 and 364 in mg AA/g protein (n = 3). Letters indicate significant difference ($p \leq 0.05$) in contents of each free AA in a PBD over the storage period, for better readability, letters were not included for free AAs with no significant change ($p > 0.05$). Not detected (ND).

Free AA	Day	Oat	Oat/Hemp	Almond	Roasted Almond	Pea	Soy	Soy/Rice
Histidine	0	0.18 ± 0.00	0.05 ± 0.00	ND	ND	0.22 ± 0.05 [a]	0.04 ± 0.00	0.17 ± 0.01
	175	0.27 ± 0.05	0.06 ± 0.00	ND	ND	0.37 ± 0.07 [ab]	0.07 ± 0.01	0.23 ± 0.05
	364	0.15 ± 0.00	0.05 ± 0.00	ND	ND	0.48 ± 0.11 [b]	0.14 ± 0.00	0.16 ± 0.03
Isoleucine	0	0.25 ± 0.01	0.13 ± 0.01	0.10 ± 0.01	0.09 ± 0.00	0.64 ± 0.04 [a]	0.02 ± 0.00	0.05 ± 0.00
	175	0.27 ± 0.03	0.13 ± 0.01	0.09 ± 0.01	0.09 ± 0.00	1.14 ± 0.20 [b]	0.02 ± 0.00	0.05 ± 0.00
	364	0.19 ± 0.01	0.17 ± 0.02	0.10 ± 0.02	0.10 ± 0.01	1.30 ± 0.09 [b]	0.04 ± 0.00	0.05 ± 0.00
Leucine	0	0.21 ± 0.00	0.29 ± 0.03	0.05 ± 0.01	0.03 ± 0.00	2.20 ± 0.13 [a]	0.01 ± 0.00	0.06 ± 0.00
	175	0.24 ± 0.03	0.32 ± 0.03	0.06 ± 0.01	0.02 ± 0.00	3.60 ± 0.57 [b]	0.02 ± 0.00	0.06 ± 0.00
	364	0.09 ± 0.00	0.38 ± 0.04	0.06 ± 0.01	0.03 ± 0.00	4.00 ± 0.17 [b]	0.07 ± 0.00	0.06 ± 0.01
Lysine	0	0.13 ± 0.03	0.60 ± 0.09 [a]	0.28 ± 0.04	0.05 ± 0.00	1.42 ± 0.05 [a]	0.06 ± 0.01	0.05 ± 0.00
	175	0.34 ± 0.04	0.84 ± 0.10 [ab]	0.30 ± 0.06	0.12 ± 0.03	3.24 ± 0.35 [b]	0.08 ± 0.02	0.07 ± 0.00
	364	0.61 ± 0.02	1.19 ± 0.03 [b]	0.31 ± 0.05	0.19 ± 0.01	3.77 ± 0.35 [c]	0.04 ± 0.00	0.05 ± 0.01
Methionine	0	0.02 ± 0.00	0.05 ± 0.00 [a]	0.02 ± 0.00	0.01 ± 0.00	0.08 ± 0.01 [a]	0.01 ± 0.00	0.02 ± 0.00
	175	0.02 ± 0.00	0.08 ± 0.01 [b]	0.02 ± 0.00	0.02 ± 0.00	0.14 ± 0.01 [b]	0.01 ± 0.00	0.03 ± 0.00
	364	0.01 ± 0.00	0.08 ± 0.01 [b]	0.01 ± 0.00	0.01 ± 0.00	0.18 ± 0.01 [c]	0.02 ± 0.00	0.02 ± 0.00
Phenylalanine	0	0.16 ± 0.01	0.14 ± 0.01	0.07 ± 0.01	0.01 ± 0.01	0.76 ± 0.07 [a]	0.01 ± 0.00	0.04 ± 0.00
	175	0.24 ± 0.01	0.15 ± 0.02	0.07 ± 0.02	0.01 ± 0.00	1.45 ± 0.27 [b]	0.01 ± 0.00	0.05 ± 0.01
	364	0.24 ± 0.00	0.19 ± 0.03	0.07 ± 0.01	0.02 ± 0.00	1.54 ± 0.12 [b]	0.05 ± 0.01	0.04 ± 0.00
Threonine	0	0.40 ± 0.03 [ab]	0.17 ± 0.01	0.09 ± 0.02	0.06 ± 0.01	0.35 ± 0.07 [a]	0.01 ± 0.00	0.02 ± 0.00
	175	0.50 ± 0.08 [b]	0.19 ± 0.03	0.09 ± 0.01	0.04 ± 0.01	0.65 ± 0.09 [b]	0.01 ± 0.00	0.03 ± 0.00
	364	0.27 ± 0.01 [a]	0.25 ± 0.03	0.08 ± 0.01	0.07 ± 0.01	0.82 ± 0.04 [c]	0.01 ± 0.00	0.02 ± 0.00
Tryptophan	0	0.95 ± 0.02 [b]	0.11 ± 0.02	0.05 ± 0.01	0.03 ± 0.00	0.02 ± 0.00	0.10 ± 0.02 [a]	0.24 ± 0.01
	175	1.16 ± 0.20 [c]	0.11 ± 0.01	0.04 ± 0.01	0.03 ± 0.01	0.02 ± 0.00	0.11 ± 0.02 [a]	0.24 ± 0.02
	364	0.30 ± 0.04 [a]	0.15 ± 0.02	0.04 ± 0.01	0.02 ± 0.00	0.04 ± 0.01	0.35 ± 0.02 [b]	0.22 ± 0.01
Valine	0	0.45 ± 0.01 [ab]	0.24 ± 0.02	0.15 ± 0.01	0.08 ± 0.00	0.61 ± 0.05 [a]	0.02 ± 0.00	0.05 ± 0.00
	175	0.59 ± 0.12 [b]	0.26 ± 0.03	0.13 ± 0.02	0.08 ± 0.01	1.10 ± 0.02 [b]	0.02 ± 0.00	0.06 ± 0.00
	364	0.33 ± 0.01 [a]	0.36 ± 0.03	0.15 ± 0.04	0.11 ± 0.03	1.43 ± 0.08 [c]	0.04 ± 0.00	0.05 ± 0.01
L-Alanine	0	1.46 ± 0.03 [a]	0.90 ± 0.06	0.70 ± 0.09	0.72 ± 0.10	0.79 ± 0.04 [a]	0.11 ± 0.01	0.22 ± 0.01
	175	1.88 ± 0.33 [a]	1.07 ± 0.09	0.79 ± 0.09	0.64 ± 0.06	1.97 ± 0.16 [b]	0.15 ± 0.02	0.24 ± 0.01
	364	2.45 ± 0.09 [b]	1.28 ± 0.08	0.67 ± 0.09	0.79 ± 0.07	2.50 ± 0.12 [c]	0.11 ± 0.01	0.22 ± 0.02
Arginine	0	ND	ND	0.24 ± 0.08	ND	0.62 ± 0.18 [a]	0.23 ± 0.07	0.80 ± 0.16
	175	1.26 ± 0.22	0.42 ± 0.39	ND	ND	1.35 ± 0.26 [b]	0.27 ± 0.15	1.05 ± 0.21
	364	ND	0.08 ± 0.00	ND	ND	1.45 ± 0.06 [b]	0.63 ± 0.03	0.88 ± 0.05
Asparagine	0	12.30 ± 0.45 [a]	1.97 ± 0.17	1.22 ± 0.13	0.72 ± 0.16	0.84 ± 0.06	0.07 ± 0.00	0.43 ± 0.01
	175	15.39 ± 2.30 [b]	2.07 ± 0.13	1.30 ± 0.30	0.72 ± 0.04	1.79 ± 0.22	0.07 ± 0.02	0.45 ± 0.03
	364	9.49 ± 1.35 [a]	3.02 ± 0.09	1.19 ± 0.19	0.82 ± 0.14	1.68 ± 0.20	0.12 ± 0.01	0.41 ± 0.03
Aspartic acid	0	4.14 ± 0.01 [a]	0.73 ± 0.11	0.59 ± 0.07	0.55 ± 0.09	0.15 ± 0.03	0.15 ± 0.03	0.32 ± 0.01
	175	5.37 ± 0.86 [b]	0.82 ± 0.05	0.66 ± 0.15	0.50 ± 0.03	0.35 ± 0.04	0.18 ± 0.04	0.35 ± 0.02
	364	3.35 ± 0.16 [a]	1.01 ± 0.15	0.59 ± 0.08	0.68 ± 0.09	0.49 ± 0.02	0.26 ± 0.01	0.33 ± 0.02
Cystine	0	0.52 ± 0.02 [a]	0.53 ± 0.07	0.32 ± 0.04	0.30 ± 0.07	0.33 ± 0.02 [a]	0.03 ± 0.01	0.06 ± 0.01
	175	0.7 ± 0.16 [a]	0.64 ± 0.07	0.37 ± 0.10	0.31 ± 0.03	1.14 ± 0.17 [b]	0.04 ± 0.02	0.06 ± 0.00
	364	1.62 ± 0.29 [b]	0.86 ± 0.11	0.33 ± 0.06	0.40 ± 0.07	1.44 ± 0.13 [b]	0.02 ± 0.00	0.05 ± 0.01
Glutamic acid	0	7.00 ± 0.01 [b]	0.89 ± 0.07	0.61 ± 0.07	0.14 ± 0.01	0.29 ± 0.01	0.16 ± 0.03	0.53 ± 0.02
	175	8.96 ± 1.43 [c]	1.09 ± 0.09	0.66 ± 0.16	0.10 ± 0.01	0.50 ± 0.03	0.21 ± 0.04	0.57 ± 0.04
	364	3.85 ± 0.25 [a]	1.35 ± 0.14	0.59 ± 0.06	0.08 ± 0.01	0.55 ± 0.00	0.39 ± 0.02	0.52 ± 0.06
Glutamine	0	ND	0.19 ± 0.03 [b]	ND	ND	0.01 ± 0.00	ND	ND
	175	ND	0.12 ± 0.02 [ab]	ND	ND	0.01 ± 0.00	ND	ND
	364	ND	0.10 ± 0.03 [a]	ND	ND	ND	ND	ND
Glycine	0	0.52 ± 0.02 [a]	0.53 ± 0.07	0.32 ± 0.04	0.30 ± 0.07	0.33 ± 0.02 [a]	0.03 ± 0.01	0.06 ± 0.01
	175	0.70 ± 0.16 [a]	0.64 ± 0.07	0.37 ± 0.10	0.31 ± 0.03	1.14 ± 0.17 [b]	0.04 ± 0.02	0.06 ± 0.00
	364	1.62 ± 0.29 [b]	0.86 ± 0.11	0.33 ± 0.06	0.40 ± 0.07	1.44 ± 0.13 [b]	0.02 ± 0.00	0.05 ± 0.01

Table 3. Cont.

Free AA	Day	Oat	Oat/Hemp	Almond	Roasted Almond	Pea	Soy	Soy/Rice
Proline	0	5.15 ± 0.15 [b]	0.67 ± 0.02	1.05 ± 0.14	0.68 ± 0.06	0.05 ± 0.01	0.03 ± 0.00	0.07 ± 0.00
	175	6.46 ± 1.01 [c]	0.75 ± 0.05	1.07 ± 0.19	0.61 ± 0.03	0.09 ± 0.01	0.03 ± 0.01	0.08 ± 0.00
	364	3.99 ± 0.17 [a]	0.94 ± 0.03	1.02 ± 0.13	0.72 ± 0.04	0.15 ± 0.01	0.04 ± 0.00	0.07 ± 0.01
Serine	0	0.60 ± 0.04	0.34 ± 0.03	0.29 ± 0.00	0.26 ± 0.06	1.03 ± 0.05 [a]	0.02 ± 0.00	0.06 ± 0.01
	175	0.77 ± 0.09	0.41 ± 0.04	0.32 ± 0.08	0.23 ± 0.01	2.89 ± 0.36 [b]	0.03 ± 0.01	0.06 ± 0.01
	364	0.95 ± 0.07	0.58 ± 0.05	0.28 ± 0.06	0.31 ± 0.04	3.37 ± 0.38 [c]	0.04 ± 0.01	0.06 ± 0.01
Tyrosine	0	0.44 ± 0.02 [b]	0.09 ± 0.01	0.04 ± 0.01	0.03 ± 0.01	0.18 ± 0.03 [a]	0.01 ± 0.00	0.05 ± 0.01
	175	0.55 ± 0.09 [b]	0.10 ± 0.02	0.05 ± 0.00	0.03 ± 0.00	0.47 ± 0.09 [b]	0.01 ± 0.00	0.06 ± 0.00
	364	0.29 ± 0.00 [a]	0.14 ± 0.02	0.03 ± 0.00	0.04 ± 0.00	0.52 ± 0.02 [b]	0.02 ± 0.00	0.05 ± 0.00
Total free AA	0	35.53 ± 0.87 [a]	8.72 ± 0.82	6.67 ± 0.83	4.46 ± 0.68	10.67 ± 0.91 [a]	1.22 ± 0.19	3.69 ± 0.29
	175	46.31 ± 7.26 [b]	10.30 ± 1.24	6.83 ± 1.37	4.21 ± 0.29	22.40 ± 2.94 [b]	1.52 ± 0.39	4.19 ± 0.44
	364	29.44 ± 2.61 [a]	13.10 ± 1.00	6.27 ± 0.89	5.19 ± 0.57	25.86 ± 1.92 [b]	2.54 ± 0.14	3.64 ± 0.33
mg Free AA/100 g PBD	0	16.20 ± 0.40	8.11 ± 0.76	2.94 ± 0.36	2.00 ± 0.30	18.16 ± 1.55	5.20 ± 0.80	9.66 ± 0.76
	175	21.12 ± 3.31	9.58 ± 1.15	3.01 ± 0.60	1.89 ± 0.13	38.13 ± 5.00	6.41 ± 1.63	10.96 ± 1.16
	364	13.43 ± 1.19	12.18 ± 0.93	2.77 ± 0.39	2.33 ± 0.26	44.01 ± 3.27	10.70 ± 0.60	9.51 ± 0.86

To investigate the specific AAs that may be released over time, the levels of free AA were quantified in the stored PBD samples on sampling days 0, 175 and 364 (Table 3) using quantitative triple Q MS-based analysis. The summarized data show the sum of all free AAs in the samples. The initial contents ranged from 1.2 to 35.6 mg free AA/g protein, with min. and max. representing soy and oat, respectively, whereas the final contents ranged from 2.5 to 29.2 mg free AA/g protein, with min. and max. representing soy and oat, respectively. Total free AA in oat PBD was significantly higher on day 175 compared to levels on both days 0 and 364, whereas for pea PBD, free AA levels were significantly higher on days 175 and 364 compared to day 0. The remaining PBDs did not differ in the content of total free AAs across the three time points. For almond and roasted almond, this pattern was also reflected for individual AAs, as these PBDs did not differ between any time points among all quantified AAs. The soy drinks also showed the same levels of free AAs at all time points, except for tryptophan, which significantly increased on day 364. Oat PBD showed significantly different levels of free AAs for most of the quantified AAs over the storage period. L-alanine, cystine and glycine significantly increased over the storage time in oat, whereas aspartic acid, asparagine, threonine and valine decreased between days 175 and 364; glutamic acid, proline, tryptophan and tyrosine decreased over the whole storage period. In the pea drink, L-alanine, lysine, methionine, serine, threonine and valine increased significantly at all three time points, whereas glycine, isoleucine, leucine, phenylalanine, tyrosine and arginine significantly increased between days 0 and 175 but not thereafter. For the oat/hemp drink, the contents of free lysine and methionine were significantly increased on day 364 compared to day 0, and the content of free glutamine was significantly decreased on day 364 compared to day 0, whereas for all other free AAs in oat/hemp, the contents did not significantly differ.

4. Discussion

This paper aimed to examine the storage stability of commercially available PBDs from the Danish market, concerning the proteolysis and generation of free amino acids. Therefore, parameters related to protein solubility and proteolysis, including the level and composition of free AAs, color and particle size, were measured in different types of PBDs stored for one year. Previous studies found that particularly the UHT treatment of bovine milk resulted in reduced storage stability due to heat- and processing-induced reactions, like the Maillard reaction and/or dehydroalanine pathway and proteolysis, potentially contributing to age gelation [12,13]. After UHT treatment, both milk and PBD are sterile and aseptically packed; however, some enzymes are heat-resistant and may withstand the UHT treatment [13]. In bovine milk, some indigenous enzymes are found to be heat-stable, e.g., those from the plasmin system, but enzymes from psychotropic bacteria can also be found in UHT milk from contamination of the raw milk [12]; hence, it has been investigated if these mechanisms could be present and quantifiable in PBDs.

As outlined in Table 1, PBDs have some inherent differences e.g., in protein contents, but these did not seem to be primary drivers for the subsequent differences in the solubility and sedimentation of protein. All PBDs (except the soy/rice PBD), representing big variations in protein contents, displayed the presence of sedimentable proteinaceous material prior to storage. The soy/rice PBD, containing the second-highest content of total protein among all PBDs studied here, was the most stable in relation to protein solubility, both before and after storage. Perhaps surprisingly, no significant decreases in total protein contents in the supernatants before and after storage were observed for any PBDs, indicating that there was no significant protein sedimentation during storage. In contrast, there were indications of increased protein solubility with time, especially for pea. This is known to be a problem in some PBDs, in that sediment is observed to develop and may be considered a quality problem for consumers [19]. Soy PBDs contained the highest protein concentrations and almond PBDs the lowest, but according to the measurements of total protein after centrifugation on day 0 and day 364, these drinks seemed stable during storage, as the total protein contents of supernatants before and after storage were not significantly different. The results here indicate that any sediment developing may be more related to non-proteinaceous material like fibres.

The particle size was influenced over time i.e., oat and pea PBD had increasing particle size towards the end of the experiment, whereas the other PBDs showed decreasing values. A previous study concluded that smaller particle sizes were associated with more stable PBDs, and larger particle sizes produced less stable drinks [20]. This relationship was not observed here using protein solubility as a parameter for stability, i.e., for pea, the particle size increased over time, but the protein solubility also increased. It is likely that in PBDs, there is no direct correlation between protein solubility and particle size.

Heat treatment through e.g., roasting of the starting material, may affect the color of PBDs, which is also apparent from Table 2, comparing almond and roasted almond, where the b* values of roasted almond are much higher compared to almond. A previous study quantified compounds related to the Maillard reaction in the same two almond drinks and found no significant difference in the levels of those compounds in non-stored PBDs; however, it is still necessary to quantify the compounds in stored PBDs [8].

Looking at parameters for proteolysis, based on the OPA assay, the level of free N-terminals in the pea PBD increased continuously over time, whereas the oat PBD increased rapidly towards the end of the storage period, and the other drinks remained relatively constant (Figure 2). For oat, there seemed to be an accelerated effect between days 175 and 364, as the level of free N-terminals was increased on days 301 and 364 compared to the remaining storage period. However, this was not reflected in total free AAs, which were highest on day 175 compared to lower total free AAs on day 364. The determination of free N-terminals via the OPA assay comprises levels of free N-terminals in intact plant proteins, generated peptides and free AAs, as well as lysine side chains. Changes in lysine side chains in proteins and peptides could also potentially contribute to the changes observed, i.e., like engaging in Maillard or dehydroalanine pathway reactions [11,12]. The level measured by the OPA assay is therefore, a balance between any proteolysis occurring, creating peptides and free AAs on the side, and then the stability of both proteins and peptides in relation to engaging in Maillard or dehydroalanine pathway reactions, as well as the generated free AAs engaging in, e.g., Maillard reactions via Strecker aldehydes [21]. This may also explain why lower levels of free AAs were seen in the oat PBD at the same time as higher levels of free N-terminals [21]. Furthermore, it is known from the processing of oat drinks that the oat slurry needs to be hydrolysed by alpha-amylases to prevent gelatinization after cooling down [22]. The utilized alpha-amylase needs to be both heat-stable in order to function efficiently during the processing of the PBDs and of high purity to avoid the presence of proteolytic side activity.

High and increasing levels of free N-terminals in pea PBD, as determined by the OPA assay, were also reflected in high and increasing total free AAs, as determined by triple Q-MS. Levels of free AAs in pea significantly increased from days 0 to 175, which, for pea, confirmed proteolysis and the generation of free AAs occurred during storage. What makes the pea PBD more prone to proteolysis is unclear, but the initial high level of amino N-terminals suggests that it is related to the raw material, which, in this case, is pea protein, as listed in Supplementary Materials Table S1. During protein fractionation, proteolysis may occur since proteases and other enzymes are released in the initial steps of the purification process, which could be the reason for the high initial levels of free N-terminals in pea PBD [23]. This indicates some kind of proteolytic activity being present in the pea PBD, however, the identity of such protease is not known at present. It can be speculated whether this increase in proteolysis in the pea PBD could be the driver behind the increased solubility of total pea protein observed in the pea PBD during storage, and if the formed peptides of free AAs are more soluble than intact protein.

A previous study concluded that the formation of free AAs in UHT milk contributed to stale aromas, which could be linked to the formation of Strecker aldehydes [24]. Another study classified free AAs in food systems into groups based on how they impact flavour and found that ten AAs contribute to bitterness, i.e., arginine, histidine, tyrosine, leucine, valine, methionine, isoleucine, phenylalanine, lysine and proline [14]. In this study, we found nine out of the ten bitter free AAs significantly increased during the storage period in pea PBD, and, therefore, conducting sensory analysis on these stored PBDs could be relevant. Free serine, alanine, arginine, glycine and threonine contribute to sweet aromas [14,25], and in pea PBD, all five of these were found to have significantly higher levels after the storage period compared to the initial level; oat alanine and glycine were observed to increase significantly during storage.

Total free AA was standardized to protein content, but it would be interesting to determine the total AA compositions of proteins in the PBDs in order to better interpret differences across, e.g., oat and pea. It is, however, clear that the free AAs in the two PBDs are dominated by different free AAs, and this could affect the nutritional value of the drinks differently and probably relates to the total AA profile of the protein sources. Pea is for instance high in branched-chain AAs, like valine, leucine and isoleucine [26], and the concentrations of these free AAs are all generally high in pea PBD and increasing during storage. The underlying enzymatic or non-enzymatic drivers still need to be confirmed. Table 3 shows the level of total free AA in 100 g of PBD, i.e., the values here are not standardized to protein content but represent the absolute level, as experienced in the drinks. From these values, it is evident that pea contains the highest content of free AAs and therefore, could be most affected by their presence since the free AA will be available for further reactions, like the formation of Strecker aldehydes as part of the Maillard reaction [11].

5. Conclusions

In the present study, the possible quality deteriorations of commercially available PBDs were studied over the course of a one-year storage period, especially in relation to color and pH, particle size, proteolysis and the generation of free AAs. It was found that the parameters varied depending on plant sources. For particle size, the oat/hemp PBD had the highest particle size. Limited variations were seen in pH, color and protein solubility over time, whereas protein solubility significantly increased in the pea PBD. Oat had the highest total level of free AAs relative to protein content, whereas pea had the highest content of free AAs relative to g PBD. The pea PBD was found to have a significantly increased amount of free AAs over the storage period, in line with pea PBD also having the highest increase in the level of total free N-terminals over storage, indicating that proteolysis could have occurred. Oat significantly increased from days 0 to 175 and decreased to the initial levels again on day 364. In conclusion, it is clear that the plant source starting material is important for the specific changes occurring during storage in

PBDs. No gelation was found, but the observed changes in free AAs may affect the flavour of the products. Generally, the underlying molecular mechanisms for the observed changes should be the focus of further studies, where the changes in amino acid content could be elucidated by specific reactions, such as the Maillard reaction and protein crosslinking through the dehydroalanine pathway, and the effect of digestibility may be highly relevant.

Supplementary Materials: The following supporting information can be downloaded at: https://www.mdpi.com/article/10.3390/foods13030367/s1, Table S1: Declared contents of PBDs, name and manufacturer: Table S2: Data for each PBDs on all 11 sampling days; Table S3: Contents of free AA in PBDs at days 0, 175 and 364.

Author Contributions: Conceptualization, N.A.P., L.B.L., S.D.-H.N. and I.S.R.; methodology, N.A.P., L.B.L., T.T.L. and I.S.R.; validation, formal analysis; investigation, I.S.R., T.C., T.T.L. and M.A.-T.; writing—original draft preparation, I.S.R., N.A.P. and L.B.L.; writing—review and editing, I.S.R., N.A.P., L.B.L., T.T.L., T.C., M.A.-T. and S.D.-H.N.; supervision, L.B.L., N.A.P. and S.D.-H.N.; project administration and funding acquisition, L.B.L. All authors have read and agreed to the published version of the manuscript.

Funding: This research was funded by the Aarhus University Research Foundation.

Institutional Review Board Statement: Not applicable.

Informed Consent Statement: Not applicable.

Data Availability Statement: The original contributions presented in the study are included in the article/Supplementary Materials, further inquiries can be directed to the corresponding author.

Conflicts of Interest: The authors declare no conflicts of interest. The funders had no role in the design of the study; in the collection, analyses, or interpretation of data; in the writing of the manuscript; or in the decision to publish the results.

References

1. Rosenfeld, D.L. The psychology of vegetarianism: Recent advances and future directions. *Appetite* **2018**, *131*, 125–138. [CrossRef] [PubMed]
2. Sethi, S.; Tyagi, S.K.; Anurag, R.K. Plant-based milk alternatives an emerging segment of functional beverages: A review. *J. Food Sci. Technol.* **2016**, *53*, 3408–3423. [CrossRef] [PubMed]
3. Fuentes, C.; Fuentes, M. Making a market for alternatives: Marketing devices and the qualification of a vegan milk substitute. *J. Mark. Manag.* **2017**, *33*, 529–555. [CrossRef]
4. Ingram, C.J.E.; Mulcare, C.A.; Itan, Y.; Thomas, M.G.; Swallow, D.M. Lactose digestion and the evolutionary genetics of lactase persistence. *Hum. Genet.* **2009**, *124*, 579–591. [CrossRef] [PubMed]
5. Patra, T.; Rinnan, Å.; Olsen, K. The physical stability of plant-based drinks and the analysis methods thereof. *Food Hydrocoll.* **2021**, *118*, 106770. [CrossRef]
6. Thorning, T.K.; Raben, A.; Tholstrup, T.; Soedamah-Muthu, S.S.; Givens, I.; Astrup, A. Milk and dairy products: Good or bad for human health? An assessment of the totality of scientific evidence. *Food Nutr. Res.* **2016**, *60*, 32527. [CrossRef]
7. Munekata, P.E.S.; Domínguez, R.; Budaraju, S.; Roselló-Soto, E.; Barba, F.J.; Mallikarjunan, K.; Roohinejad, S.; Lorenzo, J.M. Effect of Innovative Food Processing Technologies on the Physicochemical and Nutritional Properties and Quality of Non-Dairy Plant-Based Beverages. *Foods* **2020**, *4*, 288. [CrossRef]
8. Roland, I.S.; Aguilera-Toro, M.; Nielsen, S.D.-H.; Poulsen, N.A.; Larsen, L.B. Processing-Induced Markers in Proteins of Commercial Plant-Based Drinks in Relation to Compositional Aspects. *Foods* **2023**, *12*, 3282. [CrossRef] [PubMed]
9. Schilling, W.; Lesak, D. The making of a plant-based beverage. *Food Technol. Mag.* **2022**, *76*, 73–75.
10. Mäkinen, O.E.; Sozer, N.; Ercili-Cura, D.; Poutanen, K. Protein from Oat: Structure, process, functionality, and nutrition. In *Sustainable Protein Sources*; Academic Press: Cambridge, MA, USA, 2016; pp. 105–119.
11. Nielsen, S.D.; Knudsen, L.J.; Baekgaard, L.T.; Rauh, V.; Larsen, L.B. Influence of Lactose on the Maillard Reaction and Dehydroalanine-Mediated Protein Cross-Linking in Casein and Whey. *Foods* **2022**, *11*, 897. [CrossRef]
12. Aguilera-Toro, M.; Poulsen, N.A.; Akkerman, M.; Rauh, V.; Larsen, L.B.; Nielsen, S.D. Development in Maillard Reaction and Dehydroalanine Pathway Markers during Storage of UHT Milk Representing Differences in Casein Micelle Size and Sedimentation. *Foods* **2022**, *11*, 1525. [CrossRef] [PubMed]
13. Anema, S.G. Age gelation, sedimentation, and creaming in UHT milk: A review. *Compr. Rev. Food Sci. Food Saf.* **2019**, *18*, 140–166. [CrossRef] [PubMed]
14. Kim, Y.; Kim, E.Y.; Son, H.J.; Lee, J.J.; Choi, Y.H.; Rhyu, M.R. Identification of a key, umami -active fraction in modernized Korean soy sauce and the impact thereof on bitter-masking. *Food Chem.* **2017**, *233*, 256–262. [CrossRef]

15. McGuire, R.G. Reporting of Objective Color Measurements. *HortScience* **1992**, *27*, 1254–1255. [CrossRef]
16. Jones, D.B. *Factors for Converting Percentages of Nitrogen in Foods and Feeds into Percentages of Protein*; US Department of Agriculture: Washington, DC, USA, 1931; Volume 183, pp. 1–22.
17. Salazar, C.; Armenta, J.M.; Shulaev, V. An UPLC-ESI-MS/MS Assay Using 6-Aminoquinolyl-N-Hydroxysuccinimidyl Carbamate Derivatization for Targeted Amino Acid Analysis: Application to Screening of Arabidopsis thaliana Mutants. *Metabolites* **2012**, *2*, 398–428. [CrossRef] [PubMed]
18. Diep, T.T.; Pook, C.; Yoo, M.J.Y. Physicochemical properties and proximate composition of tamarillo (*Solanum betaceum* Cav.) fruits from New Zealand. *J. Food Compos. Anal.* **2020**, *92*, 103563. [CrossRef]
19. Giacalone, D.; Clausen, M.P.; Jaeger, S.R. Understanding barriers to consumption of plant-based foods and beverages: Insights from sensory and consumer science. *Curr. Opin. Food Sci.* **2022**, *48*, 100919. [CrossRef]
20. Durand, A.; Franks, G.V.; Hosken, R.W. Particle sizes and stability of UHT bovine, cereal and grain milks. *Food Hydrocoll.* **2003**, *17*, 671–678. [CrossRef]
21. Jensen, S.; Jansson, T.; Eggers, N. Storage-induced changes in the sensory characteristics and volatiles of conventional and lactose-hydrolyzed UHT processed milk. *Eur. Food Res. Technol.* **2015**, *240*, 1247–1257. [CrossRef]
22. Deswal, A.; Deora, N.S.; Mishra, H.N. Optimization of Enzymatic Production Process of Oat Milk Using Response Surface Methodology. *Food Bioprocess Technol.* **2013**, *7*, 610–618. [CrossRef]
23. Ryan, B.J.; Henehan, G.T. Avoiding Proteolysis During Protein Purification. *Methods Mol. Biol.* **2017**, *1485*, 53–69. [CrossRef]
24. Nielsen, S.D.; Jansson, T.; Le, T.T.; Jensen, S.; Eggers, N.; Rauh, V.; Sundekilde, U.K.; Sørensen, J.; Andersen, H.J.; Bertram, H.C. Correlation between sensory properties and peptides derived from hydrolysed-lactose uht milk during storage. *Int. Dairy. J.* **2017**, *68*, 23–31. [CrossRef]
25. Chen, D.-W.; Zhang, M. Non-volatile taste active compounds in the meat of Chinese mitten crab (*Eriocheir sinensis*). *Food Chem.* **2007**, *104*, 1200–1205. [CrossRef]
26. Gorissen, S.H.M.; Crombag, J.J.R.; Senden, J.M.G.; Waterval, W.A.H.; Bierau, J.; Verdijk, L.B.; van Loon, L.J.C. Protein content and amino acid composition of commercially available plant-based protein isolates. *Amino Acids* **2018**, *50*, 1685–1695. [CrossRef] [PubMed]

foods

MDPI

Article

Enhancing the Techno-Functional Properties of Lentil Protein Isolate Dispersions Using In-Line High-Shear Rotor-Stator Mixing

Nicolas Malterre [1], Francesca Bot [2], Emilie Lerda [1], Elke K. Arendt [1,3], Emanuele Zannini [1,4] and James A. O'Mahony [1,*]

[1] School of Food and Nutritional Sciences, University College Cork, T12 Y337 Cork, Ireland; nicolas.malterre@ucc.ie (N.M.); emilielerda@gmail.com (E.L.)
[2] Department of Food and Drug, University of Parma, 43124 Parma, Italy; francesca.bot@unipr.it
[3] APC Microbiome Institute Ireland, University College Cork, T12 Y337 Cork, Ireland
[4] Department of Environmental Biology, "Sapienza" University of Rome, 00185 Rome, Italy
* Correspondence: sa.omahony@ucc.ie; Tel.: +353-21-4903625

Abstract: In response to global challenges such as climate change and food insecurity, plant proteins have gained interest. Among these, lentils have emerged as a promising source of proteins due to their good nutritional profile and sustainability considerations. However, their widespread use in food products has been impeded by limited solubility. This study aimed to investigate the potential of high-shear mixing, a resource-efficient technique, to enhance lentil protein solubility and its functional properties. Red lentil protein isolate powders were rehydrated and subjected to a semi-continuous in-line high-shear treatment at 10,200 rpm for a timespan ranging from 0 to 15 min. The results highlighted a significant ($p < 0.05$) increase in solubility from 46.87 to 68.42% after 15 min of shearing and a reduction in particle size as a result of the intense shearing and disruption provided by the rotor and forced passage through the perforations of the stator. The volume-weighted mean diameter decreased from 5.13 to 1.72 μm after 15 min of shearing, also highlighted by the confocal micrographs which confirmed the breakdown of larger particles into smaller and more uniform particles. Rheological analysis indicated consistent Newtonian behaviour across all dispersions, with apparent viscosities ranging from 1.69 to 1.78 mPa.s. Surface hydrophobicity increased significantly ($p < 0.05$), from 830 to 1245, indicating exposure of otherwise buried hydrophobic groups. Furthermore, colloidal stability of the dispersion was improved, with separation rates decreasing from 71.23 to 24.16%$\cdot h^{-1}$. The significant enhancements in solubility, particle size reduction, and colloidal stability, highlight the potential of in-line high-shear mixing in improving the functional properties of lentil protein isolates for formulating sustainable food products with enhanced techno-functional properties.

Keywords: lentil proteins; plant proteins; in-line high-shear mixer; solubility; functional properties

Citation: Malterre, N.; Bot, F.; Lerda, E.; Arendt, E.K.; Zannini, E.; O'Mahony, J.A. Enhancing the Techno-Functional Properties of Lentil Protein Isolate Dispersions Using In-Line High-Shear Rotor-Stator Mixing. *Foods* **2024**, *13*, 283. https://doi.org/10.3390/foods13020283

Academic Editors: Norbert Raak, Ruifen Li, Laura Roman and Xiaona Guo

Received: 29 October 2023
Revised: 22 December 2023
Accepted: 10 January 2024
Published: 16 January 2024

1. Introduction

The need to tackle global challenges, ranging from climate change to food insecurity and chronic diseases, as well as the growing demand for healthy and sustainable food protein ingredients have recently driven interest in plant proteins [1–3]. Multiple protein sources including legumes, pseudocereals, cereals, tubers, grains, nuts, and seeds offer diverse nutritional profiles and health benefits [3,4]. Among these, pulses and more particularly lentils emerge as an interesting animal protein alternative, as they are rich in essential amino acids, dietary fibres, vitamins, and minerals as well as being affordable, sustainable and abundant raw materials [5]. Despite their positive attributes, the widespread utilisation of lentil proteins in food products has been impeded by their limited solubility [5,6]. Indeed, solubility is a fundamental functional property of proteins that plays a crucial role in processes such as gelation, foaming and emulsification where soluble proteins allow for the stabilisation of oil globules by adsorption at the oil/water interface [7–9]. On the other hand, the low solubility of protein-based ingredients, impacted by extraction processing parameters (e.g., pH changes, drying temperatures), can lead to processing challenges such as clumping, phase separation, sedimentation, and overall poor physical stability [6,10]. To enhance the functional properties of plant-based proteins, a range of strategies have been investigated, including chemical and enzymatic modifications, as well as physical treatments [11,12]. High-pressure homogenisation (HPH), a well-regarded technique known for its ability to modify protein structures and enhance their functional properties, has been widely used in recent years [13–16]. Several studies have shown that HPH could be used to improve the functional properties of plant-protein ingredients, particularly their solubility. More specifically, HPH treatment in the range of 70–150 MPa could lead to a 20 to 30% increase in solubility [17–19]. Similarly, Saricaoglu (2020) [20] has shown that HPH treatment in the pressure range of 0–150 MPa could increase the solubility of the protein ingredients from 32 to 46%. Other studies have investigated the effect of HPH on protein ingredients for the formulation of lentil protein stabilised emulsions, such as Jeske et al. (2019) [21] and Primozic et al. (2018) [22], who studied the suitability of HPH (34 and 103 MPa) to improve lentil protein isolate (LPI) for use in nanoemulsions. The results showed that HPH treatment allowed for a reduction in particle size and hydrophobicity. However, while effective, HPH comes with challenges as it is resource-intensive, both in terms of capital cost and energy consumption [23,24].

On the other hand, high-shear mixers, which are composed of a rotor-stator allowing for a shearing rate in the range of 20,000–100,000 s^{-1}, are highly efficient in increasing powder solubility in a cost-effective manner [25,26]. This physical treatment has been used extensively in the dairy industry during batch compounding for the disintegration and integration of powder particles into various aqueous phases. Indeed, O'Sullivan et al. (2017) [27] compared different methods of powder rehydration and found that the use of an in-line high-shear mixer allowed for the best reduction in particle size of the dispersion from 20 to 10 μm. Furthermore, extensive literature is available on the use of in-line shear mixers for the formulation of stable emulsions [28–31]. Scholz and Keck (2015) [32] have shown that an in-line high-shear rotor-stator mixer could allow for the formulation of stable emulsions with droplet size slightly larger than the one obtained by high-pressure homogenization; however, no evidence of differences in the functionalities of the emulsions were found.

Even though the use of in-line high-shear mixers in the dairy industry for the disruption of dairy powder particles and formulation of stable emulsions has been well documented, to the author's knowledge, there is no information available on the influence of in-line high-shear mixers on plant protein solubility and functional properties. Consequently, this study aims to investigate the potential of high-shear mixing to enhance the functional properties of lentil protein dispersions. The results of this study will support the development of cost-effective, sustainable processing solutions for the use of plant-based protein ingredients in food systems including infant formulae, milk, and cheese alternatives.

2. Materials and Methods

2.1. Preparation of Lentil Protein Isolate Dispersions

The lentil protein dispersions were prepared by rehydrating red lentil protein isolate powder (72.15% protein, *w*/*w*), provided by Döhler GmbH (Darmstadt, Germany), at a concentration of 5% powder (*w*/*v*) (corresponding to 3.61% protein (*w*/*v*)) in pre-heated deionised water (70 °C), stirred magnetically at 300 rpm for 1 h at 20 °C. While the protein content of the lentil protein isolate powder was not as high as existing protein isolates generally used in the formulation of infant nutritional products, it is in line with previous studies on plant-based nutritional products [21,33,34]. The main non-protein constituents of such ingredients are typically fat, starch and fibre, with the starch and fibre components, in particular, having the potential to contribute to functional properties (see Section 3.2 for further details). The dispersions were then adjusted to pH 6.8 [33,35] and allowed to rehydrate fully at 4 °C for 18 h, after which the samples were equilibrated at 20 °C and pH was readjusted, if necessary, to pH 6.8. This pH was chosen to align with that typical of infant formula as employed in previous research from our group [33].

2.2. High-Shear Mixing Treatment

The lentil protein dispersion was subjected to a semi-continuous in-line high-shear treatment using a Silverson laboratory high-shear mixer (L5M-A; Silverson, East Longmeadow, MA, USA), equipped with an in-line mixing assembly to allow for recirculation. A peristaltic pump (Watson Marlow, Marlow, UK) was used to feed the shearing head. A valve was set on the outlet of the shearing head to apply 0.5 bar of back pressure and the outlet was directed to the feeding tank. Samples were collected from the outlet after 0 (no shearing), 1, 3, 5 and 15 min of shearing at 10,200 rpm. A control sample, which did not pass through the shearing system was included in the analysis.

2.3. Particle Size Distribution

A Mastersizer 3000 (Malvern Instruments, Worcestershire, UK) static light scattering device fitted with an automated Hydro MV wet and dry cell disperser was used to evaluate the particle size distribution (PSD) of the dispersion and powder sample forms respectively. The protein's refractive index was set at 1.338, the dispersant's refractive index at 1.33, and the adsorption index was 0.001. The dispersions were diluted in ultrapure water until 10% laser obscuration was achieved (at 20 °C). The volume-weighted mean particle size (D[4,3]), specific surface area, and the particle sizes below which 10 and 90% of the sample volume are detected (Dv(10) and Dv(90)) are presented as the results.

2.4. Confocal Laser Scanning Microscopy

Microstructural analysis of the dispersions was performed using an OLYMPUS FV1000 confocal laser scanning biological microscope (Olympus Corporation, Tokyo, Japan). The dispersions were prepared as described by Malterre et al. (2023) [34] with the following modifications, whereby 1 mL of the dispersion was mixed in 4 mL of low gelling temperature agarose (Sigma-Aldrich, St. Louis, MO, USA) solution (1.5%, *w*/*v*) at 37 °C to fix the dispersion and allow for higher quality imaging. As described by Grasso et al. (2021) [36], 20 µL of Fast Green FCF aqueous solution (200 µL of 0.1 g/L) was added to 1 mL of the dispersion-agarose sample, which was then incubated at 20 °C until gelation. Fast Green FCF was excited at 633 nm, and representative images, performed using a 20× objective lens, were reported whereby protein was stained red.

2.5. Rheological Analysis

A modular compact rheometer MCR 102e (Anton Paar, Graz, Austria) fitted with a concentric cylinder geometry was used to assess the rheological characteristics of the dispersions. The rotor's diameter was 26.7 mm, the cell's internal diameter was 28.9 mm, and the distance between the two parts at the geometrical base was 2 mm. After loading the sample (20 mL) into the cylinder and preheating it to 20 °C, the shear rate was increased to 100 s^{-1} over 30 s and maintained for 30 s. A constant temperature of 20 °C was applied throughout the measurement, and viscosity was continuously recorded. The Ostwald-de Waele model (Equation (1)) was used to describe the rheological properties of the protein dispersions while ensuring that $R^2 \geq 0.95$.

$$\tau = K\gamma^{n} \tag{1}$$

where τ is the shear stress (mPa), γ is the shear rate (s^{-1}), K is the flow consistency coefficient (mPa.s^n) and n is the flow behaviour index (dimensionless).

2.6. Protein Solubility

An RC5C plus centrifuge (Thermo ScientificTM SorvallTM, Waltham, MA, USA) with a GSA rotor was used to determine the protein solubility of the dispersions [33]. The protein concentration in the supernatant of the dispersions, obtained after centrifugation at 1000× g for 20 min at 20 °C, was measured using the Kjeldahl method, with a nitrogen-to-protein conversion factor of 6.25. The results are expressed as a percentage of protein in the supernatant (soluble protein) as compared to the total protein content of the dispersions (3.61% (w/v)).

2.7. Protein Profile Analysis

The protein profile of the dispersions, and their fractions, were assessed using sodium dodecyl sulphate-polyacrylamide gel electrophoresis (SDS-PAGE) with precast gels (Mini-PROTEAN TGX, Bio-Rad Laboratories, Hercules, CA, USA), following the method developed by Laemmli (1970) [37], with minor modifications. The samples were centrifugated at 1000× g for 20 min at 20 °C using an RC5C plus centrifuge (Thermo Scientific™ Sorvall™, Waltham, MA, USA) with a GSA rotor to isolate the soluble protein in the supernatant and diluted to a protein concentration of 2 mg/mL. In method I (non-reducing conditions), 100 µL of the diluted supernatant was mixed with 100 µL of sample loading buffer (2× Laemmli sample buffer, Bio-Rad Laboratories, Hercules, CA, USA) whereas, in method II (reducing conditions), 100 µL of the diluted dispersion was mixed with 95 µL of sample loading buffer and 5 µL of β-mercaptoethanol. Each well was loaded with 7 µL of sample solution and 5 µL of SDS-PAGE broad range protein standard (Bio-Rad laboratories, Hercules, CA, USA) was loaded as a protein molecular weight standard. The gels were run in a running buffer (10× Tris/Glycine/SDS, Bio-Rad Laboratories, Hercules, CA, USA) at 160 V for 1 h, then fixed for 15 min using a solution of water:methanol:acetic acid (40:50:10), stained with Coomassie Brilliant Blue R-250 (Bio-Rad Laboratories, Hercules, CA, USA) and destained using a solution of water:methanol:acetic acid (50:40:10) until a clear background was achieved. The gels were then stored at 4 °C in a 5% (v/v) acetic acid solution overnight to reverse the shrinkage induced during the fixing and destaining steps.

2.8. Surface Hydrophobicity

Hydrophobicity of proteins in the supernatant (dispersion centrifugated at 1000× g for 20 min as described in Section 2.6) was measured using the method previously described by Nakai (2003) [38] and Goulding et al. (2021) [39], with minor modifications, using the hydrophobic probe 8-Anilino-1-naphthalenesulfonic acid (ANS). Samples were diluted to each of the following concentrations, 0.003, 0.006, 0.009, 0.012, and 0.015% (v/v) in sample buffer (0.01 M phosphate buffer, pH 7) and 100 µL of the diluted solutions were transferred to a black 96 microwell plate, after which 50 µL of ANS solution (8 mM in 0.1 M phosphate buffer) was added. The well plate was then covered with foil and gently rocked for 15 min on a plate shaker. After this, the fluorescence intensity of the samples was measured using a Varioskan Flash microplate reader (Fisher Scientific, Ireland) with an excitation wavelength of 390 nm and an emission wavelength of 470 nm. The fluorescence intensity was plotted as a function of protein concentration (%, v/v) and only linear regressions with an R^2 higher than 0.98 were considered. The slope (S0) value was used to compare the surface hydrophobicity of proteins in the different samples.

2.9. Accelerated Physical Stability

The colloidal stability of the dispersions was measured using an analytical centrifuge (LUMiSizer®, LUM GmbH, Berlin, Germany). The samples were centrifugated through two successive cycles at 800 and 3000 rpm for 10 min each, with near-infrared light illuminating the samples throughout the analysis and the intensity of the transmitted light over the entire length of the sample was measured every 10 s. The separation rate was determined by plotting the integral of the intensity of the transmitted light over time and extracting the slope (%.h^{-1}). The light transmission profiles before centrifugation, after the first cycle and after the second cycle were extracted and plotted.

2.10. Statistical Data Analysis

All samples were formulated in independent triplicate trials and all analyses were conducted in triplicate, unless otherwise stated. The data generated was subject to one-way analysis of variance (ANOVA) using R version 4.3.1 (R foundation for statistical computing, Vienna, Austria). A paired comparison test was used to determine statistically significant differences ($p < 0.05$) between mean values for different samples, at a 95% confidence level.

3. Results and Discussion

3.1. Particle Size Distribution and Microstructure

The particle size distribution (PSD) parameter, volume-weighted mean particle diameter (D[4,3]) of the lentil powders, control sample and that processed through the in-line high-shear mixer are shown in Table 1. The red lentil protein isolate powder ingredient displayed a D[4,3] value of 30.7 ± 0.06 µm in line with the literature [34,40,41]. The control red lentil protein dispersion and the sample processed through the in-line high-shear mixer for 0 min displayed similar ($p > 0.05$) D[4,3] values, of 5.13 ± 0.58 µm and 5.00 ± 0.04 µm, respectively, indicating that the pumping through the semi-continuous system and application of back pressure did not affect the particle size distribution. Interestingly, increasing the shearing time caused a significant ($p < 0.05$) decrease in the PSD parameters, with a decrement in D[4,3] to 2.81 ± 0.06, 2.17 ± 0.04 and 1.72 ± 0.01 µm after 1, 5 and 15 min of processing, respectively. This progressive reduction in particle size can be attributed to the high shearing forces induced by the rotor of the mixing head, which causes the breakage of the powder particles with the forced passage of the particles through the perforations of the stator [30,42,43].

Table 1. Particle size distribution parameters, protein solubility, surface hydrophobicity and separation rate for lentil protein isolate powder and protein dispersions (5% powder, *w/v*) of untreated (control) and treated with high-shear mixing for 0, 1, 3, 5 and 15 min.

	D[4,3] (µm)	Span (µm)	Protein Solubility (%)	Surface Hydrophobicity (-)	Separation Rate (%.h^{-1})
Powder	30.7 ± 0.06 ^{a}	1.84 ± 0.02 ^{a}	n/a	n/a	n/a
Control	5.13 ± 0.58 ^{b}	7.25 ± 2.98 ^{b}	46.87 ± 1.55 ^{a}	830 ± 123 ^{a}	71.23 ± 18.15 ^{a}
0 min	5.00 ± 0.04 ^{b}	6.96 ± 0.16 ^{b}	48.04 ± 4.03 ^{a}	852 ± 73 ^{a}	69.67 ± 18.85 ^{a}
1 min	2.81 ± 0.06 ^{c}	4.28 ± 0.22 ^{c}	57.18 ± 5.65 ^{b}	1044 ± 26 ^{b}	36.57 ± 5.19 ^{b}
3 min	2.49 ± 0.18 ^{d}	3.99 ± 0.22 ^{c}	63.69 ± 7.14 ^{c}	1116 ± 70 ^{bc}	30.81 ± 2.63 ^{bc}
5 min	2.17 ± 0.04 ^{e}	3.72 ± 0.06 ^{d}	63.94 ± 5.70 ^{c}	1245 ± 43 ^{c}	28.89 ± 3.01 ^{cd}
15 min	1.72 ± 0.01 ^{f}	3.14 ± 0.02 ^{e}	68.42 ± 8.04 ^{d}	1236 ± 41 ^{c}	24.16 ± 2.58 ^{d}

Values within a column that share a superscript are not significantly different from one another ($p < 0.05$). D[4,3] = volume-weighted mean particle diameter. Span = measurement of the width of the distribution calculated as (Dv(90) − Dv(10))/Dv(50). n/a: not applicable.

Changes in particle size are also visible on the confocal micrographs (Figure 1), as the control and 0 min dispersions displayed large particles in the range of 40–50 µm. Furthermore, micrographs of the 1 min dispersion, showed heterogeneous particles in the range of 1–5 µm and 15–25 µm. Longer processing up to 15 min showed further reductions of the population of large particles and an increase in small particles as illustrated by an increase in the intensity of red particles on the micrographs (Figure 1).

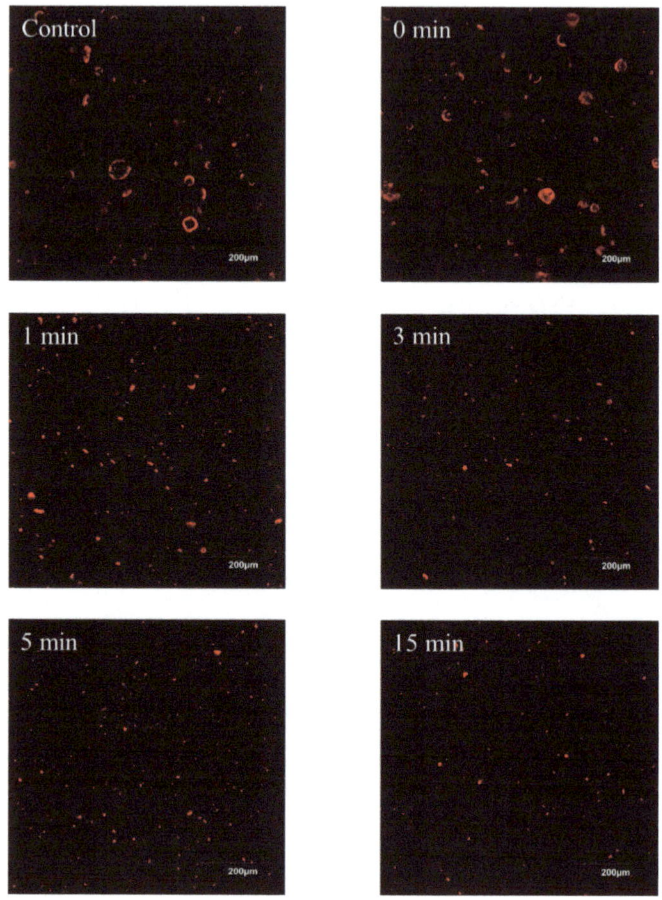

Figure 1. Confocal laser micrographs of lentil protein isolate dispersions (5% powder, *w/v*) of untreated (control) and treated with high-shear mixing for 0, 1, 3, 5, and 15 min.

The reduction in number of bigger particles can also be shown through the particle size parameters D(90) (particle size below which 90% of sample volume is found), which progressively decreased throughout the process from 12.71 $\pm$ 3.33 µm for the control dispersion to 6.83 $\pm$ 0.35 and 3.76 $\pm$ 0.07 µm after 1 and 15 min of shearing (Figure 2). On the other hand, the D(10), which represents the particle size below which 10% of the sample volume is found, had minimal reduction ($p > 0.05$) through shearing from 0.46 $\pm$ 0.05 µm for the control dispersion and 0.42 $\pm$ 0.01 µm for the dispersion treated for 1 min. This parameter then plateaued ($p > 0.05$) with increasing shearing time to reach a D(10) value of 0.35 $\pm$ 0.01 µm after 1 min. Furthermore, the particle disruption induced by the high-shearing caused a significant ($p < 0.05$) increase in specific surface area from 4.14 $\pm$ 0.46 $m^2 \cdot g^{-1}$ for the control dispersion to 5.87 $\pm$ 10 $m^2 \cdot g^{-1}$ after 15 min of shearing (Figure 2). Higher specific surface area could allow for greater exposure of each particle to water and possibly better dissolution of the powder as demonstrated by the Noyes Whitney equation, which details a link between the exposed surface and the dissolution rate of the powder [44,45]. The results are in agreement with the literature as similar results on the effect of high-shear mixing on particle disruption have been found in the literature for powders. In particular, Pacek et al. (2007) [42] and Padron and Özcan-Taskin (2018) [46], found that an in-line high-shear mixer at 8000 rpm, caused the reduction of particles of silicon from 90 μm to 9 μm after 20 min and from 100 μm to 15 μm after 30 min of shearing. The authors have also shown that in the timespan used in the present study (<15 min), in-line high-shearing did not allow for the disruption of aggregates below 1–2 μm. Similarly, O'Sullivan et al. (2017) [27] have found that the use of in-line high-shear mixer for the rehydration and dispersion of dairy powders (skim milk powder and milk protein isolate) at a concentration of 7.2% (*w/w*) caused a significant reduction of particle size from 100 to 20 μm after 15 min, while the size of the smaller particles plateaued in the range 0.1–1 μm. Overall, these results show that in-line high-shear mixing has the potential to effectively reduce the particle size of lentil protein isolates across both large and small particles/aggregates, which could have potential benefit in increasing the dissolution rate and improving overall rehydration.

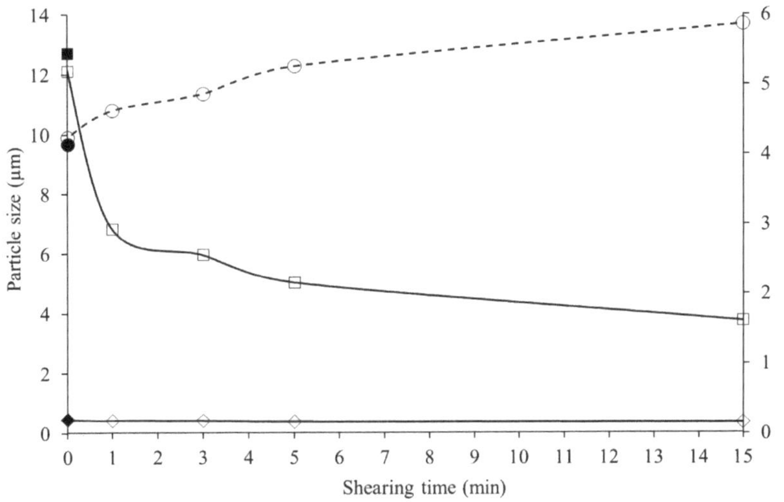

Figure 2. Particle size distribution parameters (-◇-, Dv(10), particle size below which 10% of sample volume is found), (-□- Dv(90), particle size below which 90% of sample volume is found), and (--○--, specific surface area) of protein dispersions (5% powder *w/v*) untreated (control) and treated with high-shear for 0, 1, 3, 5, and 15 min. Filled shapes (-■-, -◆- and -•-) correspond to the control dispersion.

3.2. Rheological Properties

The apparent viscosity of the dispersions as well as their flow properties n (-) and k (mPa.s^n) as defined by fitting the Ostwald de Waele model, are provided in Table 2. The apparent viscosity at 100 s^{-1} and 20 °C ranged between 1.69 ± 0.3 and 1.78 ± 0.34 mPa.s with no significant differences ($p > 0.05$) between the dispersions. Similar apparent viscosity values were measured for lentil protein dispersions with protein concentrations of 4% (*w/w*) [20]. The Ostwald de Waele model allowed for the accurate description of the rheological properties of the dispersions ($R^2 > 0.98$), and similar ($p > 0.05$) consistency coefficient ranging between $1.57 \times 10^{-3} \pm 0.81 \times 10^{-3}$ and $1.87 \times 10^{-3} \pm 0.72 \times 10^{-3}$ mPa.s were observed over the range of treatments. The flow behaviour index of all dispersions ranged between 1.00 ± 0.06 and 1.03 ± 0.09 mPa.s^n, indicating Newtonian behaviour of the dispersions. Similar results have been observed by Jeske et al. (2019) [21] who studied the HPH treatment of lentil protein isolate emulsions at a concentration of 3% (*w/w*). Furthermore, the impact of starch and fibre content on the rheological properties of plant protein emulsions was not evaluated as part of this study. According to Dikeman and Fahey (2006) [47] changes in starch or fibre content could modify the viscosity of the samples and therefore have an impact on the shearing and mechanical forces induced by high-shear mixing; therefore, future studies should investigate the impact of starch and fibre content on the techno-functional properties of lentil protein isolate dispersions for use in food systems.

Table 2. Apparent viscosity (η at 100 s^{-1}, 20 °C), flow behaviour (n) and consistency coefficient (K) of protein dispersions (5%, *w/v*, powder) untreated (control) and treated with high-shear treatment for 0, 1, 3, 5 and 15 min.

Treatment	η (mPa.s)	n (-)	K ($mPa.s^n$)
Control	1.74 ± 0.35 [a]	1.02 ± 0.06 [a]	$1.73 \times 10^{-3} \pm 0.78 \times 10^{-3}$ [a]
0 min	1.75 ± 0.33 [a]	1.02 ± 0.07 [a]	$1.77 \times 10^{-3} \pm 0.76 \times 10^{-3}$ [a]
1 min	1.69 ± 0.3 [a]	1.02 ± 0.07 [a]	$1.57 \times 10^{-3} \pm 0.81 \times 10^{-3}$ [a]
3 min	1.73 ± 0.26 [a]	1.00 ± 0.06 [a]	$1.70 \times 10^{-3} \pm 0.60 \times 10^{-3}$ [a]
5 min	1.72 ± 0.32 [a]	1.03 ± 0.09 [a]	$1.67 \times 10^{-3} \pm 0.75 \times 10^{-3}$ [a]
15 min	1.78 ± 0.34 [a]	1.00 ± 0.05 [a]	$1.87 \times 10^{-3} \pm 0.72 \times 10^{-3}$ [a]

Values within a column that share a superscript are not significantly different from one another ($p > 0.05$).

3.3. Protein Solubility

Protein solubility is an essential functional property which strongly influences other protein functional properties, such as emulsifying, foaming or gelling properties [7,9]. The control dispersion had a protein solubility of 46.87 ± 1.55%, similar ($p > 0.05$) to the dispersion treated for 0 min displaying a solubility of 48.04 ± 4.03% (Table 1), as expected for plant protein ingredients which have low solubility as compared to dairy isolates (>90% in similar conditions) [6]. The high shearing generated through the treatment allowed for a 20% increase in solubility when treated for 1 min (57.18 ± 5.65%), and continuously increased up to a value of 68.42 ± 8.04% for the dispersion treated for 15 min, allowing for 42% increase in solubility. The increase in solubility can be attributed to the particle breakage caused by the intense shearing as observed in Table 2. Similar results have been observed by other authors who showed that mechanical shearing up to 8000 rpm can increase the solubility of protein ingredients by breaking down the protein aggregates and facilitating their dispersion and dissolution [27,48].

The breakages and disruptions induced by the in-line high-shear treatment, reducing the particle size, could have allowed for better dispersion of the agglomerated powder particles Schuck et al. (2012) [49], increasing the rehydration and solubility of the ingredients as shown by other authors who worked on the improvement of protein ingredients through mechanical processing such as the well-documented use of HPH for the improvement of plant protein functionalities [18,50–54].

A better understanding of the protein solubilisation as affected by processing time was obtained by analysing the protein profile of the soluble fractions of the dispersions (Figure 3a,b). The supernatant of lentil protein isolate dispersions had proteins with molecular weight of 50 kDa, which may correspond to vicilin subunits, and the bands at 37 and 25 kDa may correspond to the acidic and basic subunits of legumin. Under reducing conditions, legumin dissociated into its acidic and basic subunits due to the dissociation of disulphide bonds, resulting in a decrease in the intensity of the bands at 50 and 37 kDa and an increase in the intensity of the bands at 20–25 kDa. Similar results have been observed by Jarpa-Parra et al. (2015) [55], Barbana and Boye (2013) [56], Joshi et al. (2012) [57], and Alonso-Miravalles et al. (2019) [41] The protein profile of all the samples were similar to the one of the control sample, indicating that in-line high-shear treatment did not affect the profile of solubilised proteins.

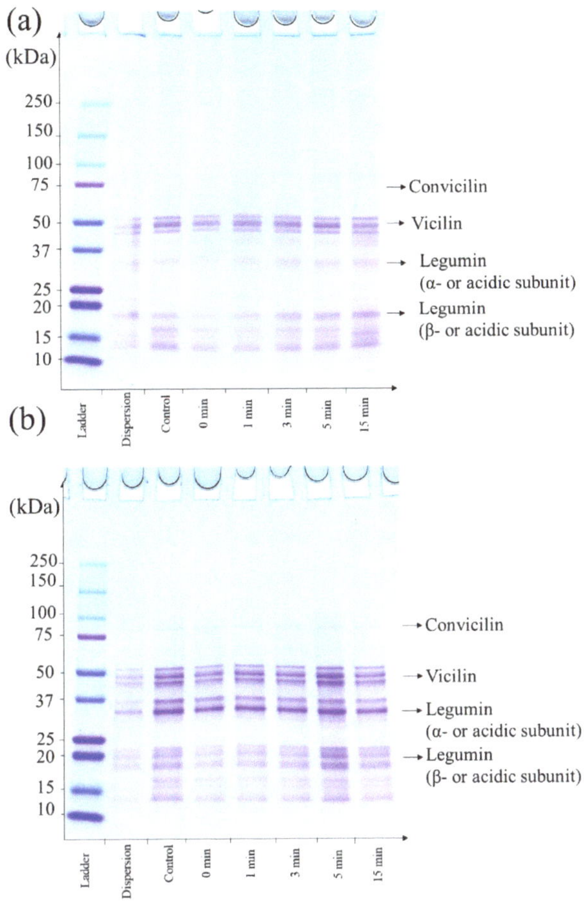

Figure 3. Representative sodium dodecyl sulphate-polyacrylamide gel electrophoretograms (SDS-PAGE) of lentil protein isolate dispersions (5% w/v) untreated, control supernatant untreated and supernatant treated with high-shear-mixing for 0, 1, 3, 5 and 15 min under non reducing (**a**) and reducing (**b**) conditions.

3.4. Surface Hydrophobicity

Understanding hydrophobic interactions is crucial for the development of food products containing high concentrations of protein ingredients, as they have a strong contribution to stability and overall interfacial properties. The control sample had the lowest value of 830 ± 123, similar ($p > 0.05$) to the sample treated for 0 min, revealing a negligible effect of the pumping in the semi-continuous system studied (Table 1). Treating the sample at 10,200 rpm for 1 min allowed for a significant ($p < 0.05$) increase of the surface hydrophobicity to 1044 ± 26, which reached the maximum value of 1245 ± 43 for the sample treated for 5 min. This increase in surface hydrophobicity could be attributed to exposure of hydrophobic residues, otherwise buried, as a result of mechanical disruption [39,52]. The results agree with the literature, where the application of mechanical forces (i.e., shear, cavitation, turbulences) caused an increase in surface hydrophobicity and thus an improvement of plant protein functionalities (i.e., emulsifying properties) [52,58].

3.5. Colloidal Stability

The physical stability of the dispersions was analysed by measuring the transmission of light through the samples during centrifugation. In Figure 4, representative light transmission profiles for each of the dispersions treated at 10,200 rpm from 0 to 15 min are shown. The control and the sample treated for 0 min showed low physical stabilities with separation rates of 71.23 ± 18.15%.h^{-1} and 69.67 ± 18.85%.h^{-1}, after the second centrifugating cycle, respectively. In addition, at low centrifugation speed (i.e., 800 rpm), it can be observed that the treated dispersions (1, 3, 5 and 15 min) did not undergo any changes through the first cycle, indicating relative physical stability of the large, dispersed particle as also underlined by Crowley et al. (2019) [59]. When increasing the speed to 3000 rpm during the second cycle, complete sedimentation and destabilisation of the samples can be measured. In particular, the separation rate of the dispersion decreased sharply ($p < 0.05$) with the treatment of 1 min (36.57 ± 5.19%.h^{-1}) indicating a slower movement of particles in the samples as a result of particle disruption and particle size reduction, which resulted in more colloidally stable dispersions. The separation rate decreased progressively to reach a value of 24.16 ± 2.58%.h^{-1} for the dispersion treated for 15 min. The results are in agreement with the literature, particularly Moll et al. (2021) [60], who found that the reduction in particle size caused by physical treatment with microfluidisation allowed for an increase in colloidal stability and Jeske et al. (2019) [21] using HPH at 18 MPa observed a reduction in separation rate from 48 to 22%.h^{-1}.

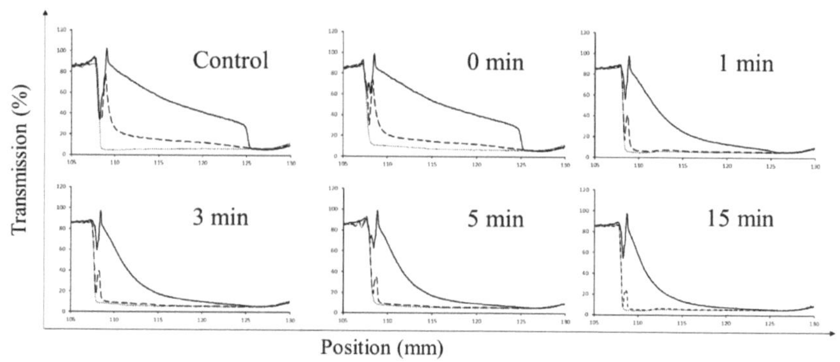

Figure 4. Accelerated physical stability profiles of lentil protein isolate dispersions (5% w/v powder) untreated (control) and treated with high-shear treatment for 0, 1, 3, 5 and 15 min. Original light transmission of the dispersions (time 0 min) (. . . .) together with the profile after the first cycle (10 min at 800 rpm) (- -) and the profile after the second cycle (10 min at 3000 rpm) (-) are shown.

4. Conclusions

The effect of in-line high-shear mixing at 10,200 rpm on the techno-functional properties of lentil protein isolate was studied during treatment times ranging between 0 and 15 min. It was shown that the high-shear and long time provided by the mixer allowed for a 42% improvement in solubility of the protein ingredients from 46.87 ± 1.55% of soluble protein at 0 min up to 68.42 ± 8.04% when treated for 15 min. This increase in solubility has been attributed to the high shearing induced at the rotor and the passage through the small perforations of the stator, with a concomitant significant ($p < 0.05$) reduction in particle size from 5.13 ± 0.58 µm (at 0 min) to 1.72 ± 0.01 µm after 15 min. Furthermore, the particle disruption caused an increase in hydrophobicity as hydrophobic groups exposed to the surface have been observed, which can be associated with an improvement of the emulsifying properties of the lentil protein isolate. The lentil protein isolate dispersions after in-line high-shear treatment had improved colloidal stability with values of 69.67%.h^{-1} at 0 min, down to 35.77%.h^{-1} after 1 min of treatment and further treatment for 15 min reduced it to 24.16%.h^{-1}. These results show that in-line high-shear mixing has the potential to improve the functional properties of lentil protein isolate dispersions for the formulation of more sustainable food products, including infant formulae with enhanced techno-functional properties.

Author Contributions: N.M., J.A.O. and F.B. conceived and designed the experiments; N.M. and E.L. performed the experiments and collated the data; N.M. analysed the data and prepared the manuscript; N.M., F.B., E.K.A., E.Z. and J.A.O. reviewed the manuscript. All authors have read and agreed to the published version of the manuscript.

Funding: The work for this publication has been undertaken as part of the SMART PROTEIN project. This project has received funding from the European Union's Horizon 2020 research and innovation program under grant agreement no. 862957.

Data Availability Statement: Data will be made available on request.

Acknowledgments: The authors would like to thank our colleagues Avril McCord, David Waldron and Johann Scollard for their technical assistance and expertise that greatly assisted the work.

Conflicts of Interest: The authors declare no conflicts of interest.

References

1. *World Population Prospects 2022: Summary of Results*; United Nations: New York, NY, USA, 2022; ISBN 978-92-1-148373-4.
2. Henchion, M.; Hayes, M.; Mullen, A.; Fenelon, M.; Tiwari, B. Future Protein Supply and Demand: Strategies and Factors Influencing a Sustainable Equilibrium. *Foods* **2017**, *6*, 53. [CrossRef] [PubMed]
3. Day, L.; Cakebread, J.A.; Loveday, S.M. Food Proteins from Animals and Plants: Differences in the Nutritional and Functional Properties. *Trends Food Sci. Technol.* **2022**, *119*, 428–442. [CrossRef]
4. Shaghaghian, S.; McClements, D.J.; Khalesi, M.; Garcia-Vaquero, M.; Mirzapour-Kouhdasht, A. Digestibility and Bioavailability of Plant-Based Proteins Intended for Use in Meat Analogues: A Review. *Trends Food Sci. Technol.* **2022**, *129*, 646–656. [CrossRef]
5. Can Karaca, A.; Low, N.; Nickerson, M. Emulsifying Properties of Chickpea, Faba Bean, Lentil and Pea Proteins Produced by Isoelectric Precipitation and Salt Extraction. *Food Res. Int.* **2011**, *44*, 2742–2750. [CrossRef]
6. Grossmann, L.; McClements, D.J. Current Insights into Protein Solubility: A Review of Its Importance for Alternative Proteins. *Food Hydrocoll.* **2023**, *137*, 108416. [CrossRef]
7. Zhao, H.; Shen, C.; Wu, Z.; Zhang, Z.; Xu, C. Comparison of Wheat, Soybean, Rice, and Pea Protein Properties for Effective Applications in Food Products. *J. Food Biochem.* **2020**, *44*, e13157. [CrossRef] [PubMed]
8. Ebert, S.; Gibis, M.; Terjung, N.; Weiss, J. Survey of Aqueous Solubility, Appearance, and pH of Plant Protein Powders from Carbohydrate and Vegetable Oil Production. *LWT* **2020**, *133*, 110078. [CrossRef]
9. Day, L. Proteins from Land Plants—Potential Resources for Human Nutrition and Food Security. *Trends Food Sci. Technol.* **2013**, *32*, 25–42. [CrossRef]
10. Grossmann, L.; Weiss, J. Alternative Protein Sources as Technofunctional Food Ingredients. *Annu. Rev. Food Sci. Technol.* **2021**, *12*, 93–117. [CrossRef]
11. Bou, R.; Navarro-Vozmediano, P.; Domínguez, R.; López-Gómez, M.; Pinent, M.; Ribas-Agustí, A.; Benedito, J.J.; Lorenzo, J.M.; Terra, X.; García-Pérez, J.V.; et al. Application of Emerging Technologies to Obtain Legume Protein Isolates with Improved Techno-Functional Properties and Health Effects. *Compr. Rev. Food Sci. Food Saf.* **2022**, *21*, 2200–2232. [CrossRef]
12. Mirmoghtadaie, L.; Shojaee Aliabadi, S.; Hosseini, S.M. Recent Approaches in Physical Modification of Protein Functionality. *Food Chem.* **2016**, *199*, 619–627. [CrossRef] [PubMed]
13. Gharibzahedi, S.M.T.; El Kantar, S.; Pasdar, N.; Altintas, Z.; Koubaa, M. Chapter 7—Emerging Technologies for Processing of Plant Proteins. In *Processing Technologies and Food Protein Digestion*; Bhat, Z.F., Morton, J.D., Bekhit, A.E.-D.A., Suleria, H.A.R., Eds.; Academic Press: Cambridge, MA, USA, 2023; pp. 175–208, ISBN 978-0-323-95052-7.
14. Sá, A.G.A.; Laurindo, J.B.; Moreno, Y.M.F.; Carciofi, B.A.M. Influence of Emerging Technologies on the Utilization of Plant Proteins. *Front. Nutr.* **2022**, *9*, 809058. [CrossRef] [PubMed]
15. Avelar, Z.; Vicente, A.A.; Saraiva, J.A.; Rodrigues, R.M. The Role of Emergent Processing Technologies in Tailoring Plant Protein Functionality: New Insights. *Trends Food Sci. Technol.* **2021**, *113*, 219–231. [CrossRef]
16. Tobin, J.; Heffernan, S.P.; Mulvihill, D.M.; Huppertz, T.; Kelly, A.L. Applications of High-Pressure Homogenization and Microfluidization for Milk and Dairy Products. In *Emerging Dairy Processing Technologies*; Datta, N., Tomasula, P.M., Eds.; John Wiley & Sons, Ltd.: Chichester, UK, 2015; pp. 93–114, ISBN 978-1-118-56047-1.
17. Keuleyan, E.; Gélébart, P.; Beaumal, V.; Kermarrec, A.; Ribourg-Birault, L.; Le Gall, S.; Meynier, A.; Riaublanc, A.; Berton-Carabin, C. Pea and Lupin Protein Ingredients: New Insights into Endogenous Lipids and the Key Effect of High-Pressure Homogenization on Their Aqueous Suspensions. *Food Hydrocoll.* **2023**, *141*, 108671. [CrossRef]
18. Zhao, S.; Huang, Y.; McClements, D.J.; Liu, X.; Wang, P.; Liu, F. Improving Pea Protein Functionality by Combining High-Pressure Homogenization with an Ultrasound-Assisted Maillard Reaction. *Food Hydrocoll.* **2022**, *126*, 107441. [CrossRef]

19. Moll, P.; Salminen, H.; Griesshaber, E.; Schmitt, C.; Weiss, J. Homogenization Improves Foaming Properties of Insoluble Pea Proteins. *J. Food Sci.* **2022**, *87*, 4622–4635. [CrossRef]
20. Saricaoglu, F.T. Application of High-Pressure Homogenization (HPH) to Modify Functional, Structural and Rheological Properties of Lentil (*Lens culinaris*) Proteins. *Int. J. Biol. Macromol.* **2020**, *144*, 760–769. [CrossRef]
21. Jeske, S.; Bez, J.; Arendt, E.K.; Zannini, E. Formation, Stability, and Sensory Characteristics of a Lentil-Based Milk Substitute as Affected by Homogenisation and Pasteurisation. *Eur. Food Res. Technol.* **2019**, *245*, 1519–1531. [CrossRef]
22. Primozic, M.; Duchek, A.; Nickerson, M.; Ghosh, S. Formation, Stability and in Vitro Digestibility of Nanoemulsions Stabilized by High-Pressure Homogenized Lentil Proteins Isolate. *Food Hydrocoll.* **2018**, *77*, 126–141. [CrossRef]
23. Fayaz, G.; Soleimanian, Y.; Mhamadi, M.; Turgeon, S.L.; Khalloufi, S. The Applications of Conventional and Innovative Mechanical Technologies to Tailor Structural and Functional Features of Dietary Fibers from Plant Wastes: A Review. *Compr. Rev. Food Sci. Food Saf.* **2022**, *21*, 2149–2199. [CrossRef]
24. Calligaris, S.; Plazzotta, S.; Bot, F.; Grasselli, S.; Malchiodi, A.; Anese, M. Nanoemulsion Preparation by Combining High Pressure Homogenization and High Power Ultrasound at Low Energy Densities. *Food Res. Int.* **2016**, *83*, 25–30. [CrossRef]
25. O'Sullivan, J.J.; Drapala, K.P.; Kelly, A.L.; O'Mahony, J.A. The Use of Inline High-Shear Rotor-Stator Mixing for Preparation of High-Solids Milk Protein-Stabilised Oil-in-Water Emulsions with Different Protein:Fat Ratios. *J. Food Eng.* **2018**, *222*, 218–225. [CrossRef]
26. Hall, S.; Cooke, M.; El-Hamouz, A.; Kowalski, A.J. Droplet Break-up by in-Line Silverson Rotor–Stator Mixer. *Chem. Eng. Sci.* **2011**, *66*, 2068–2079. [CrossRef]
27. O'Sullivan, J.J.; Schmidmeier, C.; Drapala, K.P.; O'Mahony, J.A.; Kelly, A.L. Monitoring of Pilot-Scale Induction Processes for Dairy Powders Using Inline and Offline Approaches. *J. Food Eng.* **2017**, *197*, 9–16. [CrossRef]
28. Van Der Schaaf, U.S.; Karbstein, H.P. Fabrication of Nanoemulsions by Rotor-Stator Emulsification. In *Nanoemulsions*; Elsevier: Amsterdam, The Netherlands, 2018; pp. 141–174, ISBN 978-0-12-811838-2.
29. Schuch, A.; Deiters, P.; Henne, J.; Köhler, K.; Schuchmann, H.P. Production of W/O/W (Water-in-Oil-in-Water) Multiple Emulsions: Droplet Breakup and Release of Water. *J. Colloid Interface Sci.* **2013**, *402*, 157–164. [CrossRef] [PubMed]
30. Pacek, A.W.; Hall, S.; Cooke, M.; Kowalski, A.J. Emulsification in Rotor–Stator Mixers. In *Emulsion Formation and Stability*; Tadros, T.F., Ed.; Wiley: Hoboken, NJ, USA, 2013; pp. 127–167, ISBN 978-3-527-31991-6.
31. Rueger, P.E.; Calabrese, R.V. Dispersion of Water into Oil in a Rotor–Stator Mixer. Part 2: Effect of Phase Fraction. *Chem. Eng. Res. Des.* **2013**, *91*, 2134–2141. [CrossRef]
32. Scholz, P.; Keck, C.M. Nanoemulsions Produced by Rotor–Stator High Speed Stirring. *Int. J. Pharm.* **2015**, *482*, 110–117. [CrossRef] [PubMed]
33. Alonso-Miravalles, L. Physicochemical Characterisation of Plant-Based Protein Ingredients for the Development of Infant Nutritional Formulations. Ph.D. Thesis, University College Cork, Cork, Ireland, 2020.
34. Malterre, N.; Bot, F.; O'Mahony, J.A. Formulation and Physical Stability of High Total Solids Lentil Protein-Stabilised Emulsions for Use in Plant Protein-Based Young Child Formulae. *Foods* **2023**, *12*, 1741. [CrossRef]
35. De Figueiredo Furtado, G.; da Silva Carvalho, A.G.; Hubinger, M.D. Model Infant Formulas: Influence of Types of Whey Proteins and Oil Composition on Emulsion and Powder Properties. *J. Food Eng.* **2021**, *292*, 110256. [CrossRef]
36. Grasso, N.; Roos, Y.H.; Crowley, S.V.; Arendt, E.K.; O'Mahony, J.A. Composition and Physicochemical Properties of Commercial Plant-Based Block-Style Products as Alternatives to Cheese. *Future Foods* **2021**, *4*, 100048. [CrossRef]
37. Laemmli, U.K. Cleavage of Structural Proteins during the Assembly of the Head of Bacteriophage T4. *Nature* **1970**, *227*, 680–685. [CrossRef] [PubMed]
38. Nakai, S. Measurement of Protein Hydrophobicity. *Curr. Protoc. Food Anal. Chem.* **2003**, *9*, B5.2.1–B5.2.13. [CrossRef]
39. Goulding, D.A.; O'Regan, J.; Bovetto, L.; O'Brien, N.M.; O'Mahony, J.A. Influence of Thermal Processing on the Physicochemical Properties of Bovine Lactoferrin. *Int. Dairy J.* **2021**, *119*, 105001. [CrossRef]
40. Vogelsang-O'Dwyer, M.; Sahin, A.W.; Bot, F.; O'Mahony, J.A.; Bez, J.; Arendt, E.K.; Zannini, E. Enzymatic Hydrolysis of Lentil Protein Concentrate for Modification of Physicochemical and Techno-Functional Properties. *Eur. Food Res. Technol.* **2023**, *249*, 573–586. [CrossRef]
41. Alonso-Miravalles, L.; Jeske, S.; Bez, J.; Detzel, A.; Busch, M.; Krueger, M.; Wriessnegger, C.L.; O'Mahony, J.A.; Zannini, E.; Arendt, E.K. Membrane Filtration and Isoelectric Precipitation Technological Approaches for the Preparation of Novel, Functional and Sustainable Protein Isolate from Lentils. *Eur. Food Res. Technol.* **2019**, *245*, 1855–1869. [CrossRef]
42. Pacek, A.W.; Ding, P.; Utomo, A.T. Effect of Energy Density, pH and Temperature on de-Aggregation in Nano-Particles/Water Suspensions in High Shear Mixer. *Powder Technol.* **2007**, *173*, 203–210. [CrossRef]
43. Ding, P.; Orwa, M.G.; Pacek, A.W. De-Agglomeration of Hydrophobic and Hydrophilic Silica Nano-Powders in a High Shear Mixer. *Powder Technol.* **2009**, *195*, 221–226. [CrossRef]
44. Dokoumetzidis, A.; Macheras, P. A Century of Dissolution Research: From Noyes and Whitney to the Biopharmaceutics Classification System. *Int. J. Pharm.* **2006**, *321*, 1–11. [CrossRef]
45. Noyes, A.A.; Whitney, W.R. The rate of solution of solid substances in their own solutions. *J. Am. Chem. Soc.* **1897**, *19*, 930–934. [CrossRef]
46. Padron, G.A.; Özcan-Taşkın, N.G. Particle De-Agglomeration with an in-Line Rotor-Stator Mixer at Different Solids Loadings and Viscosities. *Chem. Eng. Res. Des.* **2018**, *132*, 913–921. [CrossRef]

47. Dikeman, C.L.; Fahey, G.C. Viscosity as Related to Dietary Fiber: A Review. *Crit. Rev. Food Sci. Nutr.* **2006**, *46*, 649–663. [CrossRef] [PubMed]
48. Pereira, R.N.; Rodrigues, R.M.; Ramos, Ó.L.; Xavier Malcata, F.; Teixeira, J.A.; Vicente, A.A. Production of Whey Protein-Based Aggregates Under Ohmic Heating. *Food Bioprocess. Technol.* **2016**, *9*, 576–587. [CrossRef]
49. Schuck, P.; Dolivet, A.; Jeantet, R. *Analytical Methods for Food and Dairy Powders*; Wiley-Blackwell: Chichester, UK, 2012; ISBN 978-0-470-65598-6.
50. Melchior, S.; Moretton, M.; Calligaris, S.; Manzocco, L.; Nicoli, M.C. High Pressure Homogenization Shapes the Techno-Functionalities and Digestibility of Pea Proteins. *Food Bioprod. Process.* **2022**, *131*, 77–85. [CrossRef]
51. Yang, J.; Liu, G.; Zeng, H.; Chen, L. Effects of High Pressure Homogenization on Faba Bean Protein Aggregation in Relation to Solubility and Interfacial Properties. *Food Hydrocoll.* **2018**, *83*, 275–286. [CrossRef]
52. Yu, S.; Liu, J.-J.; Yun, E.J.; Kwak, S.; Kim, K.H.; Jin, Y.-S. Production of a Human Milk Oligosaccharide 2′-Fucosyllactose by Metabolically Engineered Saccharomyces Cerevisiae. *Microb. Cell Fact.* **2018**, *17*, 101. [CrossRef] [PubMed]
53. Song, X.; Zhou, C.; Fu, F.; Chen, Z.; Wu, Q. Effect of High-Pressure Homogenization on Particle Size and Film Properties of Soy Protein Isolate. *Ind. Crops Prod.* **2013**, *43*, 538–544. [CrossRef]
54. Bader, S.; Bez, J.; Eisner, P. Can Protein Functionalities Be Enhanced by High-Pressure Homogenization?—A Study on Functional Properties of Lupin Proteins. *Procedia Food Sci.* **2011**, *1*, 1359–1366. [CrossRef]
55. Jarpa-Parra, M.; Bamdad, F.; Tian, Z.; Zeng, H.; Temelli, F.; Chen, L. Impact of pH on Molecular Structure and Surface Properties of Lentil Legumin-like Protein and Its Application as Foam Stabilizer. *Colloids Surf. B Biointerfaces* **2015**, *132*, 45–53. [CrossRef]
56. Barbana, C.; Boye, J.I. In Vitro Protein Digestibility and Physico-Chemical Properties of Flours and Protein Concentrates from Two Varieties of Lentil (*Lens culinaris*). *Food Funct.* **2013**, *4*, 310–321. [CrossRef]
57. Joshi, M.; Adhikari, B.; Aldred, P.; Panozzo, J.F.; Kasapis, S.; Barrow, C.J. Interfacial and Emulsifying Properties of Lentil Protein Isolate. *Food Chem.* **2012**, *134*, 1343–1353. [CrossRef]
58. Xu, Y.; Wang, G.; Wang, X.; Yu, J.; Wang, J.; Zhang, Z.; Li, R. Effects of Homogenization on the Molecular Flexibility and Emulsifying Properties of Soy Protein Isolate. *Food Sci. Biotechnol.* **2018**, *27*, 1293–1299. [CrossRef] [PubMed]
59. Crowley, S.V.; Kelly, A.L.; O'Mahony, J.A.; Lucey, J.A. Colloidal Properties of Protein Complexes Formed in β-Casein Concentrate Solutions as Influenced by Heating and Cooling in the Presence of Different Solutes. *Colloids Surf. B Biointerfaces* **2019**, *174*, 343–351. [CrossRef] [PubMed]
60. Moll, P.; Salminen, H.; Schmitt, C.; Weiss, J. Impact of Microfluidization on Colloidal Properties of Insoluble Pea Protein Fractions. *Eur. Food Res. Technol.* **2021**, *247*, 545–554. [CrossRef]

foods

MDPI

Article

Dietary Supplementation with Popped Amaranth Modulates the Gut Microbiota in Low Height-for-Age Children: A Nonrandomized Pilot Trial

Oscar de Jesús Calva-Cruz [1,†], Cesaré Ovando-Vázquez [2,†], Antonio De León-Rodríguez [1], Fabiola Veana [3], Eduardo Espitia-Rangel [4], Samuel Treviño [5] and Ana Paulina Barba-de la Rosa [1,*]

1 Molecular Biology Division, Instituto Potosino de Investigación Científica y Tecnológica, San Luis Potosí 78216, Mexico; oscar.calva@ipicyt.edu.mx (O.d.J.C.-C.); aleonr@ipicyt.edu.mx (A.D.L.-R.)
2 CONACYT-Centro Nacional de Supercómputo, Instituto Potosino de Investigación Científica y Tecnológica, San Luis Potosí 78216, Mexico; cesare.ovando@ipicyt.edu.mx
3 Tecnológico Nacional de México, Instituto Tecnológico de Ciudad Valles, Ciudad Valles 79010, Mexico; fabiola.veana@tecvalles.mx
4 Instituto Nacional de Investigaciones Forestales, Agrícolas y Pecuarias, Texcoco 56250, Mexico; espitia.eduardo@inifap.gob.mx
5 Facultad de Ciencias Químicas, Benemérita Universidad Autónoma de Puebla, Av. San Claudio S/N, Ciudad Universitaria, Puebla 72000, Mexico; samuel_trevino@hotmail.com
* Correspondence: apbarba@ipicyt.edu.mx
† These authors contributed equally to this work.

Citation: Calva-Cruz, O.d.J.; Ovando-Vázquez, C.; De León-Rodríguez, A.; Veana, F.; Espitia-Rangel, E.; Treviño, S.; Barba-de la Rosa, A.P. Dietary Supplementation with Popped Amaranth Modulates the Gut Microbiota in Low Height-for-Age Children: A Nonrandomized Pilot Trial. *Foods* **2023**, *12*, 2760. https://doi.org/10.3390/foods12142760

Academic Editors: Norbert Raak, Ruifen Li and Laura Roman

Received: 15 June 2023
Revised: 17 July 2023
Accepted: 18 July 2023
Published: 20 July 2023

Abstract: Amaranth has been recognized as a nutraceutical food because it contains high-quality proteins due to its adequate amino acid composition that covers the recommended requirements for children and adults. Since pre-Hispanic times, amaranth has been consumed as popped grain; the popping process improves its nutritive quality and improves its digestibility. Popped amaranth consumption has been associated with the recovery of malnourished children. However, there is no information on the impact that popped amaranth consumption has on gut microbiota composition. A non-randomized pilot trial was conducted to evaluate the changes in composition, structure, and function of the gut microbiota of stunted children who received four grams of popped amaranth daily for three months. Stool and serum were collected at the beginning and at the end of the trial. Short-chain fatty acids (SCFA) were quantified, and gut bacterial composition was analyzed by *16S rRNA* gene sequencing. Biometry and hematology results showed that children had no pathology other than low height-for-age. A decrease in the relative abundance of *Alistipes putredinis*, *Bacteroides coprocola*, and *Bacteroides stercoris* bacteria related to inflammation and colitis, and an increase in the relative abundance of *Akkermansia muciniphila* and *Streptococcus thermophiles* bacteria associated with health and longevity, was observed. The results demonstrate that popped amaranth is a nutritious food that helps to combat childhood malnutrition through gut microbiota modulation.

Keywords: 16S rRNA sequencing; diet; popped amaranth; gut microbiota; short-chain fatty acids

1. Introduction

Malnutrition is a term that encompasses many different manifestations of inadequate nutrition, including both undernutrition and obesity, which are characterized by an imbalance in energy intake and energy expenditure [1]. Undernutrition, defined as a deficiency of calories or a shortage of one or more essential nutrients, is a significant pediatric health problem and a pressing and overwhelming global health issue contributing to nearly half of all deaths in children under five years of age [2], mostly occurring in low- and middle-income countries where, at the same time, rates of childhood overweight and obesity are rising [3]. Assessment of a child's health has been based on measurements of the body mass index (weight [kg]/height^2 [m^2]), but to recognize clinical signs of certain undernutrition problems, weight, length, height, or z-score values have been used to classify the nutritional status of children. In this sense, undernutrition is classified as stunting when there is low height-for-age, e.g., <-2 standard deviations (SD); wasting, when there is low weight-for-age (<-2 SD), and underweight, when there is low weight-for-height (<-2 SD). Malnutrition is also classified as moderate when it falls between -2 and -3 SD and severe when it falls below -3 SD [3]. According to the z-score, 165 million children under five years of age are stunted and 50 million children are wasting [4]. Being classified as stunted is a key risk factor for diminished survival, poor child and adult health, decreased learning capacity, and lost future productivity [5], and is considered as a multifactorial disease caused not only by inadequate protein/energy intake, but also other factors, such as poor sanitation, no access to clean drinking water, and inadequate psychosocial stimulation. All these factors cause disruption of the normal gut microbial ecosystem known as dysbiosis, which results in a predisposition to common infections, impaired immunity, and worsening malnutrition [6,7].

Growing evidence has indicated that dietary patterns and adequate food processing conditions are factors that affect food digestibility and functionality properties, and influence the composition, structure, and function of gut microbiota, and, therefore, human health [8,9]. It is also known that the quality and quantity of protein intake affects the metabolites produced by the gut microbiota [9–12]. Carbohydrates (resistant starch and dietary fiber) that escape digestion and absorption in the small intestine are also used by the microbiota and, through saccharolytic fermentation, lead to the generation of short-chain fatty acids (SCFA). SCFA, mainly acetate, propionate, and butyrate, are primary fermentation end products and represent the major flow of carbon from the diet through the microbiome to the host. SCFA have a positive influence on gut integrity and nutritional health by improving the energy yield, modulation of colonic pH, production of vitamins, and the stimulation of gut homeostasis, including anti-pathogenic activities [7].

Amaranth has been recognized as a nutraceutical food because it contains higher amounts of proteins (compared with traditional cereals, such as corn, wheat, and rice), but, most importantly, because of its high nutritive value due to its adequate amino acid composition, which covers the requirements recommended for children and adults [13]. Amaranth grains contain several encrypted peptides with antidiabetic, antihypertensive, and antioxidant functions [14]. They are rich in lipids containing several sterols, including tocopherols, and they are also a good source of squalene, a key metabolite in the sterol pathway. In relation to micronutrients, amaranth grains are a good source of minerals, including phosphorus, potassium, magnesium, calcium, iron, zinc, manganese, and selenium, and are also rich in vitamins (B2, B6, and E), niacin, and thiamine [13]. Since pre-Hispanic times, popped amaranth grain has been consumed, which was obtained by placing the grain on a clay pot heated with firewood, resulting in a pre-cooked food with a nutty flavor [15]. Currently, popped amaranth is obtained in a gas-fluidized bed with very short residence times, so popped amaranth could be considered as a minimally processed healthy food snack due to its high quality and quantity of proteins [16]. As for other plant-based food products, heat treatment processing does not alter amaranth grain properties but popping increases the digestibility of its nutrients [17], enhances its antioxidative properties [18,19], and reduces adverse antinutritional compounds, such as tannins, lectins, and trypsin inhibitors [17]. The consumption of popped amaranth grain has been associated with health benefits in humans, including recovery of severely malnourished children [17,20,21].

Recent reports have shown that the administration of quinoa protein and its hydrolysates was able to restore the gut microbiota in a mice model of colorectal cancer to similar to that of the control group [22]. Amaranth as quinoa is known as a super-grain because of its high nutritive quality; however, despite the extensive evidence that supports the beneficial effects on health of amaranth consumption, to date, there have been no studies that have analyzed the effects of popped amaranth consumption on gut microbiota composition. Therefore, the present study sought to utilize a nonrandomized pilot trial design to explore the effect of popped amaranth consumption on changes in the structure and abundance of the gut microbiota of children classified as low height-for-age (i.e., stunted children). The results obtained show that popped amaranth consumption helps to combat child malnutrition through gut microbiota modulation.

2. Materials and Methods

2.1. Recruitment of Participants

Children living in San Antonio Huichimal, Lima, and the Vista Hermosa rural area of Tenek, Ciudad Valles, San Luis Potosi, S.L.P., Mexico, were recruited for this research through screening based on inclusion and exclusion criteria. Sick children or those children with a reported disease were excluded. The selected children had not taken antibiotics in the last three months prior to the study. Body weight was measured using a digital weight scale with the infant wearing a light cloth and no shoes (accuracy: 0.1 kg). Meanwhile, body height was measured using a 2-meter-long microtoise without shoes (accuracy: 0.1 cm). Children were eligible for inclusion if they were aged between 6 and 7 years. They were grouped into two groups: a control group (Ctrl), including those that presented normal height-for-age with a mean of HAZ = -0.03 ± 0.5 (25 children), and a stunted or low height-for-age (HAZ < -2 SD) group (9 children).

To identify factors related to the children's health status, a family survey was carried out. The survey included questions such as the type of home construction, access to basic services, type of food, type of child's diet, type of house kitchen, and drinking water services.

2.2. Research Design

The study was approved by the ethics committees of the Health Services of San Luis Potosi (approval reference: SLP/006-2018), and DIF-Ciudad Valles, San Luis Potosí. This research was conducted following the applicable regulations and guidelines of the Helsinki Declaration, revised in 2000. Informed consent was signed by the participant children, as well as by the children's parents/legal guardians. The trial was conducted from 13 March to 12 January 2021. Stunted children consumed four grams of popped amaranth daily for three months. Serum and stool samples from all participant children were obtained before and after the trial. Blood samples were used for chemistry and liver function analysis, while stool samples were used for SCFA analysis and gut microbiota composition. The research design is shown in Figure 1.

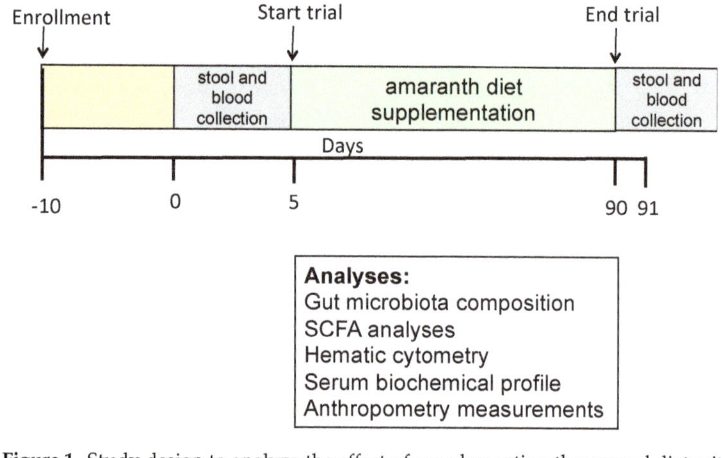

Figure 1. Study design to analyze the effect of supplementing the normal diet with amaranth on gut microbiota of children between 6 to 7 years old with moderate acute undernutrition living in rural areas.

2.3. Preparation of Amaranth Popped Grain

Popped seeds were obtained by heating the seeds in an industrial hot air fluidized bed machine popper (Amaranta^R, San Miguel de proyectos Agropecuarios, Hidalgo, México). The popped amaranth complied with the Official Mexican Standards (NOM-051-SCFI/SSA1-2010) for food and drinks for human consumption. The protein, fat, fiber, and ash contents were determined by standard methods [23]. Total carbohydrate was calculated by subtracting protein, fat, ash, and fiber from 100 [24].

2.4. Research Outcome

During the intervention period, the daily popped amaranth intake was supervised. Blood and stool samples were taken before and after the study. The primary outcomes were gut microbiota composition and SCFA profile, while a secondary outcomes was the hematic biochemical profile.

2.5. Serum Collection, Hematic Cytometry, and Biochemical Profile Analysis

Blood samples (10 mL) were taken early in the morning under fasting conditions and following the established Official Mexican Norm (NOM-016S-SA3-2012). Children were requested to not take any food or water before blood collection. Blood was collected in BD Vacutainer® Blood Collection tubes without anticoagulant. Samples were centrifuged at 400× g for 10 min; the serum was separated and frozen at −70 °C until analysis. Hematic cytometry was performed in an automatic system Kontrolab Model 5H+ (KontroLab Co., Ltd., Giudornia, Roma, Italy). Then, from each sample, a drop of blood was spread on a clear glass slide and stained with a May—Grünwald–Giemsa stain to differentiate the types of white blood cells in the sample. An aleotory count of 100 white blood cells in a bright-field microscope Axio Imager-A2m (Carl Zeiss Microscopy, White Plains, NY, USA) at 100× magnification was carried out. Serum concentrations of glucose, triglycerides, total cholesterol, urea, acid uric, creatinine, total protein, albumin, aspartate aminotransferase (AST), alanine aminotransferase (ALT), gamma-glutamyl transferase (GGT), alkaline phosphatase (ALKP), and lactate dehydrogenase (LDH) were determined by spectrophotometry using commercial kits (SPINREACT, Girona, Spain) and read on a MultiskanGO microplate spectrophotometer (ThermoFischer Scientific, Waltham, MA, USA).

2.6. Stool Sample Collection

For the collection of stool samples, a clean potty was placed in the toilet; immediately after defecation, stool samples were collected aseptically in a sterile stool container. The children were instructed to urinate first to avoid urine contamination; they were supported by their parents and trained personnel were in charge for stool collection. Samples were transported to the laboratory using ice packs, aliquoted, and immediately stored at −70 °C until further processing.

2.6.1. Measurement of Fecal Short-Chain Fatty Acids (SCFA)

An aliquot of stool samples (250 mg) was homogenized in one mL of H_2SO_4 (0.5 mmol/L) and mixed at 1400 rpm for 3 min (Thermomixer, Eppendorf, Hamburg, Germany). The homogenized sample was incubated for 20 min in an ice-water bath and centrifuged at 4 °C and 4800× g for 15 min. The supernatant was recovered and subsequently centrifuged at 4 °C and 13,000× g for 15 min. This procedure was repeated two times for clarification. The sample was filtered through a 0.22 µm Millipore filter (Merck, Darmstadt, Germany) before injection into the chromatographic system. SCFAs were analyzed using an Agilent 1100 (Hewlett Packard, Santa Clara, CA, USA) and a Rezex ROA LC Column 150 × 7.8 mm (Phenomenex Inc., Torrance, CA, USA). The mobile phase was composed of H_2SO_4 (0.5 mmol/L). The column temperature was 60 °C, the flow rate was 0.5 mL/min, and measurement was performed using an RID-10A RI detector. Calibration curves were calculated from 0.18 to 1.8 mg/mL for acetic acid, 0.08 to 0.8 g/L for propionic acid, and 0.11 to 1.1 g/L for butyric acid.

2.6.2. DNA Extraction and Integrity Verification

Stool samples (250 μg) were diluted in 1300 μL of saline solution (0.85%), removing coarse remains. A quantity of 600 μL of the decanted sample was taken and resuspended in a new tube. The suspension was centrifuged at 10,000× *g* for 10 min at 4 °C. The obtained pellet was resuspended in 1 mL of cold PBS and centrifuged at 700× *g*, 4 °C for 1 min. The supernatant was collected and centrifuged at 9000× *g* for 5 min at 4 °C. The resulting pellet was used for DNA extraction following the specifications of the DNeasy UltraClean Microbial Kit from QIAGEN (Hilden, Germany). DNA was quantified in a NanoDrop One (ThermoFischer Scientific, Waltham, MA, USA), and integrity was verified by visualization on a 1% agarose gel in a Tris-Borate-EDTA (TBE) buffer for 60 min at 70 volts. DNA extracts were stored at −80 °C until sequenced.

2.7. 16S rRNA Gene Sequencing and Bioinformatics Analyses

For microbiome analysis, the V4–V5 region of the *16S ribosomal RNA* gene was amplified once using the universal bacterial primers 515FB: 5′-GTGYCAGCMGCCGCGGTAA-3′ and 926R: 5′-CCGYCAATTYMTTTRAGTTT-3′. A total of 51 Samples were sequenced on the Illumina MiSeq platform (Illumina, San Diego, CA, USA) using 300 + 300 bp paired-end according to the protocol described elsewhere [25]. Samples were sequenced at the Integrated Microbiome Resource (IMR) (Dalhousie University, Halifax, NS, Canada).

2.7.1. Amplicon Sequence Variant Inference

The R package DADA2 v1.16.0 [26] was used to process the 16S sequencing data. We trimmed 20 nt to the right side of the forward fragments and 60 nt to the right side of the reverse fragments using the *filterAndTrim ()* function with default parameters. We used the *learnErrors()* function with default parameters to learn the error rates. We applied the *dada()* function to infer the sample composition using default parameters, the filtered fragments, and the calculated error rates. We merged the fragments with the *mergePairs()* function and removed the chimeras using the *removeBimeraDenovo()* function with default parameters. The taxonomy for each sequence was assigned using the *assignTaxonomy()* function and the silva_nr99_v138 database. The amplicon sequence variant (ASV) quantification and the phylogenetic assignment were obtained.

2.7.2. Diversity Quantification and Functional Prediction

Alpha diversity, Shannon, inverse of Simpson, Fisher, Chao1, and ACE indices were calculated with the *diversity()* function from the R package vegan v2.5.6 [27]. To determine the structural variation of microbial communities, beta diversity and NMD plots were generated with custom R scripts and using the BrayCurtis dissimilarity calculated with the *ordinate()* function from the vegan package.

The functional prediction of the 16S sequencing data was performed with Tax4Fun2 [28]. The *RunRefBlast()* and *makeFunctionalPrediction()* functions were used with default parameters to predict the functional profiles of the ASV quantification results. The reference used in both steps *runRefBlast* and *makeFunctionalPrediction* was RF99NR. The pathway predictions table (pathway scores per library) was used to perform a sparse partial least squares discriminant analysis (sPLS-DA). The function *spls-da()* from the R package mixOmics with default parameters and $ncomp = 2$ was used [29]. The plots to represent the results of the discriminative analysis (DA) were generated with custom R scripts.

Using significant taxa and routes, taxa-pathways networks were created (permutation test spls-da, 1000 permutations, p-value < 0.1). The Spearman correlation between two nodes (taxa or pathways) was used to identify their connection using the igraph R package [30]. Only correlations greater than the 30th percentile were employed to create the networks.

2.8. Statistical Analyses

The statistical analysis of results was performed on Prism 8.0 for Mac (GraphPad Software, San Diego, CA, USA). The normal distribution was tested by D'Agostino–Pearson and Shapiro–Wilk tests. One-way ANOVA was performed for data with a normal distribution, followed by Bonferroni's post hoc test. A Kruskall–Wallis test, followed by Dunn's post hoc test, was carried out if data were not normally distributed. A *p*-value less than 0.05 ($p < 0.05$) was considered as statistically significant.

3. Results and Discussion

3.1. Popped Amaranth as Minimally Processed Food

Popping is a traditional process that has been carried out in Mexico since pre-Hispanic times. At that time popping was obtained by heating the amaranth grains for 15–25 s in covered clay pots at 175 to 195 °C, which resulted in grain expansion [31,32]. Currently, popped amaranth is obtained in an industrial hot air fluidized bed machine at 300 to 330 °C with 9 s of residence; hence, the popped amaranth could be considered a minimally processed food. Under these conditions, popped amaranth showed high protein content (15.8%), fat (6.7%), fiber (3.8%), and ashes (2.2%) (Supplementary Materials, Table S1).

3.2. Children's Selection and Sociodemographic Living Conditions Questionaries

Children showing undernutrition and classified as stunted based on the height-for-age parameters [3] were selected and named as the undernutrition group (UN, $n = 9$). The UN group daily consumed four grams of popped amaranth and were named the UNA group, but during the trial and COVID-19 pandemic situation, two children abandoned the study (UNA, $n = 7$). A control group (Ctrl, $n = 21$) represented by children of normal height-for-age living in the same conditions as the UN group was also selected.

The main results of the sociodemographic questionaries showed that factors associated with children's health status were related to living conditions, such as the use of latrines (not properly a bathroom), firewood used as fuel in kitchens, and the dormitory was the same place as the kitchen. The age of the mother at the birth of the child ranged from 20–24 years and most had high school as the highest level of studies. Although, in general, all the children had access to basic food (beans, rice, tortillas), undernourished children showed low appetite and normally ate only once a day, while the control children ate three times a day. Some other problems, such as parasitic infections, were detected in the children [33].

3.3. Children's Serum Biochemical Analysis

The hematic cytometry results for the participating children are shown in the Supplementary Materials, Table S2. The baseline values for participants reflected that they did not have any other pathology than low height-for-age in the UN group. However, although all parameters fell within biological limits, it was observed that hematocrit and mean globular volume (MCV) values had a tendency to increase in the UNA children, compared with the control group. Low values of MCV are related to protein-energy malnutrition or deficiency of iron and folate [34]. Amaranth is a rich source of protein and contains high amounts of minerals, such as iron, calcium, and magnesium [13]. This could be one of the reasons for the observed trend of an increase in these values and reported child health improvement.

The serum biochemical profile did not show any pathology. The renal function, tested by creatinine, uric acid, and urea, was normal in all participants. The biomarkers for liver function (ALT, TGO) showed normal levels. No significant changes were observed in cholesterol, triglycerides, and high-density lipoprotein (HDL) amongst the groups (Supplementary Materials, Table S3).

3.4. Modulation of Gut Microbiota in the Undernutrition Children's Group after Amaranth Consumption

Sequencing of the *16S ribosomal ARN* gene (*16S rRNA*) has generated great interest in the study of bacterial communities. Amplicon sequencing variants (ASVs) have been proposed as an alternative to operational taxonomic units (OTUs) to analyze microbial communities. The use of ASVs has grown in popularity because they reflect a more refined level of taxonomy; however, the use of ASVs must be undertaken carefully in order to avoid the risk of dividing a single bacterial genome into separate clusters [35]. Once the sequences were refined, a total of 90,904 ASVs were obtained for the microbiome of the control children, 105,312 ASVs for the underweight children, and 91,526 for the underweight children after amaranth consumption.

The alpha diversity, which reflects the gut microbial richness was assessed based on the observed species and the Shannon index [36]. Although no significant differences amongst groups were observed, a tendency toward lower values was observed in the UN group; however, after amaranth consumption, the UNA group showed a tendency to increase to similar values to those of the Ctrl group (Figure 2A,B). Simpson's inverse index, which indicates the diversity or dominance of species in the sample, also showed a tendency to decrease in the UN group, which suggests that the bacterial community was smaller in this group. However, this value also increased after amaranth consumption for the UNA group (Figure 2C), and the same tendency was observed for Chao1 (Figure 2D), and for the Fisher, ACE, and Simpson values (Supplementary Materials, Figure S1).

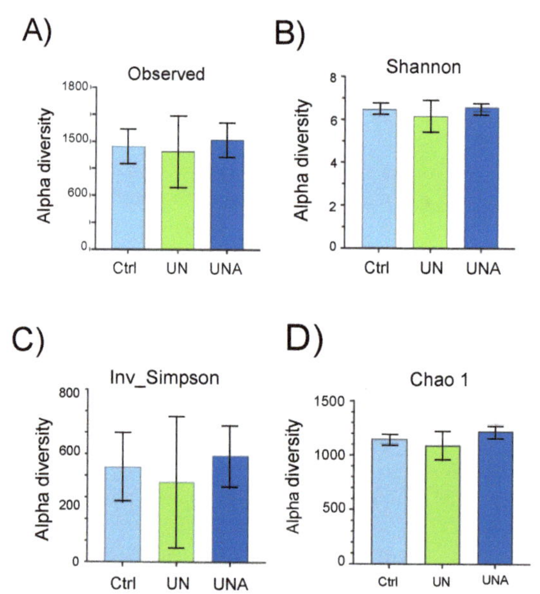

Figure 2. The effect of amaranth consumption on the gut microbiota composition of children living in rural areas. Alpha-diversity expressed as (**A**) observed richness, (**B**) Shannon index, (**C**) Inv-Simpson, and (**D**) Chao 1 indexes. Boxes express the IQR (interquartile range), bars indicate the minimum and maximum values. A Kruskall–Wallis test and Dunn's post hoc analysis ($p < 0.05$) were performed. No significant differences were found amongst the groups and different indexes. Ctrl = control group of children with normal height-for-age; UN = undernutrition group with low height-for-age; UNA = UN group after three months of amaranth consumption.

Previous reports have indicated that children with undernutrition show lower gut microbiota diversity [12,37,38], which agrees with the observed results in the present work. It was also reported that the Shannon index showed a tendency to decrease in malnourished children with protein malnutrition (Kwashiorkor), although no significant values were reported [39]. It is known that results in humans are sometimes not statistically significant because of the intrinsic variation among people, but they are clinically relevant as they reflect a clear trend [40]. The beta diversity reflects the differences in the gut microbial composition of different groups and is represented by non-metric multidimensional scaling (NMDS) or PCoA, based on the Bray–Curtis dissimilarity. As shown in Figure 3, PCoA analysis showed a difference in the gut microbiota between the Ctrl and UN groups; however, after amaranth consumption, the UNA group tended to be more like the Ctrl group.

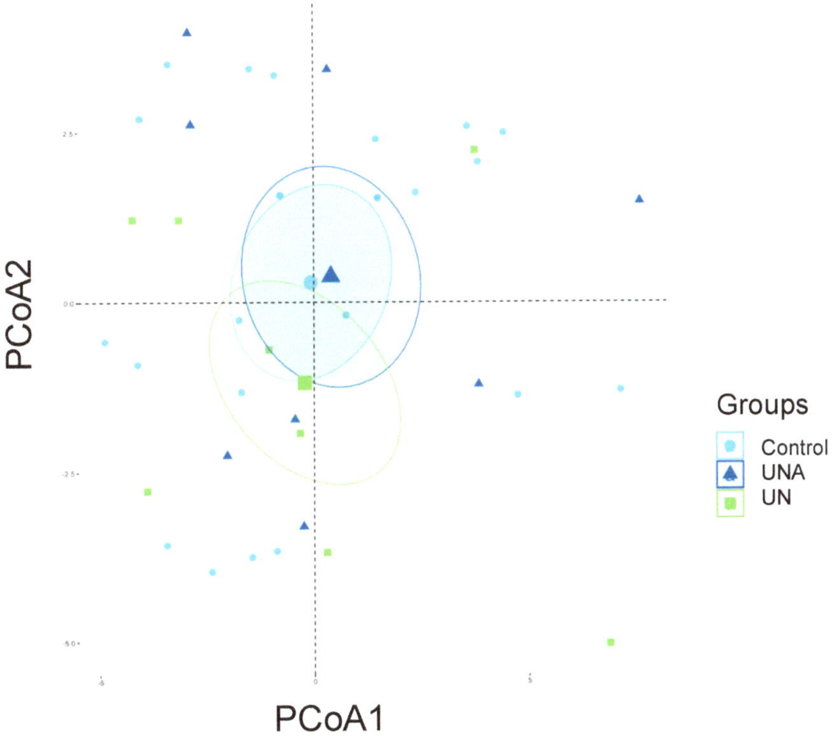

Figure 3. The effect of amaranth consumption on the gut microbiota composition of children living in rural areas. Beta-diversity measured using principal coordinate analysis (PCoA) based on the Bray Curtis dissimilarity ($p < 0.05$). Ctrl = control group of children with normal height-for-age; UN = undernutrition group with low height-for-age; UNA = UN group after three months of amaranth consumption.

The major phylum in the gut microbiota of children in the Ctrl group was the Actinobacteriota (28.8%), followed by the Bacteroidota and the Firmicutes (40.2%), Proteobacteria (12.8%), and Verrucomicrobiota (16.8%). It was observed that the UN group showed a significant increase in Actinobacteriota (50.9%) with loss of the Verrucomicrobiota phylum, which agrees with data reported by Kamil et al. [41]. After the trial, the UNA group showed a significant decrease in the abundance of Actinobacteria (20%) with an increase in Bacteriodotes and Firmicutes. Interestingly, Verrumicrobiota were recovered after amaranth consumption with similar values as the control group (13.4%) (Figure 4A). It has been reported that the Proteobacteria is the phylum that contributes to dysbiosis [42]; however, in this work, this phylum did not show significant changes in abundance amongst groups (12.8 to 15.4%) (Figure 4A).

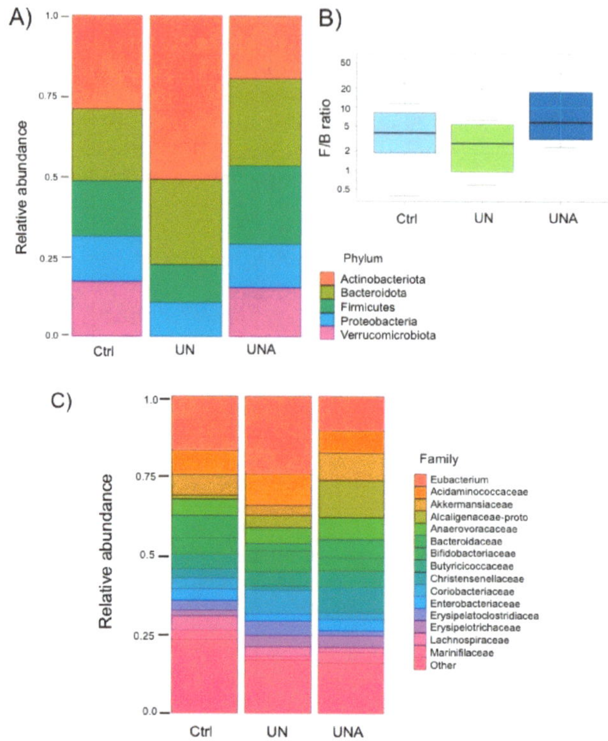

Figure 4. (**A**) Relative abundance in gut microbiota at phylum level; (**B**) Ratio of Firmicutes to Bacteroidetes (B/F). Significant differences were calculated using the non-parametric Wilcoxon test and non-significant differences were detected. (**C**) Relative abundance in gut microbiota at family level. Ctrl = control group of children with normal height-for-age; UN = undernutrition group with low height-for-age; UNA = undernutrition group after three months of amaranth consumption.

It is considered that the Firmicutes and Bacteroidetes ratio (F/B) is important for determining health status, and together represent the dominant phyla in the gut microbiota. Change in the F/B ratio has been associated with several diseases; however, it is important to be aware that this ratio could be affected by the change in abundance of all the other phyla, as observed in the present results. Our data show that, after amaranth consumption, these two phyla reached up to 51.6% in the balance abundance of F/B (Figure 4B). It has been reported that a diet high in sugar and low in fiber in children with undernutrition might cause a high F/B ratio [43]. An abundance of Firmicutes has been associated with the modulation of calorie absorption effectiveness by increasing the number of lipid droplets [44], and a high F/B ratio has been associated with the incidence of obesity [45]. Lower abundance of Firmicutes with higher abundance of Bacteroidetes has been associated with inflammatory bowel disease. It is inferred that the balance of these two phyla is important for maintaining normal gut homeostasis [46].

At the family level, a high abundance of Eubacteria was observed in the UN group with a significant decrease after amaranth consumption (24.8% and 11.1%). By contrast, the Akkermansiaceae family was found at very low abundance (3.3%) in the UN group; however, this family increased after amaranth consumption (9.1%) to an even higher abundance than the control group (1.0%). The Alcaligenaceae family was significantly increased in abundance after amaranth consumption (11.7%), and the Bacteroidaceae family, that was almost depleted in the UN group (2.6%), increased after amaranth consumption to similar values to the control group (5.9 and 7.6%, respectively). The Christensenellaceae family was also found in high abundance after amaranth consumption (7.8%), while the Coriobacteriaceae decreased from 7.2% in the undernourished group to up to 2.0% after amaranth consumption (Figure 4C).

3.5. Modulation of Gut Microbiota at Family-Genus Level was Observed in Undernourished Children after Amaranth Consumption

As described above, although the F/B ratio provides an estimation of a person's health, care should be taken because this ratio could be affected by changes in the other phyla, as well as the type of genus that changes in each phylum. After amaranth consumption most of the species' changes in relative abundance were observed in the Firmicutes phylum.

3.5.1. Changes in Firmicutes Phylum at Genus/Species Level

It has been reported that species of the Lachnospiraceae family belonging to the Eubacteriales order occur in over-abundance in undernourished children, with some of the members of this family serving as metabolic regulators in undernourished individuals [43]. Our results also showed high abundance of this family in the undernutrition group (Figure 4), and changes in the relative abundance of *Eubacterium hallii*, *Ruminococcus gauvreauii*, *Blautia faecis*, *Blautia obeum*, CAG-56, *Fusicanilobacter*, *Roseburia hominis*, and *Roseburia intestinalis*, amongst others, were observed after amaranth consumption (Figure 5). *E. hallii* was observed to decrease in the undernutrition group, but after amaranth consumption there was a relative increase. *E. hallii* is an anaerobic bacterium that belongs to the group of butyrate producers [47], and its administration in mice with diabetes changed the function of the intestinal microbiome, improving the metabolic phenotype [48].

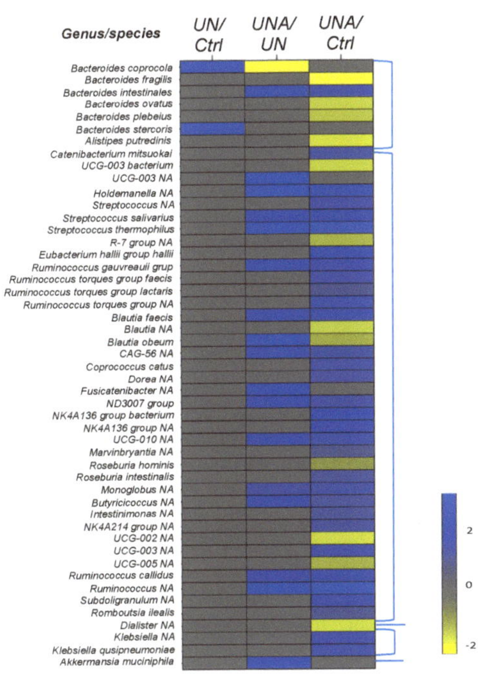

Figure 5. Heatmap of differential ASVs observed between: UN/Ctrl, undernutrition group (U/N) vs. control group (Ctrl); UNA/UN, undernutrition group after amaranth consumption (UNA) vs. undernutrition group at the beginning of assay (UN); UNA/Ctrl, undernutrition group after amaranth consumption (UNA) vs. control group (Ctrl). Values were determined using LogFC, at $p < 0.05$ and FDR < 0.1.

At the genus level, *Blautia* has attracted particular interest since its establishment in the gastrointestinal tract has been related to the alleviation of inflammatory and metabolic diseases as well as to antimicrobial activity. This bacterium has also been reported to play a role in the crosstalk with other intestinal microorganisms [49]. *Blautia* actively participates in intestinal immunomodulation in human health [38,48], improves serum HDL-C, and a relative increase in its abundance was reported after a diet intervention for obesity [50]. Although *B. obeum* has been related to obesity [46], its abundance in undernourished children after amaranth consumption was at the same level as in control children (Figure 5).

Roseburia hominis, and *Roseburia intestinalis* were identified; both bacteria showed low changes in their relative abundances after amaranth consumption, while the first decreased, the second increased. The *Roseburia* genus is described as comprising beneficial organisms as they are butyrate-producing bacteria and both species have been found to be depleted in the stools of undernourished children [37]. *R. hominis* was positively related to colonic mucosal melatonin levels, a pineal hormone that can maintain circadian rhythms and regulate immune, antioxidant, and anti-inflammatory functions, alleviating the symptoms of digestive disorders, such as irritable bowel syndrome and ulcerative colitis [51]. It is also known that *R. intestinales* prevents inflammation and maintains energy homeostasis [52].

Within the Erysipelotrichaceae family, *Holdemanella*, was found to increase relatively after amaranth consumption (Figure 5). In malnutrition, it has been shown that *Holdomanella* was depleted [53]. *Holdomanella* has been isolated from the feces of healthy metabolic volunteers and has been linked to the amelioration of hyperglycemia, improvement in oral glucose tolerance, and restoration of gluconeogenesis and insulin signaling in the liver of obese mice [54].

The Streptococcaceae family was represented by a relative increase in the abundance of *S. salivarus* and *S. thermophilus* after amaranth consumption (Figure 5). *S. thermophilus* is a well-known probiotic and is commonly found in yogurts [55]. *S. salivarius* is a commensal in humans and is phylogenetically close to *S. thermophilus* [56]. Within the Oscillospiraceae family, *Subdoligranulum* showed a high relative abundance after amaranth consumption. *Subdoligranulum* is a butyrate-producing bacterium with high potential as a probiotic to handle metabolic diseases. It also affects the activity of acetyl-CoA acetyltransferase, acetyl/propionyl-CoA carboxylase, and butanol dehydrogenase, which contribute to butyric acid production [57]. Butyrate has several beneficial properties that are essential to maintain gastrointestinal health. Therefore, the butyrate-producing bacteria represent a niche-specific next-generation of probiotics with *Butyricicoccus* species being one of the potential bacteria [58]. *Butyricicoccus* have also been shown to have potential as a therapeutic target for food allergy [59]. *Butyricicoccus* showed an increase in relative abundance in undernourished children after amaranth consumption.

3.5.2. Changes in the Bacteroidetes Phylum at Genus/Species Level

Within the Bacteroidota phylum, a relative increase in *Bacteroides coprocola* and *Bacteroides stercoris* was observed in the undernutrition group; however, after amaranth consumption their relative abundance decreased as well as that of *Alistipes putredinis* (Figure 5). Different strains of *B. coprocola* have been reported, some are related to type 2 diabetes, some have been identified in patients with hypertension, and some others have been linked to patients with attention-deficit/hyperactivity disorder [60]; thus, it is suggested to be cautious about its presence or absence [61,62]. Although *B. stercoris* is an SCFA-producing bacterium, some studies have linked its presence to ulcerative colitis as well as to that of *Alistipes putredinis* [63].

A decrease in the relative abundance of *B. fragilis*, *B. ovatus*, and *B. plebeius* was observed after amaranth consumption when compared with the control group (Figure 5). *B. fragilis* is a commensal organism but has also been reported as an opportunistic pathogen in clinical infections, and as being responsible for a variety of diseases involving disruption of the intestinal barrier [64]. Other studies have reported that in mice the non-toxigenic *B. fragilis* strain may mediate beneficial interactions with the host by directing the immune response and suppressing intestinal inflammation [65]; thus, a deep characterization should be carried out to identify the *B. fragilis* species detected in undernourished children. Some of the *B. ovatus* strains have been related to the induction of gut IgA and maintenance of tissue homeostasis [66]. *B. plebeius* was found to be a dominant bacterium in children with multisystemic inflammatory syndrome [67]. Only *B. intestinalis* showed a high relative abundance after amaranth consumption (Figure 5). This bacterium has been identified as a type of polyamine-producing bacteria, such as putrescine, spermidine, and spermine, all of which are organic cations required for animal cell growth and differentiation. They are also involved in various steps of DNA, RNA, and protein synthesis [68].

3.5.3. Changes in the Verrucomicrobiota Phylum

Amaranth consumption induced a relative increase in abundance of *Akkermansia muciniphila* (Figure 5). *A. muciniphila*, is a colonic mucin-degrading bacterium that has been linked to intestinal health and immune signaling and has been associated with reduction in the incidence of diabetes when administered as a probiotic [69]. It was reported that *A. muciniphila* improves metabolic disorders in dietary obese mice and was found to increase after resistant starch intake. It has been proposed as a new probiotic with the capacity to promote healthy longevity [70]. *A. muciniphila* can metabolize the mucosal layer as a source of carbon and nitrogen to produce acetate and propionate, playing an important role in maintaining the gut barrier, especially when enteral nutrition intake is low, such as during long-term fasting and in malnutrition [71].

The diversity and richness of the mouse microbiota changed when mice were fed fermented soybeans, resulting in enhancement of *Lactobacillus*, *Butyricicoccus*, members of the Lachnospiraceae family, and *A. muciniphila* [72]. Our results showed that these species were also found in increased abundance in undernourished children after amaranth consumption.

3.6. Amaranth Consumption Promotes Gut Microbiota-Dependent Metabolites

Popped amaranth grains, in addition to high-quality proteins, are rich in carbohydrates that escape digestion (resistant starch), as well as dietary fiber from testa [13]. The gut microbiota is predominantly involved in the fermentation of indigestible carbohydrates into SCFA, which exert multiple effects on energy homeostasis and are crucial for intestinal health. The most abundant SCFA are acetate, butyrate, and propionate, comprising > 95% of the SCFA content [73]. Investigations have shown that SCFA concentrations in children with severe acute and moderate undernutrition are low, especially propionic and butyric acid, and these SCFA increase during recovery along with the gut microbiota number. In this study, it was observed that, after amaranth consumption, the levels of acetic, butyric, and propionic acids showed a tendency to increase, although only propionic acid was significant as well as the sum of these three SCFAs (Figure 6). *E. hallii*, *Ruminococcus*, *Blautia*, *R. hominis*, and *Butyricicoccus*, known as butyrate-producer species [40], were all observed in increased relative abundance after amaranth consumption. *A. muciniphila* and *Subdoligranulum* sp. have been described as acetate- and propionate-producer bacteria, respectively [57,71]—both were also found in increased abundance after amaranth consumption. Altogether, these data suggest that SFCA levels are restored after amaranth consumption, correlating with an increased abundance of SCFA-producer bacteria.

The effect of proteins in the gut on the microbiota metabolism is less well understood since the high variability of proteins due to their amino acid composition and the presence or absence of essential amino acids induces differences in protein digestion by the microbiota generating a great diversity of metabolites, many of which could be detrimental to host health [74]. Recent work using quinoa protein isolates showed that they had a positive effect on the modulation of mice gut microbiota [22], similar to the effect observed in the present work; however, previous works using mice as models, when fed soy proteins showed that, in addition to a tendency to fat accumulation, the effect on the modulation of the gut microbiota was not the same compared to feeding of amaranth proteins [75].

Further work should be carried out in order to understand the effect of dietary protein on the protein–gut–host–health axis.

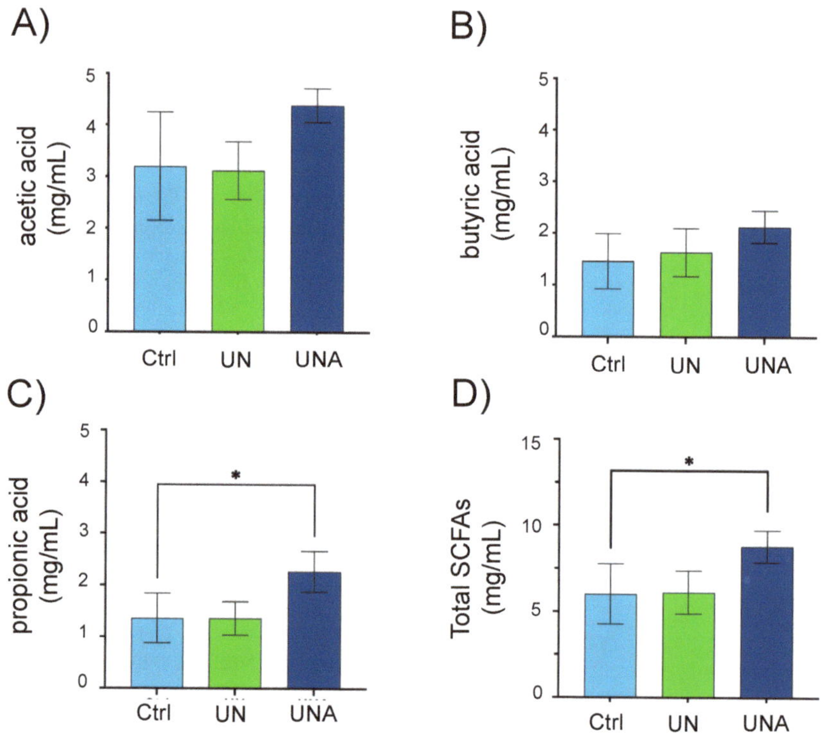

Figure 6. Short-chain fatty acid (SCFA) levels in children's feces samples. One-way ANOVA was carried out followed by Tukey's post hoc test. (**A**) acetic acid; (**B**) butyric acid; (**C**) propionic acid; (**D**) Total SCFA. Data is presented as mean of mg/mL ± standard deviation. Asterisk shows significant differences at $p < 0.05$. Ctrl = control group of children normal weight-for-age; UN = undernutrition children group with low height-for-age; UNA = undernutrition children group after three months of amaranth consumption.

3.7. Functional Prediction of Bacterial Taxa

The metabolic pathway predictions of gut microbiota composition in the undernutrition group compared to the control group (UN/Ctrl) (Figure 7) showed increased starch and sucrose metabolism, which decreased after amaranth consumption (UNA/Ctrl). The biosynthesis of secondary metabolism was increased after amaranth consumption, as well as butanoate, pyruvate, and propanoate metabolism, which are the routes to produce butyric and propionic acid, associated with the tendency to increase of these metabolites after amaranth consumption (Figure 7).

Purine and amino sugar and nucleotide metabolism, as well as the pathways related to vitamin metabolism, such as biotin (vitamin H), nicotinate and nicotinamide (B3), and riboflavin (B2), showed a tendency to increase after amaranth consumption. Increase in metabolic pathways related to sphingolipid and arachidonic acid metabolism was observed after amaranth consumption. The sphingolipids are considered to be a class of bioactive lipids that play key roles in the regulation of several cellular processes, while arachidonic acid forms part of cell membranes and is important for human health and tissue homeostasis. A correlation between Ruminococcaceae UCG 009 and arachidonic acid has been reported [76]. Several *Ruminocococcus* species were detected here after amaranth consumption (Figure 5).

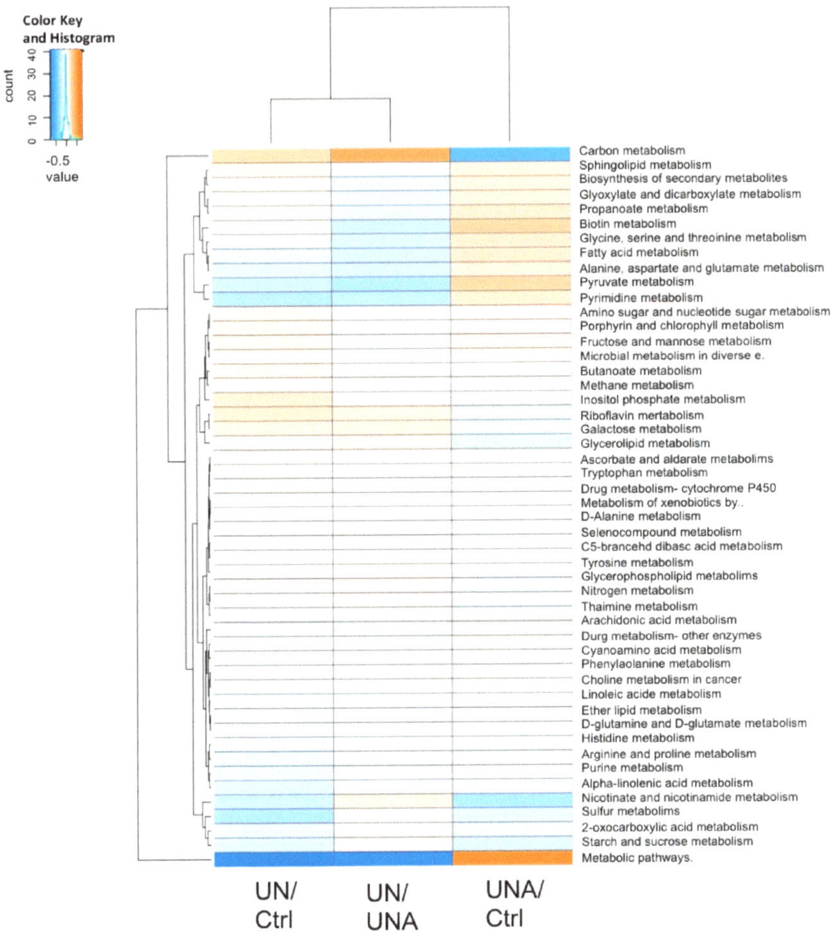

Figure 7. Heatmap of predicted functions of the gut microbiome by Tax4fun2 evaluated according to the differences in the Kyoto Encyclopedia of Genes and Genomes (KEGG) pathway. Differences in the KEGG pathway: undernutrition group (UN) vs. control children (Ctrl); undernutrition group after amaranth consumption (UNA) vs. undernutrition group at the beginning of assay (UN); undernutrition group before amaranth consumption (UN) vs. control group (Ctrl).

A network of correlations with metabolic activities was obtained and as shown in Figure 8, *Akkermansia* appeared to be an important species in controlling *Ruminococcus*, *Bacteroides*, and *Blautia*. There was also a positive relation with *Klebsiella* and *E. hallii*. *Akkermansia* also showed a positive correlation with species from the Oscillospiracea family, such as UCG-005, 003, and 002. The UCG strains regulate Christensenellaceae R-7 and *Butyrivibrio*, which is an important species in lipid metabolism. These groups of bacteria have a role in the regulation of nucleotide and terpenoid metabolism, transport, signal transduction, and replication and repair.

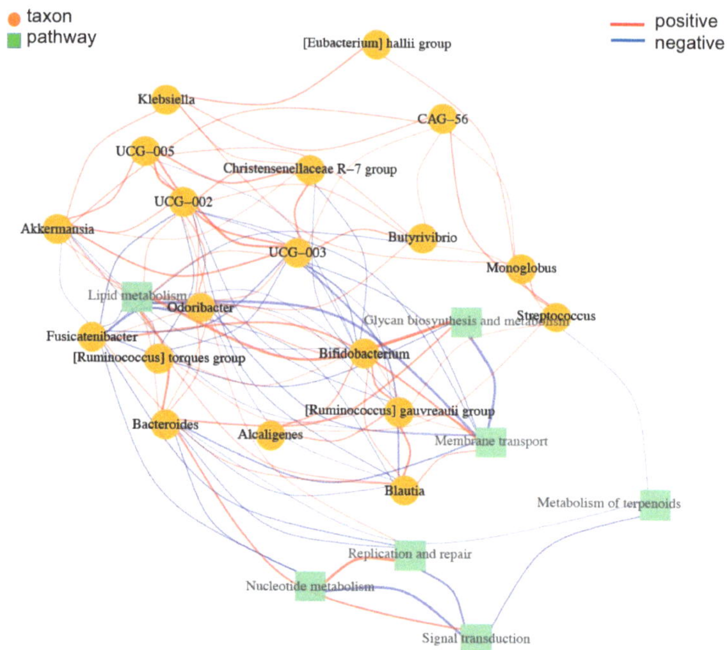

Figure 8. Association between taxa and pathways. Taxa are represented by yellow circles and pathways by green squares. The network's nodes are taxa and pathways. The association between the nodes and edges was calculated using the Spearman correlation methodology. Only correlations of absolute co-ordinates > 0.5 were considered. The nodes correlations are represented by the vertex, where red are positive and blue negative correlations.

4. Conclusions

Popped amaranth has been promoted as a food source with high potential to combat protein-energy malnutrition. Our results show that popped amaranth intervention induces an increased relative abundance of *A. muciniphila*, a bacterium related to intestinal health and host longevity, as well as *Subdoligranulum*, which is considered to be a new class of probiotics. We also observed decrease in the relative abundance of *B. coprocola* and *B. stercoris*, both related to inflammation and colitis. In summary, the present work highlights the potential uses of popped amaranth as a source of plant-based proteins, which requires minimum processing to achieve its biological functions in health. However, eradication of malnutrition does not occur only through dietary changes, it is also important to improve the infrastructural conditions and education in rural areas. Further work in a larger randomized controlled trial will be important to confirm the beneficial effects of popped amaranth consumption.

Supplementary Materials: The following supporting information can be downloaded at: https://www.mdpi.com/article/10.3390/foods12142760/s1, Supplementary Table S1: Proximal composition of popped amaranth (*Amaranthus hyphochondriacus*). Supplementary Table S2: Hematic cytometry of participant children with low height-for age according to HAZ index. Supplementary Table S3: Serum biochemical profile of participant children. Supplementary Figure S1: Alpha-diversity according to Fisher, ACE, and Simpson values. Supplementary Figure S2: Distribution of gut microbiota at class and order levels.

Author Contributions: O.d.J.C.-C. and A.P.B.-d.l.R.: conceptualization, methodology, and investigation; C.O.-V.: bioinformatics, data curation, and writing review; O.d.J.C.-C. and A.D.L.-R.: SCFA analysis; O.d.J.C.-C. and F.V.: investigation, blood and feces samples; E.E.-R.: amaranth popped preparation; O.d.J.C.-C. and S.T.: hematological and biometry analysis; A.P.B.-d.l.R.: supervision, funding acquisition, project administration, writing, review and editing. A.D.L.-R.: funding. All authors have read and agreed to the published version of the manuscript.

Funding: This work was supported by CONAHCYT Grant from the "Fondo Sectorial de Investigación en Salud y Seguridad Social" (No. A3-S-37825).

Data Availability Statement: The data presented in this study are available on request from the corresponding author. The data are not publicly available due is in process to uploaded into public database.

Acknowledgments: The authors acknowledge the Instituto Potosino de Investigación Científica y Tecnológica (IPICYT) and the National Supercomputing Center (CNS) computing resources provided to develop this project under the project grant TKII-R2018- COV1. COV acknowledges the *Investigadores por México* CONACYT program. Thanks to the rural schools located at Ciudad Valles, San Luis Potosí and to all participant children and their parents. We also thank L. Medina-Contreras and C. I. Cano-Gómez for their technical assistance.

Conflicts of Interest: The authors declare no conflict of interest.

References

1. Gordon, J.I.; Dewey, K.G.; Mills, D.A.; Medzhitov, R.M. The Human Gut Microbiota and Undernutrition. *Sci. Transl. Med.* **2012**, *4*, 137–149. [CrossRef] [PubMed]
2. Blanton, L.V.; Barratt, M.J.; Charbonneau, M.R.; Ahmed, T.; Gordon, J.I. Childhood undernutrition, the gut microbiota, and microbiota-directed therapeutics. *Science* **2016**, *352*, 1533. [CrossRef] [PubMed]
3. WHO/UNICEF. Levels and Trends in Child Malnutrition: UNICEF/WHO/The World Bank Group Joint Child Malnutrition Estimates: Key Findings of the 2021 Edition. *Retrieved October 2022.* Available online: https://www.who.int/publications/i/item/9789240025257 (accessed on 4 March 2023).
4. Prendergast, A.J.; Humphrey, J.H. The stunting syndrome in developing countries. *Paediatr. Int. Child Health* **2014**, *34*, 250–265. [CrossRef]
5. Iddrisu, I.; Monteagudo-Mera, A.; Poveda, C.; Pyle, S.; Shahzad, M.; Andrews, S.; Walton, G.E. Malnutrition and Gut Microbiota in Children. *Nutrients* **2021**, *13*, 2727. [CrossRef] [PubMed]
6. García-Mantrana, I.; Martínez-Costa, C.; Salminen, S.; Isolauri, E.; Collado, M.C. The Microbiota and Malnutrition: Impact of Nutritional Status During Early Life. *Annu. Rev. Nutr.* **2019**, *39*, 267–290. [CrossRef]
7. Calder, N.; Walsh, K.; Olupot-Olupot, P.; Ssenyondo, T.; Muhindo, R.; Mpoya, A.; Brignardello, J.; Wang, X.; McKay, E.; Morrison, D.; et al. Modifying gut integrity and microbiome in children with severe acute malnutrition using legume-based feeds (MIMBLE): A pilot trial. *Cell Rep. Med.* **2021**, *2*, 100280. [CrossRef]
8. Nova, E.; Gómez-Martinez, S.; González-Soltero, R. The Influence of Dietary Factors on the Gut Microbiota. *Microorganisms* **2022**, *10*, 1368. [CrossRef]
9. Wu, S.; Bhat, Z.F.; Gounder, R.S.; Ahmed, I.A.M.; Al-Juhaimi, F.Y.; Ding, Y.; Bekhit, A.E.D. Effect of Dietary Protein and Processing on Gut Microbiota—A Systematic Review. *Nutrients* **2022**, *14*, 453. [CrossRef]
10. Berding, K.; Vlckova, K.; Marx, W.; Schellekens, H.; Stanton, C.; Clarke, G.; Jacka, F.; Dinan, T.G.; Cryan, J.F. Diet and the Microbiota–Gut–Brain Axis: Sowing the Seeds of Good Mental Health. *Adv. Nutr. Int. Rev. J.* **2021**, *12*, 1239–1285. [CrossRef]
11. Aloo, S.O.; Oh, D.-H. The Functional Interplay between Gut Microbiota, Protein Hydrolysates/Bioactive Peptides, and Obesity: A Critical Review on the Study Advances. *Antioxidants* **2022**, *11*, 333. [CrossRef]
12. Gatya, M.; Fibri, D.L.N.; Utami, T.; Suroto, D.A.; Rahayu, E.S. Gut Microbiota Composition in Undernourished Children Associated with Diet and Sociodemographic Factors: A Case–Control Study in Indonesia. *Microorganisms* **2022**, *10*, 1748. [CrossRef] [PubMed]
13. Coelho, L.M.; Silva, P.M.; Martins, J.T.; Pinheiro, A.C.; Vicente, A.A. Emerging opportunities in exploring the nutritional/functional value of amaranth. *Food Funct.* **2018**, *9*, 5499–5512. [CrossRef] [PubMed]
14. Silva-Sánchez, C.; de la Rosa, A.P.B.; León-Galván, M.F.; de Lumen, B.O.; de León-Rodríguez, A.; de Mejía, E.G. Bioactive Peptides in Amaranth (*Amaranthus hypochondriacus*) Seed. *J. Agric. Food Chem.* **2008**, *56*, 1233–1240. [CrossRef]
15. Kaufam, C.S.; Weber, L.E. Grain amaranth. In *Advances in New Crops*; Janick, J., Simon, J.E., Eds.; Timber Press: Portland, OR, USA, 1990; pp. 127–139.
16. Solanki, C.; Indore, N.; Saha, D.; Kudos, S.K.A. Effect of popping methods on popping characteristic of amaranth grain. *Int. J. Chem. Stud.* **2018**, *6*, 2779–2782.
17. Valadez-Vega, C.; Lugo-Magaña, O.; Figueroa-Hernández, C.; Bautista, M.; Betanzos-Cabrera, G.; Bernardino-Nicanor, A.; González-Amaro, R.M.; Alonso-Villegas, R.; Morales-González, J.A.; González-Cruz, L. Effects of Germination and Popping on the Anti-Nutritional Compounds and the Digestibility of *Amaranthus hypochondriacus* Seeds. *Foods* **2022**, *11*, 2075. [CrossRef] [PubMed]

18. Paśko, P.; Barton, H.J.; Fołta, M.; Gwizdz, J. Evaluation of antioxidant activity of amaranth (*Amaranthus cruentus*) grain and by-products (flour, popping, cereal). *Roczniki Państwowego Zakładu Higieny* **2007**, *58*, 35–40. [PubMed]
19. Muyonga, J.H.; Andabati, B.; Ssepuuya, G. Effect of heat processing on selected grain amaranth physicochemical properties. *Food Sci. Nutr.* **2013**, *2*, 9–16. [CrossRef]
20. Aderibigbe, O.R.; Ezekiel, O.O.; Owolade, S.O.; Korese, J.K.; Sturm, B.; Hensel, O. Exploring the potentials of underutilized grain amaranth (*Amaranthus* spp.) along the value chain for food and nutrition security: A review. *Crit. Rev. Food Sci. Nutr.* **2020**, *62*, 656–669. [CrossRef]
21. Grundy, M.M.; Momanyi, D.K.; Holland, C.; Kawaka, F.; Tan, S.; Salim, M.; Boyd, B.J.; Bajka, B.; Mulet-Cabero, A.-I.; Bishop, J.; et al. Effects of grain source and processing methods on the nutritional profile and digestibility of grain amaranth. *J. Funct. Foods* **2020**, *72*, 104065. [CrossRef]
22. Fan, X.; Guo, H.; Teng, C.; Yang, X.; Qin, P.; Richel, A.; Zhang, L.; Blecker, C.; Ren, G. Supplementation of quinoa peptides alleviates colorectal cancer and restores gut microbiota in AOM/DSS-treated mice. *Food Chem.* **2023**, *408*, 135196. [CrossRef]
23. AOAC. *Official Methods of Analysis*, 18th ed.; Association of Official Analytical Chemists: Washington, DC, USA, 2007.
24. Gao, L.; Zhang, L.; Liu, H.; Hu, J. In vitro gastrointestinal digestion of whole grain noodles supplemented with soluble dietary fiber and their effects on children fecal microbiota. *Food Biosci.* **2023**, *53*, 102600. [CrossRef]
25. Comeau, A.M.; Li, W.K.W.; Tremblay, J.; Carmack, E.C.; Lovejoy, C. Arctic Ocean Microbial Community Structure before and after the 2007 Record Sea Ice Minimum. *PLoS ONE* **2011**, *6*, e27492. [CrossRef] [PubMed]
26. Callahan, B.J.; Mcmurdie, P.J.; Rosen, M.J.; Han, A.W.; Johnson, A.J.A.; Holmes, S.P. DADA$_2$: High-resolution sample inference from Illumina amplicon data. *Nat. Methods* **2016**, *13*, 581–583. [CrossRef] [PubMed]
27. Oksanen, J.; Blanchet, F.G.; Kindt, R.; Legendre, P.; Minchin, P.R.; O´Hara, R.B.; Solymos, P.; Weedon, J.; Vegan: Community Ecology Package. R Package 2015, version 2.6-4. *Jari Oksanen (maintainer), CRAN repository.* Available online: https://github.com/vegandevs/vegan/issues (accessed on 4 March 2023).
28. Wemheuer, F.; Taylor, J.A.; Daniel, R.; Johnston, E.; Meinicke, P.; Thomas, T.; Wemheuer, B. Tax4Fun2: Prediction of habitat-specific functional profiles and functional redundancy based on 16S rRNA gene sequences. *Environ. Microbiome* **2020**, *15*, 11. [CrossRef]
29. Rohart, F.; Gautier, B.; Singh, A.; Lê Cao, K.-A. mixOmics: An R package for 'omics feature selection and multiple data integration. *PLoS Comput. Biol.* **2017**, *13*, e1005752. [CrossRef]
30. Csardi, G.; Nepusz, T. The igraph software package for complex network research. *Int. J. Complex Syst.* **2006**, *1695*, 1–9. Available online: http://igraph.sf.net (accessed on 4 March 2023).
31. Morales, E.; Lembcke, J.; Graham, G.G. Nutritional Value for Young Children of Grain Amaranth and Maize-Amaranth Mixtures: Effect of Processing. *J. Nutr.* **1988**, *118*, 78–85. [CrossRef]
32. Velasco-Lozano, A.M. Los cuerpos divinos: El amarnato, comida ritual y cotidiana. *Amarnato Fuente de Alegría Arqueología Mexicana* **2016**, *23*, 26–33.
33. Cano-Gómez, C.I.; Veana, F.; Calva-Cruz, O.J.; Espitia-Rangel, E.; Barba-de la Rosa, A.P. Influence of amaranth consumption on the anthropometric and hematological profile of school-age children with low height for age. *RIIIT* **2023**, *11*, 90–101.
34. Getawa, S.; Getaneh, Z.; Melku, M. Hematological Abnormalities and Associated Factors Among Undernourished Under-Five Children Attending University of Gondar Specialized Referral Hospital, Northwest Ethiopia. *J. Blood Med.* **2020**, *11*, 465–478. [CrossRef]
35. Schloss, P.D. Amplicon Sequence Variants Artificially Split Bacterial Genomes into Separate Clusters. *Msphere* **2021**, *6*, e0019121. [CrossRef] [PubMed]
36. Qi, C.; Cai, Y.; Qian, K.; Li, X.; Ren, J.; Wang, P.; Fu, T.; Zhao, T.; Cheng, L.; Shi, L.; et al. gutMDisorder v2.0: A comprehensive database for dysbiosis of gut microbiota in phenotypes and interventions. *Nucleic Acids Res.* **2022**, *51*, D717–D722. [CrossRef]
37. Monira, S.; Nakamura, S.; Gotoh, K.; Izutsu, K.; Watanabe, H.; Alam, N.H.; Endtz, H.P.; Cravioto, A.; Ali, S.I.; Nakaya, T.; et al. Gut Microbiota of Healthy and Malnourished Children in Bangladesh. *Front. Microbiol.* **2011**, *2*, 228. [CrossRef] [PubMed]
38. Kamil, R.Z.; Murdiati, A.; Juffrie, M.; Nakayama, J.; Rahayu, E.S. Gut microbiota and shot-chain fatty acid profile between normal and moderate malnutrition children in Yogyakarta, Indonesia. *Microorganisms* **2021**, *9*, 127. [CrossRef] [PubMed]
39. Subramanian, S.; Huq, S.; Yatsunenko, T.; Haque, R.; Mahfuz, M.; Alam, M.A.; Benezra, A.; DeStefano, J.; Meier, M.F.; Muegge, B.D.; et al. Persistent gut microbiota immaturity in malnourished Bangladeshi children. *Nature* **2014**, *510*, 417–421. [CrossRef]
40. Kim, K.; Veerappan, K.; Woo, N.; Park, B.; Natarajan, S.; Chung, H.; Kim, C.; Park, J. Ulmus macrocarpa Hance extract modulates intestinal microbiota in healthy adults: A randomized, placebo-controlled clinical trial. *J. Microbiol.* **2021**, *59*, 1150–1156. [CrossRef]
41. Kamil, R.Z.; Murdiati, A.; Juffrie, M.; Rahayu, E.S. Gut Microbiota Modulation of Moderate Undernutrition in Infants through Gummy *Lactobacillus plantarum* Dad-13 Consumption: A Randomized Double-Blind Controlled Trial. *Nutrients* **2022**, *14*, 1049. [CrossRef]
42. Shin, N.-R.; Whon, T.W.; Bae, J.-W. Proteobacteria: Microbial signature of dysbiosis in gut microbiota. *Trends Biotechnol.* **2015**, *33*, 496–503. [CrossRef]
43. Saleh, M.; Rüschenbaum, S.; Welsch, C.; Zeuzem, S.; Moradpour, D.; Gouttenoire, J.; Lange, C.M. Glycogen Synthase Kinase 3β Enhances Hepatitis C Virus Replication by Supporting miR-122. *Front. Microbiol.* **2018**, *9*, 2949. [CrossRef]
44. Sutoyo, D.A.; Atmaka, D.R.; Sidabutar, L.M.G.B. Dietary Factors Affecting Firmicutes and Bacteroidetes Ratio in Solving Obesity Problem: A Literature Review. *Media Gizi Indones.* **2020**, *15*, 94–109. [CrossRef]

45. Bervoets, L.; Van Hoorenbeeck, K.; Kortleven, I.; Van Noten, C.; Hens, N.; Vael, C.; Goossens, H.; Desager, K.N.; Vankerckhoven, V. Differences in gut microbiota composition between obese and lean children: A cross-sectional study. *Gut Pathog.* **2013**, *5*, 10. [CrossRef] [PubMed]
46. Stojanov, S.; Berlec, A.; Štrukelj, B. The influence of probiotics on the Firmicutes/Bacteroides ratio in the treatment of obesity and inflammatory bowel disease. *Microorganisms* **2020**, *8*, 1715. [CrossRef] [PubMed]
47. Wong, J.M.W.; de Souza, R.; Kendall, C.W.C.; Emam, A.; Jenkins, D.J.A. Colonic Health: Fermentation and Short Chain Fatty Acids. *J. Clin. Gastroenterol.* **2006**, *40*, 235–243. [CrossRef] [PubMed]
48. Udayappan, S.; Manneras-Holm, L.; Chaplin-Scott, A.; Belzer, C.; Herrema, H.; Dallinga-Thie, G.M.; Duncan, S.H.; Stroes, E.S.G.; Groen, A.K.; Flint, H.J.; et al. Oral treatment with Eubacterium hallii improves insulin sensitivity in db/db mice. *npj Biofilms Microbiomes* **2016**, *2*, 16009. [CrossRef] [PubMed]
49. Liu, X.; Mao, B.; Gu, J.; Wu, J.; Cui, S.; Wang, G.; Zhao, J.; Zhang, H.; Chen, W. *Blautia*—A new functional genus with potential probiotic properties? *Gut Microbes* **2021**, *13*, 1–21. [CrossRef]
50. Shi, Z.; Zhu, Y.; Teng, C.; Yao, Y.; Ren, G.; Richel, A. Anti-obesity effects of α-amylase inhibitor enriched-extract from white common beans (*Phaseolus vulgaris* L.) associated with the modulation of gut microbiota composition in high-fat diet-induced obese rats. *Food Funct.* **2020**, *11*, 1624–1634. [CrossRef]
51. Song, L.; He, M.; Sun, Q.; Wang, Y.; Zhang, J.; Fang, Y.; Liu, S.; Duan, L. *Roseburia hominis* Increases Intestinal Melatonin Level by Activating p-CREB-AANAT Pathway. *Nutrients* **2021**, *14*, 117. [CrossRef]
52. Nie, K.; Ma, K.; Luo, W.; Shen, Z.; Yang, Z.; Xiao, M.; Tong, T.; Yang, Y.; Wang, X. Roseburia intestinalis: A Beneficial Gut Organism From the Discoveries in Genus and Species. *Front. Cell. Infect. Microbiol.* **2021**, *11*, 1–15. [CrossRef]
53. Million, M.; Alou, M.T.; Khelaifia, S.; Bachar, D.; Lagier, J.-C.; Dione, N.; Brah, S.; Hugon, P.; Lombard, V.; Armougom, F.; et al. Increased Gut Redox and Depletion of Anaerobic and Methanogenic Prokaryotes in Severe Acute Malnutrition. *Sci. Rep.* **2016**, *6*, 26051. [CrossRef]
54. Romaní-Pérez, M.; López-Almela, I.; Bullich-Vilarrubias, C.; Rueda-Ruzafa, L.; Del Pulgar, E.M.G.; Benítez-Páez, A.; Liebisch, G.; Lamas, J.A.; Sanz, Y. *Holdemanella biformis* improves glucose tolerance and regulates GLP-1 signaling in obese mice. *FASEB J.* **2021**, *35*, e21734. [CrossRef]
55. Zhu, Y.; Wang, H.; Hollis, J.H.; Jacques, P.F. The associations between yogurt consumption, diet quality, and metabolic profiles in children in the USA. *Eur. J. Nutr.* **2014**, *54*, 543–550. [CrossRef] [PubMed]
56. Delorme, C.; Abraham, A.-L.; Renault, P.; Guédon, E. Genomics of Streptococcus salivarius, a major human commensal. *Infect. Genet. Evol.* **2015**, *33*, 381–392. [CrossRef] [PubMed]
57. Van Hul, M.; Le Roy, T.; Prifti, E.; Dao, M.C.; Paquot, A.; Zucker, J.-D.; Delzenne, N.M.; Muccioli, G.G.; Clément, K.; Cani, P.D. From correlation to causality: The case of *Subdoligranulum*. *Gut Microbes* **2020**, *12*, 1–13. [CrossRef] [PubMed]
58. Boesmans, L.; Valles-Colomer, M.; Wang, J.; Eeckhaut, V.; Falony, G.; Ducatelle, R.; Van Immerseel, F.; Raes, J.; Verbeke, K. Butyrate Producers as Potential Next-Generation Probiotics: Safety Assessment of the Administration of Butyricicoccus pullicaecorum to Healthy Volunteers. *mSystems* **2018**, *3*, e00094-18. [CrossRef] [PubMed]
59. Wang, Y.; Wang, C.; Huang, J.; Xie, M.; Li, X.; Fu, L. *Butyricicoccus* plays a key role in mediating the antagonism between probiotic and antibiotic on food allergy. *Food Agric. Immunol.* **2019**, *30*, 446–461. [CrossRef]
60. Wang, L.-J.; Yang, C.-Y.; Chou, W.-J.; Lee, M.-J.; Chou, M.-C.; Kuo, H.-C.; Yeh, Y.-M.; Lee, S.-Y.; Huang, L.-H.; Li, S.-C. Gut microbiota and dietary patterns in children with attention-deficit/hyperactivity disorder. *Eur. Child Adolesc. Psychiatry* **2019**, *29*, 287–297. [CrossRef] [PubMed]
61. Chen, Y.; Li, Z.; Hu, S.; Zhang, J.; Wu, J.; Shao, N.; Bo, X.; Ni, M.; Ying, X. Gut metagenomes of type 2 diabetic patients have characteristic single-nucleotide polymorphism distribution in Bacteroides coprocola. *Microbiome* **2017**, *5*, 15. [CrossRef]
62. Calderón-Pérez, L.; Gosalbes, M.J.; Yuste, S.; Valls, R.M.; Pedret, A.; Llauradó, E.; Jimenez-Hernandez, N.; Artacho, A.; Pla-Pagà, L.; Companys, J.; et al. Gut metagenomic and short chain fatty acids signature in hypertension: A cross-sectional study. *Sci. Rep.* **2020**, *10*, 6436. [CrossRef]
63. Nomura, K.; Ishikawa, D.; Okahara, K.; Ito, S.; Haga, K.; Takahashi, M.; Arakawa, A.; Shibuya, T.; Osada, T.; Kuwahara-Arai, K.; et al. Bacteroidetes Species Are Correlated with Disease Activity in Ulcerative Colitis. *J. Clin. Med.* **2021**, *10*, 1749. [CrossRef]
64. Sun, Y.; Wu, D.; Zeng, W.; Chen, Y.; Guo, M.; Lu, B.; Li, H.; Sun, C.; Yang, L.; Jiang, X.; et al. The Role of Intestinal Dysbacteriosis Induced Arachidonic Acid Metabolism Disorder in Inflammaging in Atherosclerosis. *Front. Cell. Infect. Microbiol.* **2021**, *11*, 618265. [CrossRef]
65. Chu, H.; Khosravi, A.; Kusumawardhani, I.P.; Kwon, A.H.K.; Vasconcelos, A.C.; Cunha, L.D.; Mayer, A.E.; Shen, Y.; Wu, W.-L.; Kambal, A.; et al. Gene-microbiota interactions contribute to the pathogenesis of inflammatory bowel disease. *Science* **2016**, *352*, 1116–1120. [CrossRef] [PubMed]
66. Yang, C.; Mogno, I.; Contijoch, E.J.; Borgerding, J.N.; Aggarwala, V.; Li, Z.; Siu, S.; Grasset, E.K.; Helmus, D.S.; Dubinsky, M.C.; et al. Fecal IgA Levels Are Determined by Strain-Level Differences in Bacteroides ovatus and Are Modifiable by Gut Microbiota Manipulation. *Cell Host Microbe* **2020**, *27*, 467–475.e6. [CrossRef] [PubMed]
67. Suskun, C.; Kilic, O.; Ciftdogan, D.Y.; Guven, S.; Karbuz, A.; Parlakay, A.O.; Kara, Y.; Kacmaz, E.; Sahin, A.; Boga, A.; et al. Intestinal microbiota composition of children with infection with severe acute respiratory syndrome coronavirus 2 (SARS-CoV-2) and multisystem inflammatory syndrome (MIS-C). *Eur. J. Pediatr.* **2022**, *181*, 3175–3191. [CrossRef] [PubMed]

68. Bakir, M.A.; Kitahara, M.; Sakamoto, M.; Matsumoto, M.; Benno, Y. *Bacteroides intestinalis* sp. nov., isolated from human faeces. *Int. J. Syst. Evol. Microbiol.* **2006**, *56*, 151–154. [CrossRef] [PubMed]
69. Hänninen, A.; Toivonen, R.; Pöysti, S.; Belzer, C.; Plovier, H.; Ouwerkerk, J.P.; Emani, R.; Cani, P.D.; De Vos, W.M. Akkermansia muciniphila induces gut microbiota remodelling and controls islet autoimmunity in NOD mice. *Gut* **2018**, *67*, 1445–1453. [CrossRef] [PubMed]
70. Cerro, E.D.-D.; Lambea, M.; Félix, J.; Salazar, N.; Gueimonde, M.; De la Fuente, M. Daily ingestion of Akkermansia mucciniphila for one month promotes healthy aging and increases lifespan in old female mice. *Biogerontology* **2021**, *23*, 35–52. [CrossRef]
71. Derrien, M.; Van Baarlen, P.; Hooiveld, G.; Norin, E.; Müller, M.; de Vos, W.M. Modulation of Mucosal Immune Response, Tolerance, and Proliferation in Mice Colonized by the Mucin-Degrader *Akkermansia muciniphila*. *Front. Microbiol.* **2011**, *2*, 166. [CrossRef]
72. Han, W.; Zhuang, X.; Liu, Q.; Sun, B.; Miao, H.; Zhang, X. Fermented soy whey induced changes on intestinal microbiota and metabolic influence in mice. *Food Sci. Hum. Wellness* **2021**, *11*, 41–48. [CrossRef]
73. Cummings, J.H. Short chain fatty acids in the human colon. *Gut* **1981**, *22*, 763–779. [CrossRef]
74. Bartlett, A.; Kleiner, M. Dietary protein and the intestinal microbiota: An understudied relationship. *iScience* **2022**, *25*. [CrossRef]
75. Olguin-Calderon, D. Modulation of Caecal Microbiome in Obese Mice Associated with Administration of Amaranth or Soybean Protein Isolates. *Pol. J. Food Nutr. Sci.* **2018**, *69*, 35–44. [CrossRef]
76. Sun, Y.; O'Riordan, M.X.D. Regulation of Bacterial Pathogenesis by Intestinal Short-Chain Fatty Acids. *Adv. Appl. Microbiol.* **2013**, *85*, 93–118. [PubMed]

foods
MDPI

Article

Rework Potential of Soy and Pea Protein Isolates in High-Moisture Extrusion

Silvia J. E. Snel [1,2], Yasmine Amroussi [1], Atze Jan van der Goot [2] and Michael Beyrer [1,*]

[1] Institute of Life Technologies, University of Applied Sciences and Arts Western Switzerland, 1950 Sion, Switzerland
[2] Food Process Engineering, Agrotechnology and Food Sciences Group, Wageningen University & Research, 6708 PD Wageningen, The Netherlands
* Correspondence: michael.beyrer@hes-so.com

Abstract: High-moisture extrusion (HME) is an effective process to make fibrous products that can be used as meat analogues. In this study, the effect of extrusion of already extruded products (i.e., re-extrusion) was tested with the aim to explore the potential of rework in HME. The rework of material is important because it is a route to reduce waste, which is always produced, for example during the start or at the end of a production run. Pea and soy protein isolates (PPI and SPI) were first extruded, then freeze-dried and ground, and extruded again. The visual and textural properties of the fibrous products were evaluated. Also, the rheological properties, solubility, and water-holding capacity (WHC) of the ingredients and the products after the first and second extrusion were quantified. The obtained freeze-dried powders after the first HME cycle had a reduction in solubility of 15% for PPI and 74% for SPI. Furthermore, WHC was reduced by 65% and 17% for PPI and SPI, respectively. After the second HME cycle, the reduction in solubility and WHC was augmented to 22% and 90% for PPI, and 79% and 63% for SPI. No effect on stock and loss moduli after heating and cooling were found, even after two HME cycles. SPI fibrous products did not differ in cutting strength, anisotropy index, or visual appearance after re-extrusion. Only, a decrease in hardness was detected, from 62.0 N to 51.1 N. For PPI, re-extrusion did reduce the cutting force and hardness but not the anisotropy index. It was concluded that even though HME induces a loss of solubility and WHC, this did not affect the fibrous texture formation of the protein. This means that the texture formed during HME does not depend on the process history and that rework is thus possible for fibrous products.

Keywords: meat analogues; plant protein; rheology; Lissajous plot; solubility; anisotropy

Citation: Snel, S.J.E.; Amroussi, Y.; van der Goot, A.J.; Beyrer, M. Rework Potential of Soy and Pea Protein Isolates in High-Moisture Extrusion. *Foods* **2023**, *12*, 2543. https://doi.org/10.3390/foods12132543

Academic Editor: Norbert Raak

Received: 11 May 2023
Revised: 20 June 2023
Accepted: 27 June 2023
Published: 29 June 2023

1. Introduction

High-moisture extrusion (HME) is a common method that has been used for several decades to produce fibrous products that can be used as meat analogues [1]. Soy and pea protein are often used ingredients in high-moisture extrusion to make fibrous products [2,3]. The HME process could generate waste streams during the start or end of the extrusion run. Waste produced during processing is one of the main factors that contribute to global food loss [4]. It is thus of importance to find solutions to reduce losses, for example, by reworking waste streams. HME consists of a mixing and hydration step, a thermomechanical processing step, and finally a cooling step [5]. Most proposed mechanisms behind texture formation consider the deformation of a flow in the extruder cooling die, being either laminar flow [6], phase deformation [7], or elongational flow [8]. Apart from the mechanisms considering deformation, some authors have discussed the importance of protein–protein interactions and shown that aggregates had formed after extrusion of pea and soy [9,10]. With dead-stop HME, Liu and Hsieh [11] showed that covalent binding took place in the extruder barrel before entering the cooling die. These protein–protein interactions could reduce the solubility of the protein, as has been shown for pea and soy [9,11].

However, van der Sman and van der Goot [12] argued that the high temperatures and high shear stresses during HME induce a transient protein network and that instead of measuring relative contributions of the different bonds, rheology is a more accurate method to characterize the protein melt. A closed cavity rheometer (CCR) was recently introduced to mimic extrusion-like conditions and to quantify the effects of processing on ingredient properties [13]. The CCR is able to quantify rheological properties of dense protein dispersions at high temperature and pressure through oscillatory deformation. It has been further suggested that large amplitude oscillatory shear (LAOS) in combination with Lissajous curves could reveal relevant information on the melt properties [14]. With the use of LAOS and the dissipation ratio, it was found that the elasticity of pea protein isolate (PPI) and soy protein isolate (SPI) increased after a heat treatment [15].

Noguchi [16] reported that it was possible to extrude soy protein in three to four steps by grinding the extrudate and feeding it again to the extruder, and a similar fibrous product was obtained. It was concluded that the extrudate was little affected by multiple extrusion steps, as no visual effects were observed. Furthermore, it was proposed that in HME 'reaction' and 'texture formation' should be considered independently [16]. In addition, the protein-protein reactions are expected to take place at a much smaller length scale compared to the formation of the fibrous texture [12]. This would mean that denaturation of the plant protein is less important during HME and not likely to be important for fibrous texture formation. We hypothesize that extrusion changes protein properties, for example, the solubility and water-holding capacity, and this could impact its rework potential. Therefore, this study aims to report how important these changes are for the rework ability of soy and pea by extruding them in a second step and to compare the product properties in terms of anisotropy, hardness, and cutting strength. Furthermore, the dry protein powders after the first and second steps of extrusion were compared to the isolates in terms of rheological properties and solubility and water-holding capacity.

2. Materials and Methods

2.1. Materials

Pea protein isolate (PPI, NUTRALYS® F85M) was obtained from Roquette Frères S.A. (Lestrem, France). Soy protein isolate (SPI, SUPRO 500E A® 8) was obtained from Solae (St. Louis, MO, USA). PPI contained at least 83 wt% protein, and SPI 90% (N × 6.25, indicated by supplier). Powders were sieved to obtain a particle size of <400 μm (PowCN-Sif X600, CapsulCN, Ruian, China) to exclude size effects when compared with the freeze-dried powder.

2.2. Design of Experiments

Figure 1 shows an overview of the performed experimental steps. First, PPI and SPI were textured into fibrous products with HME (PPI-E1, SPI-E2). The obtained extrudates were freeze-dried, ground, and used again as starting material for HME. The obtained fibrous products (PPI-E2, SPI-E2) were compared to the PPI-E1 and SPI-E1 fibrous products in terms of visual appearance and texture properties. The solubility, water-holding capacity (WHC), and rheological properties of the protein isolates and freeze-dried powders were measured. All analyses are discussed in detail in the following sections.

2.3. High Moisture Extrusion

Protein isolates and freeze-dried extrudates (Figure 1) were extruded with an Evolum 25 twin-screw extruder (Clextral, Firminy, France). The screw diameter was 25 mm and the length/diameter ratio was 40. The extruder barrel consisted of 10 sections, which were heated to 30, 50, 70, 90, 100, 120, 130, 145, 145, and 125 °C, respectively. The rotational speed of the screws was set to 300 rpm. A rotating cooling die was attached that consists of a rotating inner cylinder and a thermo-regulated, static outer cylinder [17]. The cylinder can be further divided into two sections that can differ in their rotating speed. The rotation speed of the inner cylinder was set at 75 rpm in the first section, while the second section was kept at 0.5 rpm. The temperature of the cooling die was 85 °C. A breaker plate was placed between the extruder barrel and rotating die, which had 31 holes with a diameter of 3 mm. A twin-screw gravimetric feeder type KCM (K-Tron, Niederlenz, Switzerland) was used to feed the dry ingredients into the extruder, and water was injected in the second section with a water pump (DKM). The dry feed rate and water rate were adjusted according to the moisture content of the isolates to obtain samples with a moisture content of 58% for PPI and 62% for SPI. Throughput was 18 kg h^{-1} for all conditions.

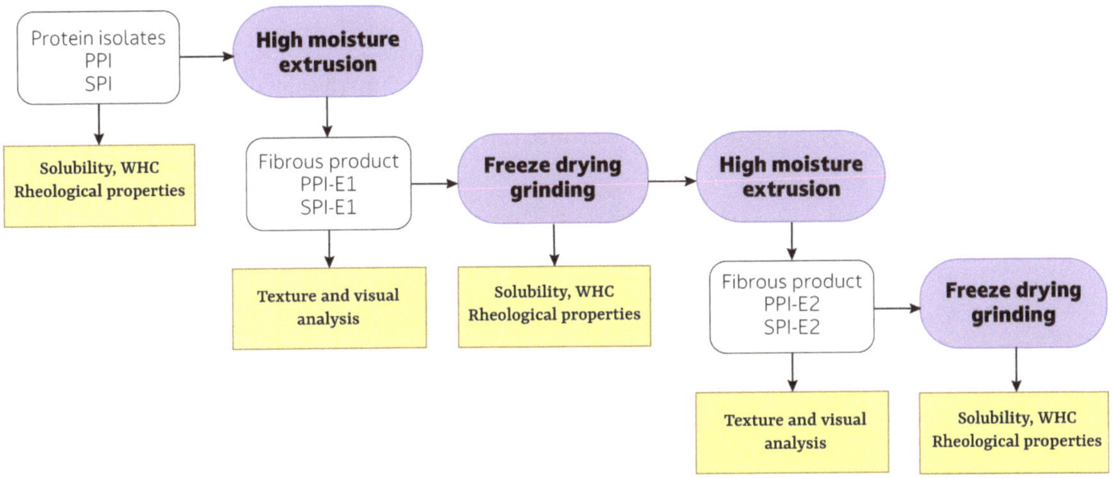

Figure 1. Schematic overview of the experimental design of experiments. PPI = pea protein isolate, SPI = soy protein isolate, WHC = water holding capacity.

2.4. Freeze-Drying of the Extrudates

Extrudates were freeze-dried and milled to facilitate feeding of those extrudates to the extruder. Freeze-drying was chosen for technical reasons and since it has limited effects on protein solubility and gelling capacity compared to other drying techniques such as oven and spray drying [18]. Extrudates were cut and collected in a sealed plastic bag. Packed extrudates were then frozen to −20 °C with a blast freezer (Electrolux, Stockholm, Sweden). Samples were then taken out of the plastic bag and transferred to a vacuum freeze-dryer (Sublimater, Zirbus, Bad Grund, Germany). The temperature of the samples was followed with temperature probes, one for each tray of samples. First, temperature was lowered to −30 °C for 3 h followed by a drying step under vacuum (0.1 mbar) for 34 h. After the samples were dehydrated, temperature and pressure were increased to 20 °C and 10 mbar. Dried samples were collected and ground with a hammer mill (Pulverizer MP, Hosokawa Alpine, Augsburg, Germany) and sieved with a vibrating sieve to obtain a particle size of <400 µm (PowCN-Sif X600, Capsulcn, Ruian, China).

2.5. Texture Analysis

Extrudates obtained from SPI and PPI were compared to extrudates made from freeze-dried extrudate powders of SPI and PPI. For simplicity, we will call the former SPI-E1 and PPI-E1, and the latter SPI-E2 and PPI-E1 (Figure 1). To prevent any other parameters from influencing the extrudates, all samples were prepared on the same day. Samples were defrosted overnight and compared visually by making a small inclination to aid in breaking the sample. Samples were opened in both parallel and perpendicular directions and photographed. Texture profile analysis and cutting tests were performed with a TA-XT2 texture analyzer and the Exponent Connect software (Stable Micro Systems, Surrey, UK). The texture analyzer was equipped with a 50 kg load cell and calibrated with a 1 kg weight.

2.5.1. Texture Profile Analysis

The larger extrudates were cut into cylinders of 20 mm in diameter and a height of 10 mm. Samples were compressed twice with a 60 mm aluminum cylinder probe to 30% of original height with a test speed of 1 mm s^{-1} and a waiting time in between the two compressions of 5 s. Samples were measured in triplicate. The peak maximum force at first compression was taken as the hardness [19].

2.5.2. Cutting Test

Cutting tests were performed in both parallel and perpendicular directions to the rotational shear flow. First, samples were cut in squares of 20 × 20 mm. Height was again 10 mm. Samples were cut up to 90% of their initial height with a Warner–Bratzler blade at a speed of 1 mm s^{-1}. Anisotropy was calculated as the ratio between cutting force in parallel and perpendicular directions :

$$AI_F = \frac{F_{\parallel}}{F_{\perp}} \tag{1}$$

2.6. Solubility and Water-Holding Capacity

The moisture contents of the powders were measured with a moisture analyzer (Mettler Toledo, Columbus, OH, USA). Solubility and water holding capacity (WHC) of the protein isolates and freeze-dried powders was measured in triplicate with a method reported by [20] with some minor alterations. A dispersion of 0.0125 g g^{-1} protein isolate in demineralized water was prepared and shaken overnight at 300 rpm at room temperature. Subsequently, the dispersions were centrifuged at 10,000× g at 21 °C for 20 min. The weight of the pellet was recorded and dried at 100 °C for at least 10 h. The solubility and WHC were calculated as follows:

$$Solubility = \frac{M_{drypowder} - M_{drypellet}}{M_{drypowder}} \tag{2}$$

$$WHC = \frac{M_{wetpellet} - M_{drypellet}}{M_{drypowder}} \tag{3}$$

in which $M_{drypowder}$ is the overall weight of the isolate, and $M_{wetpellet}$ and $M_{drypellet}$ are the weights of the pellet before and after drying.

2.7. Rheological Properties

Strain amplitude sweeps were performed with a CCR (RPA elite, TA instruments, New Castle, DE, USA). The geometry in the CCR has a radius of 2.25 mm, a maximum height of 4 mm, and a biconical opening with an angle of 3.35° to ensure homogeneous transmission of the shear stress applied. The top and bottom cones have grooves to prevent slip. The upper cone remains stationary while the lower cone oscillates. First, powders were mixed with demineralized water to obtain the same moisture content as in extrusion: 58% for PPI and 62% for SPI, calculated with the relative dry matter content of the powders. Next, approximately 5 g of the sample was placed between two plastic films in the CCR. The cavity was then sealed with a pressure of 4 bar to prevent evaporation. Samples were then heated to 30 or 145 °C for 2 min without shear. Three measuring conditions were tested: 30 °C, 145 °C, or cooled from 145 °C to 30 °C. The latter could reflect the behavior during a HME cycle, in terms of temperature profile. An amplitude strain sweep was performed from 0.1 to 1000% at 1 Hz. The yield stress and strain at the end of the linear viscoelastic (LVE) regime were defined as the point where G′ differs more than 5% from its value in the LVE regime. The flow stress and strain were defined at the crossover point, where G′ (stock modulus) is equal to G″ (loss modulus). These values were used to define if the materials were mushy (low stress and low strain), brittle (low strain and high stress), rubbery (high strain and low stress), or tough (high strain and high stress) (Schreuders et al. [21]).

2.8. Lissajous Plots

The data obtained with the strain amplitude sweeps were further analyzed with the MITlaos software (version 2.1 beta, freeware distributed from MITlaos@mit.edu). With this software, Lissajous plots were made to visualize the response of the material to the oscillatory strain for both the elastic and viscous stress. Furthermore, the dissipation ratio was calculated [21]. The energy dissipated per unit volume in a single cycle as a function of the first-order viscous Fourier coefficient (G_1'') was calculated as:

$$E_d = G_1'' \gamma_0^2 \tag{4}$$

in which E_d is the energy dissipated and γ_0 the strain amplitude. For a perfect plastic material, the energy dissipated is equal to:

$$E_d^{pp} = 4\gamma_0\, \sigma_{max} \tag{5}$$

in which σ_{max} is the maximum stress. The ratio between the actual energy dissipated and the energy dissipated by a perfect plastic then gives the energy dissipation ratio ϕ [22]:

$$\phi = \frac{E_d}{E_d^{pp}} = \frac{\pi G_1'' \gamma_0}{4\, \sigma_{max}} \tag{6}$$

For a perfect plastic material, this ratio would be 1. An elastic material would have a dissipation ratio of 0, and a purely viscous material would have a ratio of 0.8.

2.9. Standardization of Results

The relative difference between the extruded powder and extrudates was illustrated by calculating a ratio for each parameter measured. For each parameter, the obtained value was divided by the obtained value of their corresponding isolate, so SPI or PPI, or their corresponding extrudate, SPI-E1 or PPI-E1. For the corresponding sample, the value was thus taken as 1.

2.10. Statistical Analysis

Statistical analysis was performed with R prior to standardization of the results. It was assumed that the different factors tested in this study (protein type, processing, and temperature in the case of the rheological properties) did not interact. To confirm this, a two-way analysis of variance (ANOVA) or three-way ANOVA in the case of rheological properties was performed. First, normality and equal variance were tested with descriptive statistics. If the data were normally distributed, a two- or three-way ANOVA was performed to test if the observed differences between samples were significant ($\alpha = 0.05$). Multiple comparison Tukey tests were performed to indicate which treatments were significantly different from each other ($\alpha = 0.05$). If the data was not normally distributed, Welch's ANOVA and Dunn's tests were performed. All tests in this study were performed in triplicate.

3. Results

3.1. Extrudate Properties after Second Step HME

Extrudates were made with HME from protein isolates (PPI-E1, SPI-E1) and freeze-dried powders of the extrudates (PPI-E2, SPI-E2). The visual appearance, hardness, cutting strength, and anisotropy of cutting strength were compared.

The samples prepared from protein isolates and freeze-dried powders of the extrudates did not differ in their visual appearance for both SPI and PPI (Figure 2). In general, it was noted that SPI samples looked more inhomogeneous when broken in the parallel direction, while for PPI no clear difference was observed between the parallel and perpendicular direction of breaking. This observation was made for products obtained after the first and second extrusion steps.

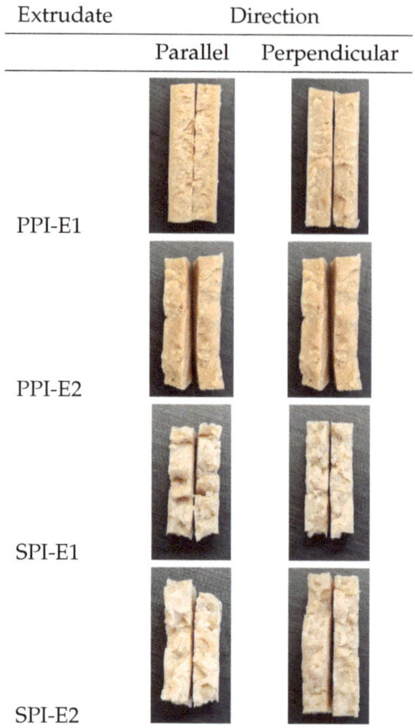

Figure 2. Visual appearance of extrudates from PPI (58% MC) and SPI (62%). A small inclination was made to allow breaking of the extrudates in parallel and perpendicular direction to the shear flow in the rotating cooling die. Samples -E1 were made from the isolates (PPI, SPI), and samples -E2 were made from the freeze-dried and ground -E1 extrudates.

The hardness, cutting force in both parallel and perpendicular directions, and the calculated anisotropy index were set to 1 for the PPI-E1 and SPI-E1 extrudates (Table 1). Then, the obtained values of the PPI-E2, and SPI-E2 samples were divided by the values obtained for PPI-E1, and SPI-E1, to show their relative difference (Figure 3). For PPI, cutting force decreased in both parallel and perpendicular directions after the second HME step, and a similar anisotropy index was thus found (Figure 3a–c). The cutting force in both directions was not altered for SPI after a second HME step, resulting in a similar anisotropy index. It can further be observed that the hardness of the PPI-E2 and SPI-E2 samples was slightly lower compared to PPI-E1 and SPI-E1 (Figure 3d).

Table 1. Hardness (N), cutting force (N) in both parallel (par) and perpendicular (per) direction, and the anisotropy index calculated as the ratio between the two cutting forces for PPI-E1 (58% MC) and SPI-E1 (62% MC) extrudates, letters indicate significant groups, $n = 3$.

Extrudate	Hardness (N)	Cutting Force Par (N)	Cutting Force per (N)	AI (-)
PPI-E1	$116 \pm 3.5\ ^{a}$	$14.3 \pm 0.2\ ^{a}$	$15.1 \pm 0.3\ ^{b}$	$1.1 \pm 0.03\ ^{b}$
SPI-E1	$62.0 \pm 2.5\ ^{b}$	$13.6 \pm 0.8\ ^{a}$	$22.1 \pm 0.4\ ^{a}$	$1.6 \pm 0.1\ ^{a}$

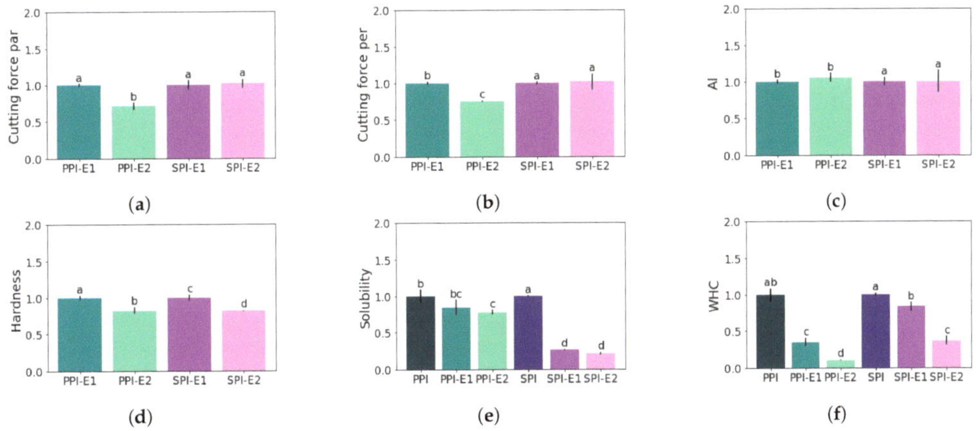

Figure 3. Relative change of extrudate parameters obtained from a cutting test (**a**–**c**), texture profile analysis (**d**), and relative solubility (**e**) and WHC (**f**) of the powders after HME after a first (-E1) and second (-E2) HME step as compared to PPI and SPI (Table 2). Values are reported as averages with standard deviations (black bars); letters indicate significant groups; $n = 3$. A relation was found between HME cycle and protein type for cutting force in both par and per direction, solubility, and WHC ($\alpha < 0.05$).

Table 2. Solubility, and WHC for PPI and SPI before and after extrusion. After extrusion, extrudates were freeze-dried, ground, and sieved (-E), this powder was extruded, freeze-dried, ground, and sieved again (-E2). Values are averages ± standard deviation, and letters indicate significant groups, $n = 3$.

Powder	Solubility (g g^{-1})	WHC (g g^{-1})
PPI	0.30 ± 0.03 ^{b}	9.6 ± 0.8 ^{ab}
SPI	0.47 ± 0.00 ^{a}	11 ± 0.2 ^{a}

3.2. Effect of HME on Protein Properties

Apart from the comparison between extrudates after a second HME step, the obtained freeze-dried powders were also analyzed. Extrudates obtained after the first (called -E1) and second (-E2) HME were freeze-dried, ground, and sieved, and their properties were compared to the protein isolates (PPI, SPI) in terms of solubility, WHC (Table 2, Figure 3e,f), and rheological properties (Table 3, Figures 4–7).

Table 3. Storage modulus (G′), loss modulus (G″), yield strain (γ_y), yield stress (σ_y), flow strain (γ_{co}), and flow stress (σ_{co}) for PPI and SPI before and after extrusion. After extrusion, extrudates were freeze-dried, ground, and sieved (-E1), this powder was extruded, freeze-dried, ground, and sieved again (-E2). G′, γ_y, and γ_{co} were determined from strain amplitude sweeps. Values are averages ± standard deviation, and letters indicate significant groups, $n = 3$.

Powder	Temperature (°C)	G′ (10^2 kPa)	G″ (10^2 kPa)	γ_y (%)	σ_y (kPa)	γ_{co} (%)	σ_{co} (kPa)
PPI	30	1.4 ± 0.3 ^{b}	0.3 ± 0.0 ^{a}	4.8 ± 0.5 ^{bc}	6.9 ± 0.5 ^{b}	137 ± 19 ^{b}	48 ± 5.1 ^{b}
	145-30	2.2 ± 0.4 ^{ab}	0.5 ± 0.1 ^{a}	3.3 ± 0.6 ^{c}	7.1 ± 1.0 ^{b}	173 ± 24 ^{ab}	62 ± 5.9 ^{b}
SPI	30	1.7 ± 0.3 ^{ab}	0.3 ± 0.1^{a}	8.0 ± 0.0 ^{a}	14 ± 2.1 ^{ab}	200 ± 0.0 ^{a}	75 ± 11 ^{ab}
	145-30	2.5 ± 0.4 ^{a}	0.5 ± 0.1 ^{a}	6.5 ± 1.2 ^{b}	16 ± 1.1 ^{a}	200 ± 0.0 ^{a}	105 ± 13 ^{a}

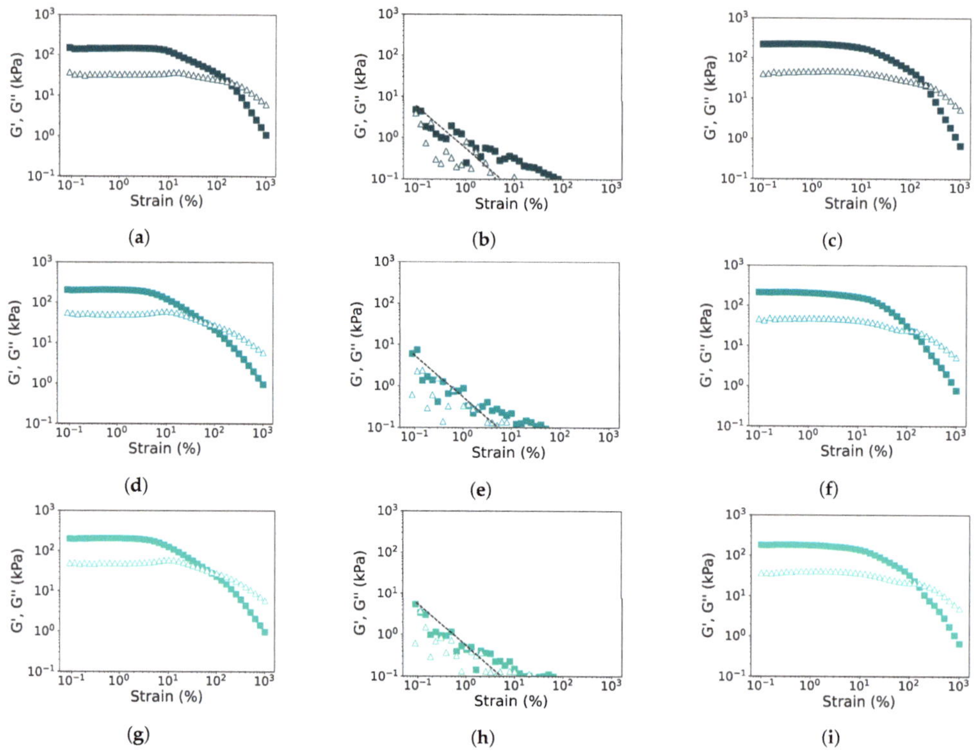

Figure 4. Strain sweeps (f = 1 Hz) of PPI (**a–c**), PPI-E1 (**d–f**), and PPI-E2 (**g–i**) at 30 °C (**a**,**d**,**g**), at 145 °C (**b**,**e**,**h**), and heated to 145 °C and cooled to 30 °C (**c**,**f**,**i**). G′: closed symbols and G″: open symbols. The black dotted line in the measurement at 145 °C represents the torque limits of the CCR.

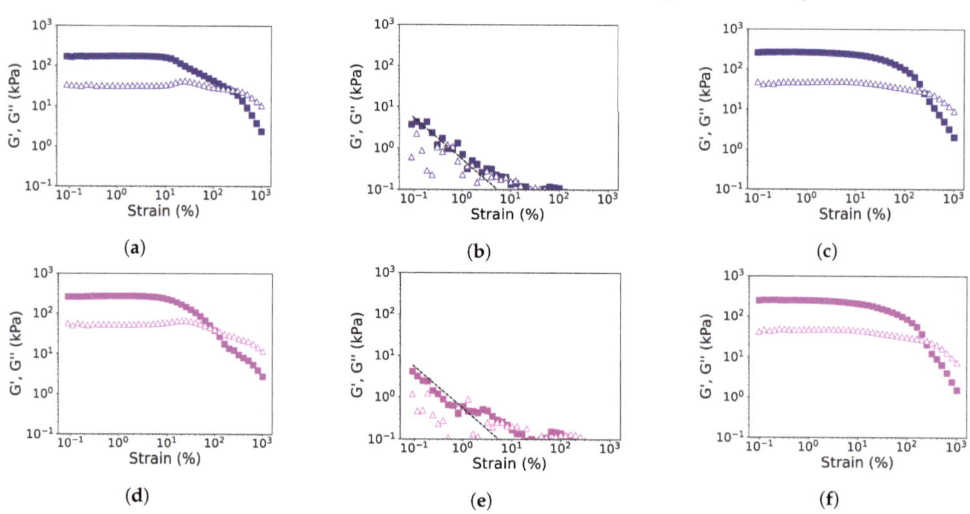

Figure 5. *Cont.*

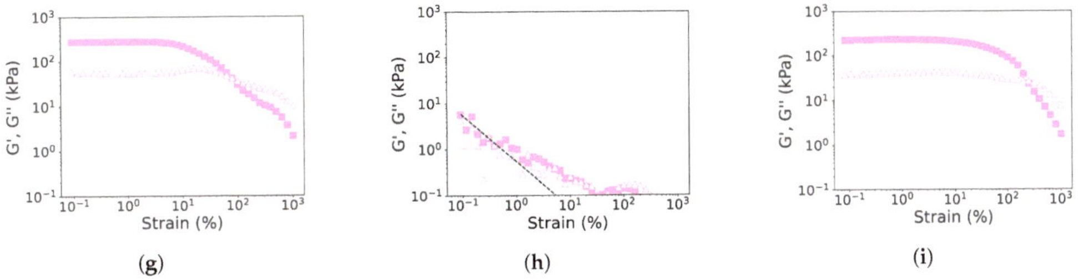

Figure 5. Strain sweeps (f = 1 Hz) of SPI (**a**–**c**), SPI-E1 (**d**–**f**), and SPI-E2 (**g**–**i**) at 30 °C (**a**,**d**,**g**), at 145 °C (**b**,**e**,**h**), and heated to 145 °C and cooled to 30 °C (**c**,**f**,**i**). G′: closed symbols and G″: open symbols. The black dotted line in the measurement at 145 °C represents the torque limits of the CCR.

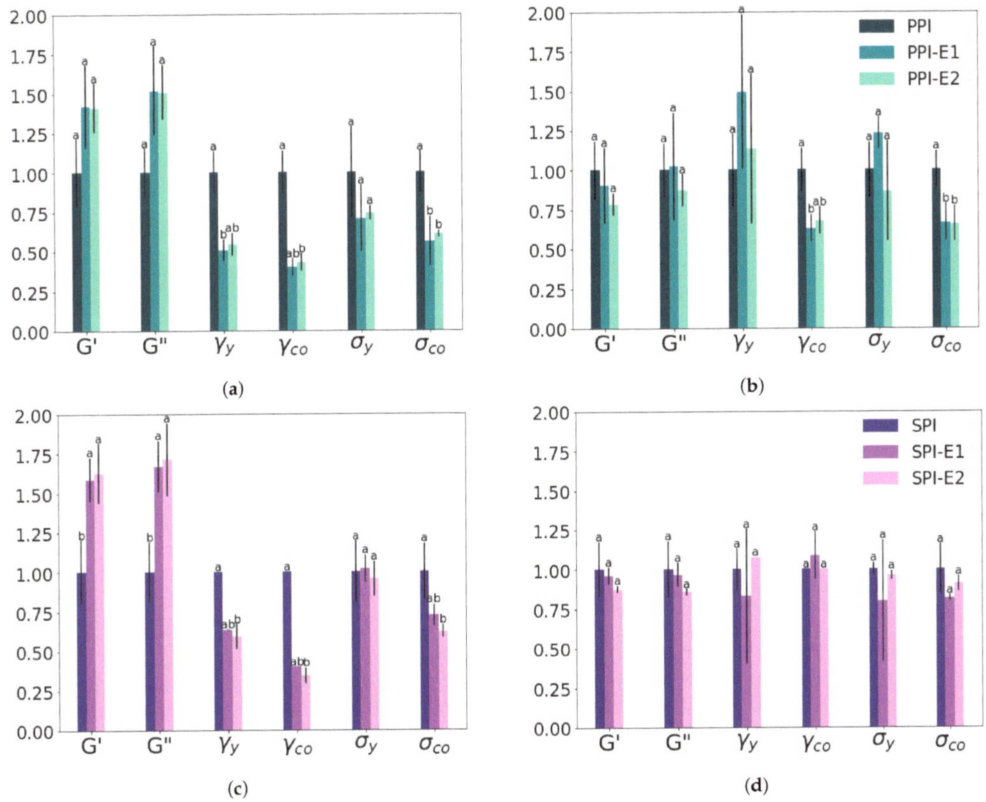

Figure 6. Relative change of rheological parameters obtained from a strain sweep after a first (-E1) and second (-E2) HME step as compared to PPI (**a**,**b**) and SPI (**c**,**d**) (Table 3). The strain sweeps were performed at 30 °C (**a**,**c**) and after heating to 145 °C and cooling to 30 °C (**c**,**d**). Values are reported as averages with standard deviations (black bars); letters indicate significant groups per variable; n = 3.

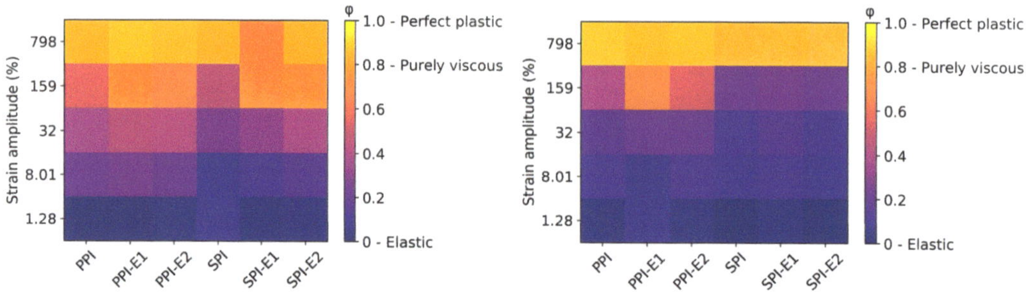

(a) (b)

Figure 7. Heat map of the dissipation ratio ϕ at different strain amplitudes for the powders before (SPI, PPI) and after one (-E1) and two (-E2) HME cycles, measured at 30 °C, (**a**) and after heating to 145 °C followed by cooling to 30 °C (**b**).

Solubility and WHC of PPI were slightly lower than the solubility and WHC of SPI (Table 2). The obtained powders from freeze-dried extrudates, PPI-E1 and SPI-E1, had a lower solubility compared to PPI and SPI (Figure 3e). For PPI, this reduction was not significant. The powders obtained from the second-step extrudates, PPI-E2 and SPI-E2 had again a lower solubility, but this additional decrease was not significantly different from the PPI-E1 and SPI-E1 samples.

The WHC of the PPI-E1 and SPI-E1 samples was significantly lower compared to PPI and SPI (Figure 3f). The second HME step significantly reduced the WHC further.

Overall, the effect of HME on solubility and WHC was similar for both PPI and SPI, although the solubility was more drastically reduced for SPI and the WHC for PPI. It seems that HME leads to additional cross-links which impact the powder properties. With the use of cross-linking and reducing agents, it was found that more cross-links lead to lower WHC for whey protein [23]. The WHC could be used to predict the maximum moisture content during HME. For example, PPI-E1 had a WHC of 3.3 ± 0.4 g g^{-1}, and extrudates had a dry matter content of 42%. The 42 g PPI-E1 could thus hold 140 g water, which translates to a maximum moisture content of 77%, which is still higher than the 58% in the extrudate. However, PPI-E2 had a WHC of 1.0 ± 0.1 g g^{-1}, which would result in a maximum moisture content of 50%. Therefore, extruding PPI-E2 in a third HME cycle might become problematic.

Rheological properties were examined with a strain amplitude sweep at 30 °C, 145 °C, and heating to 145 °C followed by cooling to 30 °C (Figures 4 and 5). The measured G' and G'' at 145 °C were found to be close to or even below the torque limits of the CCR, and thus it was not possible to determine the LVE-regime, yielding, and cross-over point for the measurements at this temperature. From the other strain amplitude sweeps, G' and G'' in the LVE-regime, yield strain and stress, and flow strain and stress were determined (Table 3). The strain and stress at the yield and flow point give an indication of the texture of the material, being either mushy, rubbery, tough, or brittle [21]. The alterations of these parameters after the first and second HME steps were again calculated relative to PPI and SPI (Figure 6). It can be seen that the HME step increased the G' and G'' measured at 30 °C for both PPI and SPI, and this remained constant after the second step. After heating to 145 °C and cooling to 30 °C, no significant changes were found for both PPI and SPI, although G' and G'' decreased slightly with each step.

The lower yield stress and strain of PPI samples compared to SPI, indicated a more mushy material for PPI compared to the brittle SPI (Table 3). Both the yield strain and flow strain decreased significantly for both PPI and SPI measured at 30 °C (Figure 6a,c). In other words, after an HME step, the isolates will yield and flow at a lower strain, and this might affect the flow behavior in the extruder barrel. After heating to 145 °C and cooling to 30 °C, PPI samples were less affected, and yield strain and stress even increased after the first

HME step (Figure 6b). SPI became slightly more brittle after the first HME cycle, seen by the lower yield strain, but after the second cycle, a similar stress and strain value was found as for the non-extruded SPI (Figure 6d). So, breakdown of the network structure seems to take place, but it can (partly) recover.

The flow stress and strain were determined at the cross-over point of the amplitude sweeps. Again, a lower flow strain and stress were observed for PPI compared to SPI, indicating that SPI is a more tough material (Table 3). The first HME step reduced the flow strain and stress for both PPI and SPI, indicating a more mushy or brittle material (Figure 6a,c). After the second HME step, the strain and stress were not reduced further. After heating to 145 °C and cooling to 30 °C, both the flow strain and stress were less affected, but still, a reduction was observed for PPI samples (Figure 6b,d). Interestingly, the cross-over stress and strain of SPI were not changed after two cycles of extrusion after heating and cooling, indicating that the SPI samples remained tough.

Lissajous curves were made from the strain amplitude sweeps (Appendix A), and dissipation ratios were calculated (Figure 7). At 30 °C, all samples were in the elastic regime at low strain amplitude and became more viscous with increasing strain amplitude. For both PPI and SPI, the point where the samples became viscous was at a lower strain amplitude after the first HME cycle. A second HME cycle did not seem to affect this further. When the samples were first heated to 145 °C and then cooled again to 30 °C, both PPI and SPI remained elastic at higher strain amplitudes compared to the unheated samples. This increase in elasticity has been found before for both pea and soy [21]. The first and second HME cycles made the PPI samples slightly more viscous at higher shear, but almost no difference was observed for the SPI samples.

4. Discussion

In this study, the rework potential of fibrous products was tested by repetitive extrusion of protein isolates. This was accomplished by comparing the relative change of the obtained fibrous products in terms of hardness, cutting strength, and anisotropy. For improved comparability, extrudates were freeze-dried and ground before the second HME step. The properties of the obtained powder after the first and second HME steps were compared to the untreated protein isolate in terms of solubility, WHC, and rheological properties.

The effect of repetitive extrusion on the extrudate properties was small, especially for SPI. Visually, no differences were found between samples after the first and second HME steps. For soy protein, this was observed before, as no visual difference was reported for defatted soy protein after three HME steps [24]. PPI extrudates had a lower anisotropy index compared to SPI, which has been observed previously [17]. For SPI samples the cutting force and anisotropy indexes were not affected by the second HME step. Only the hardness was slightly decreased. For PPI, the cutting force and hardness were decreased, however, the anisotropy index remained the same. This then confirms that soy can be extruded in several cycles without altering product properties, as was suggested before [24]. The ability to extrude SPI and PPI in several HME cycles is remarkable, as we see that ingredient properties (solubility, WHC) changed. This could be explained because some changes are less relevant; for example, the WHC was still sufficient, and other changes were partly reversible through heating. The changes in rheological properties became smaller after heating, suggesting the breaking and reforming of bonds upon heating. The hydrophobic bonds in proteins are broken at temperatures of 130–140 °C [7]. Furthermore, the high shear rates in the extruder barrel could lower the activation energy for breaking the disulfide bonds [25]. Each HME cycle reduced the solubility and the WHC of both PPI and SPI, even if the reduction in solubility was only significant after two HME cycles for PPI. The decrease in solubility after HME has been reported previously for both PPI [9] and SPI [11], although both studies used reducing buffers to break specific bonds between the proteins. Isobe and Noguchi [24] also found a reduction of soy protein solubility after each HME step. The reduced solubility and WHC were probably caused by the formation of insoluble aggregates during HME. Fang et al. [10] performed sodium dodecyl sulfate-polyacrylamide gel electrophoresis (SDS-PAGE) before and after extruding SPI and observed a reduction in the intensity of specific bands, which they attributed to aggregate formation. Similarly, Osen et al. [9] saw a reduction of specific bands in SDS-PAGE after HME of PPI and came to a similar conclusion. The decrease in WHC could also reflect a higher density of cross-links [23].

The formation of aggregates could also explain observed changes in rheological properties when measured at 30 °C. Remarkably, these changes became smaller after heat treatment was performed. Possibly, the heating and cooling step allowed the proteins to rearrange in a more optimal way. For pea protein, it has been found previously that a low cooling rate allowed pea vicilin to make more optimal interactions (O'Kane et al. [26]). O'Kane et al. [26] showed that the elastic modulus of soy protein gels followed the same trajectory during reheating and subsequent cooling between 85 and 25 °C, even at a higher cooling rate. This then further confirms the found reversibility of soy during extrusion [16]. In this study, rework was tested by freeze-drying the extrudates and feeding them as a powder to the extruder. We are confident that rework is also possible using other mild drying techniques. In the case rework is combined with native ingredients, it might be possible to add the rework without a drying step. The latter has been proven for extrudates made from soy flour [24].

5. Conclusions

SPI and PPI were extruded twice to test rework potential. First, extrudates for both protein isolates were obtained with HME and subsequently, these extrudates were freeze-dried and ground for a next HME cycle. HME led to a reduction of solubility and WHC of proteins, probably caused by protein–protein interactions and aggregate formation. A second HME cycle reduced solubility and WHC further. However, after heating, the rheological properties were not significantly different after HME. This was explained by the weakening of protein–protein interactions upon temperatures of 145 °C. In HME, the weakening of the bonds will be further achieved by the high shear stresses. We, therefore, hypothesize that protein aggregates can be formed and aligned in a repetitive, reversible manner, and fibrous textures can be formed again. This hypothesis was confirmed by the similarity in structural and textural properties of the fibrous products of the first and second HME cycle. For both PPI and SPI, the visual appearance of the extrudates was not altered. Even though the hardness and cutting strength of PPI were slightly reduced, visual aspects and anisotropy of PPI fibrous products remained similar after the second HME step. It is concluded that the reactions taking place in HME can be viewed as separate from the texture formation. SPI could be seen as a thermoplastic reversible material, and even PPI was not affected largely by HME. This thus concludes that rework is possible for both PPI and SPI.

Author Contributions: Conceptualization, S.J.E.S., A.J.v.d.G. and M.B.; methodology, S.J.E.S.; software, S.J.E.S.; validation, S.J.E.S.; formal analysis, S.J.E.S.; investigation, S.J.E.S. and Y.A.; resources, A.J.v.d.G. and M.B.; data curation, S.J.E.S.; writing—original draft preparation, S.J.E.S.; writing—review and editing, S.J.E.S., A.J.v.d.G. and M.B.; visualization, S.J.E.S.; supervision, A.J.v.d.G. and M.B.; project administration, S.J.E.S. and M.B.; funding acquisition, M.B. All authors have read and agreed to the published version of the manuscript.

Funding: This work is part of the project greenXprot, co-financed by the Swiss Innovation Agency 'Innosuisse' under contract number 36440.1 IP-LS.

Data Availability Statement: Data available on request from the corresponding author.

Acknowledgments: We thank Frédéric Maruel (School of Engineering, Hes-so VS) for his help and assistance during extrusion trials and Jos Sewalt, Miek Schlangen, and Nienke Köllmann (Food Process Engineering, Wageningen University & Research) for their help and guidance during CCR experiments.

Conflicts of Interest: The authors declare no conflict of interest.

Appendix A

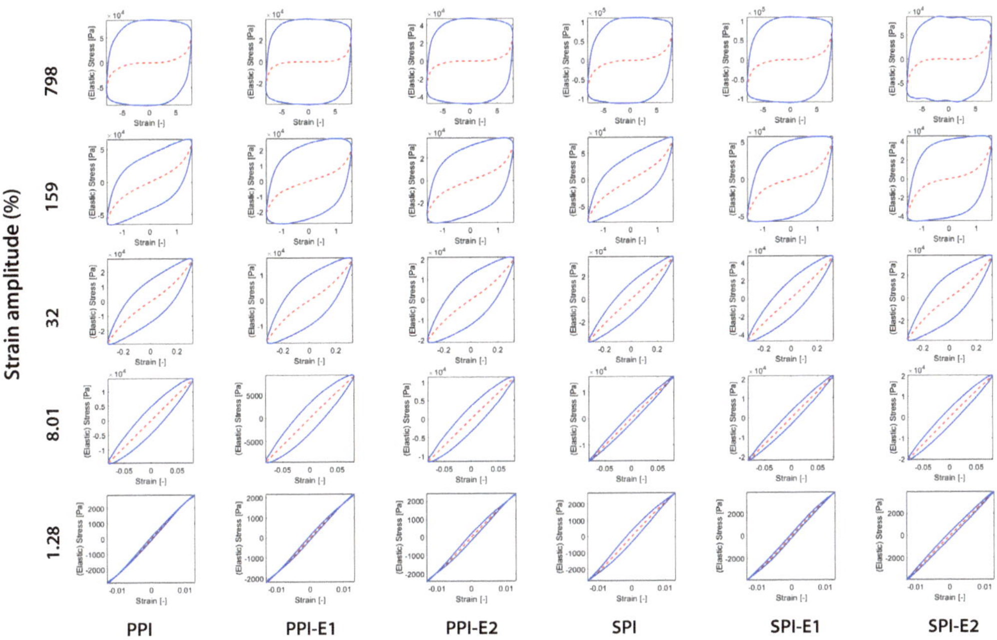

Figure A1. Elastic lissajous curves of the stress versus strain amplitude at 30 °C. Solid lines represent normalized stress and dashed lines the elastic stress.

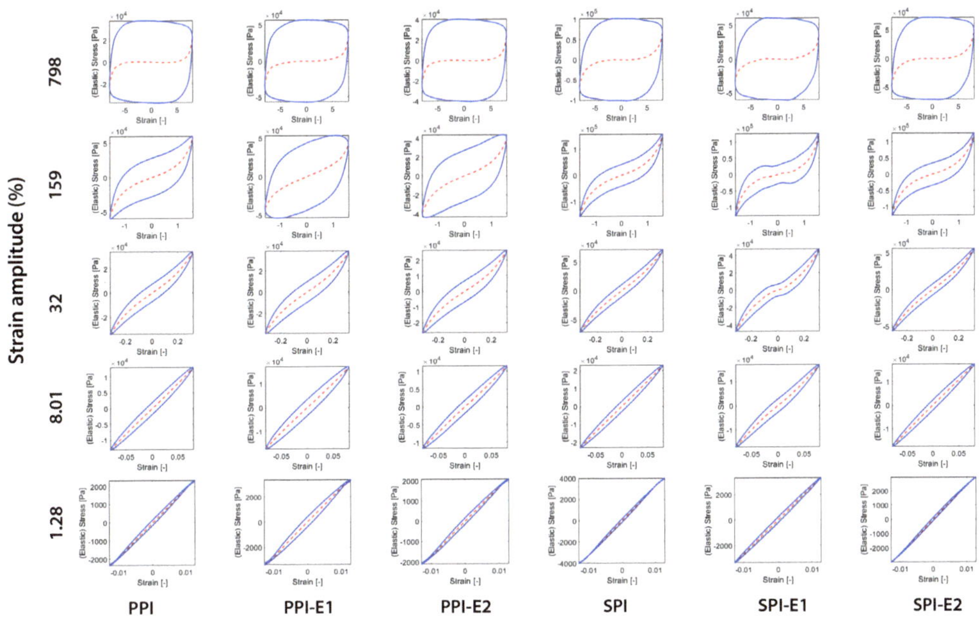

Figure A2. Elastic lissajous curves of the stress versus strain amplitude of samples heated to 145 °C and cooled to 30 °C. Solid lines represent normalized stress and dashed lines represent the elastic stress.

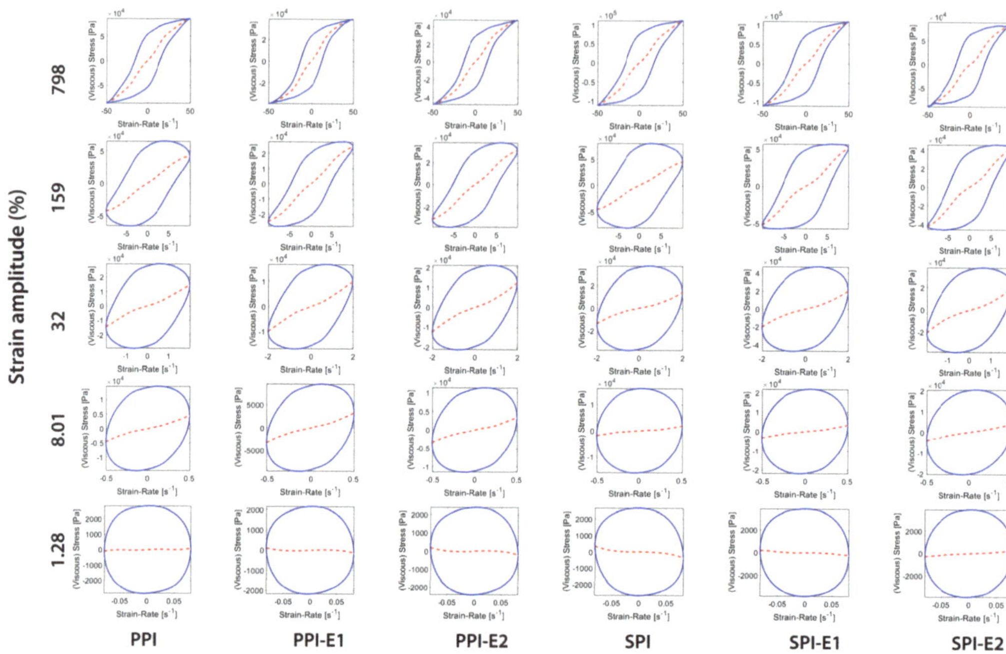

Figure A3. Viscous lissajous curves of the stress versus strain amplitude at 30 °C. Solid lines represent normalized stress and dashed lines represent the elastic stress.

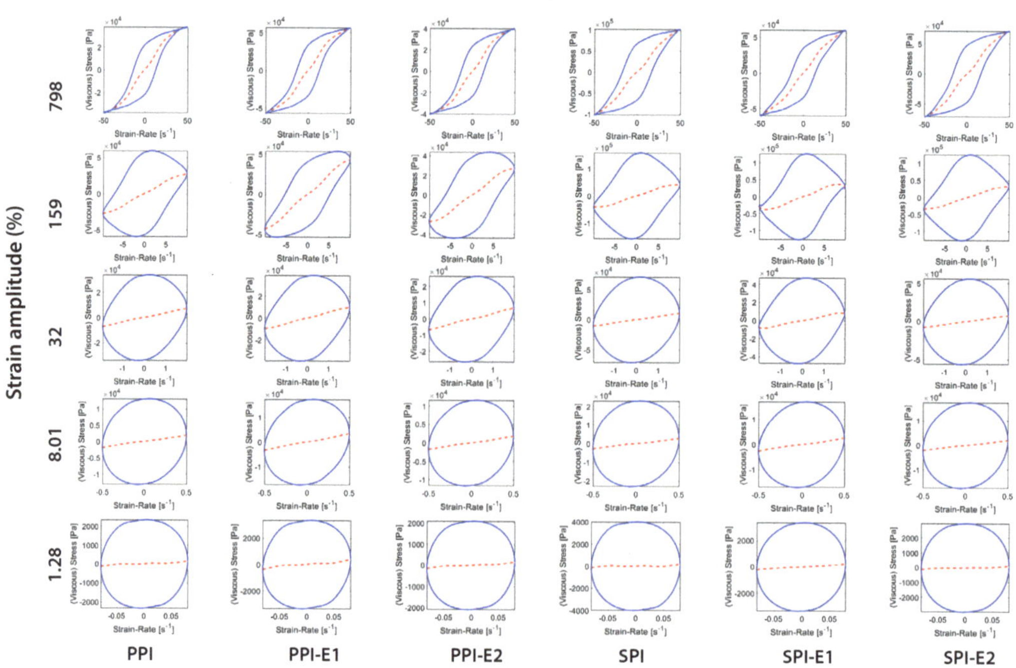

Figure A4. Viscous lissajous curves of the stress versus strain amplitude of samples heated to 145 °C and cooled to 30 °C. Solid lines represent normalized stress and dashed lines the elastic stress.

References

1. Cheftel, J.C.; Kitagawa, M.; Queguiner, C. New Protein Texturization Processes by Extrusion Cooking at High Moisture Levels. *Food Rev. Int.* **1992**, *8*, 235–275. [CrossRef]
2. Pietsch, V.L.; Bühler, J.M.; Karbstein, H.P.; Emin, M.A. High moisture extrusion of soy protein concentrate: Influence of thermomechanical treatment on protein-protein interactions and rheological properties. *J. Food Eng.* **2019**, *251*, 11–18. [CrossRef]
3. Osen, R.; Toelstede, S.; Wild, F.; Eisner, P.; Schweiggert-Weisz, U. High moisture extrusion cooking of pea protein isolates: Raw material characteristics, extruder responses, and texture properties. *J. Food Eng.* **2014**, *127*, 67–74. [CrossRef]
4. Ishangulyyev, R.; Kim, S.; Lee, S.H. Understanding food loss and waste-why are we losing and wasting food? *Foods* **2019**, *8*, 297. [CrossRef]
5. Cornet, S.H.V.; Snel, S.J.E.; Schreuders, F.K.G.; Van der Sman, R.G.M.; Beyrer, M.; Van der Goot, A.J. Thermo-mechanical processing of plant proteins using shear cell and high-moisture extrusion cooking. *Crit. Rev. Food Sci. Nutr.* **2022**, *62*, 3264–3280. [CrossRef]
6. Akdogan, H. High moisture food extrusion. *Int. J. Food Sci. Technol.* **1999**, *34*, 195–207. [CrossRef]
7. Tolstoguzov, V.B. Thermoplastic extrusion—The mechanism of the formation of extrudate structure and properties. *J. Am. Oil Chem. Soc.* **1993**, *70*, 417–424. [CrossRef]
8. Wittek, P.; Ellwanger, F.; Karbstein, H.P.; Emin, M.A. Morphology development and flow characteristics during high moisture extrusion of a plant-based meat analogue. *Foods* **2021**, *10*, 1753. [CrossRef] [PubMed]
9. Osen, R.; Toelstede, S.; Eisner, P.; Schweiggert-Weisz, U. Effect of high-moisture extrusion cooking on protein-protein interactions of pea (*Pisum sativum* L.) protein isolates. *Int. J. Food Sci. Technol.* **2015**, *50*, 1390–1396. [CrossRef]
10. Fang, Y.; Zhang, B.; Wei, Y. Effects of the specific mechanical energy on the physicochemical properties of texturized soy protein during high-moisture extrusion cooking. *J. Food Eng.* **2014**, *121*, 32–38. [CrossRef]
11. Liu, K.; Hsieh, F.H. Protein-protein interactions during high-moisture extrusion for fibrous meat analogues and comparison of protein solubility methods using different solvent systems. *J. Agric. Food Chem.* **2008**, *56*, 2681–2687. [CrossRef] [PubMed]
12. van der Sman, R.G.; van der Goot, A.J. Hypotheses concerning structuring of extruded meat analogs. *Curr. Res. Food Sci.* **2023**, *6*, 100510. [CrossRef] [PubMed]
13. Emin, M.A.; Quevedo, M.; Wilhelm, M.; Karbstein, H.P. Analysis of the reaction behavior of highly concentrated plant proteins in extrusion-like conditions. *Innov. Food Sci. Emerg. Technol.* **2017**, *44*, 15–20. [CrossRef]
14. Schreuders, F.K.G.; Sagis, L.M.C.; Bodnár, I.; Erni, P.; Boom, R.M.; van der Goot, A.J. Small and large oscillatory shear properties of concentrated proteins. *Food Hydrocoll.* **2021**, *110*, 106172. [CrossRef]
15. Schreuders, F.K.; Dekkers, B.L.; Bodnár, I.; Erni, P.; Boom, R.M.; van der Goot, A.J. Comparing structuring potential of pea and soy protein with gluten for meat analogue preparation. *J. Food Eng.* **2019**, *261*, 32–39. [CrossRef]
16. Noguchi, A. Extrusion Cooking of High-Moisture Protein foods. In *Extrusion Cooking*, 1st ed.; Mercier, C., Linko, P., Harper, J.M., Eds.; American Association of Cereal Chemists: St. Paul, MO, USA, 1989; Chapter 11, pp. 343–370.
17. Snel, S.J.; Bellwald, Y.; van der Goot, A.J.; Beyrer, M. Novel rotating die coupled to a twin-screw extruder as a new route to produce meat analogues with soy, pea and gluten. *Innov. Food Sci. Emerg. Technol.* **2022**, *81*, 103152. [CrossRef]
18. Brishti, F.H.; Chay, S.Y.; Muhammad, K.; Ismail-Fitry, M.R.; Zarei, M.; Karthikeyan, S.; Saari, N. Effects of drying techniques on the physicochemical, functional, thermal, structural and rheological properties of mung bean (*Vigna radiata*) protein isolate powder. *Food Res. Int.* **2020**, *138*, 109783. [CrossRef]
19. Meullenet, J. Relationship between sensory and instrumental texture profile attributes. *J. Sens. Stud.* **1998**, *13*, 77–93. [CrossRef]
20. Bühler, J.M.; Dekkers, B.L.; Bruins, M.E.; Goot, A.J.V.D. Modifying Faba Bean Protein Concentrate Using Dry Heat to Increase Water Holding Capacity. *Foods* **2020**, *9*, 1077.
21. Schreuders, F.K.; Sagis, L.M.; Bodnár, I.; Erni, P.; Boom, R.M.; van der Goot, A.J. Mapping the texture of plant protein blends for meat analogues. *Food Hydrocoll.* **2021**, *118*, 106753. [CrossRef]
22. Ewoldt, R.H.; Winter, P.; Maxey, J.; McKinley, G.H. Large amplitude oscillatory shear of pseudoplastic and elastoviscoplastic materials. *Rheol. Acta* **2010**, *49*, 191–212. [CrossRef]
23. Peters, J.P.; Luyten, H.; Alting, A.C.; Boom, R.M.; Van der Goot, A.J. Effect of crosslink density on the water-binding capacity of whey protein microparticles. *Food Hydrocoll.* **2015**, *44*, 277–284. [CrossRef]
24. Isobe, S.; Noguchi, A. High moisture extrusion with twin screw extruder—Fate of soy protein during the repetition of extrusion cooking. *Nippon Shokuhin Kogyo Gakkaishi.* **1987**, *34*, 456–461. [CrossRef] [PubMed]
25. Pommet, M.; Redl, A.; Helene, M.; Morel, M.H.; Domenek, S.; Guilbert, S. Thermoplastic processing of protein-based bioplastics: Chemical engineering aspects of mixing, extrusion and hot molding Thermoplastic processing of protein-based bioplastics: Chemical engineering aspects of mixing, extru-sion and hot molding. *Macromol. Symp.* **2003**, *197*, 207–218. [CrossRef]
26. O'Kane, F.E.; Happe, R.P.; Vereijken, J.M.; Gruppen, H.; Van Boekel, M.A. Heat-induced gelation of pea legumin: Comparison with soybean glycinin. *J. Agric. Food Chem.* **2004**, *52*, 5071–5078. [CrossRef] [PubMed]

foods

MDPI

Communication

Transmembrane Pressure during Micro- and Diafiltration of Milk Affects the Release of Non-Sedimentable Caseins

Norbert Raak [1,2,*], Özgenur Coşkun [1,2] and Milena Corredig [1,2]

1 Department of Food Science, Aarhus University, Agro Food Park 48, 8200 Aarhus N, Denmark; ozgenurcoskun@food.au.dk (Ö.C.); mc@food.au.dk (M.C.)

2 CiFOOD Centre for Innovative Food Research, Aarhus University, 8000 Aarhus, Denmark

* Correspondence: norbert@food.au.dk

Abstract: Membrane filtration, especially in combination with diafiltration, can affect the colloidal structure of casein micelles in milk and concentrated milks. The partial dissociation of casein proteins from the casein micelles into the serum phase has been shown to depend on diafiltration conditions. This dissociation can affect the technological functionality of the milk concentrates. The present study aimed at determining the contribution of the gel layer deposited onto the membrane during filtration in the colloidal equilibrium between soluble and micellar caseins. Skimmed milk was concentrated by microfiltration combined with diafiltration using a cross-flow spiral-wound membrane at two transmembrane pressure (TMP) levels, causing differences in the extent of the gel layer formed. Non-sedimentable casein aggregates were formed to a greater extent at a low TMP compared to a high operating TMP. This difference was attributed to the greater compression of the deposit layer during filtration at a high TMP. This study contributes new knowledge to the understanding of how to modulate the functionality of milk concentrates through the control of processing conditions.

Keywords: membrane filtration; microfiltration; casein micelles; milk protein concentrate; transmembrane pressure; whey protein depletion

Citation: Raak, N.; Coşkun, Ö.; Corredig, M. Transmembrane Pressure during Micro- and Diafiltration of Milk Affects the Release of Non-Sedimentable Caseins. *Foods* **2023**, *12*, 2234. https://doi.org/10.3390/foods12112234

Academic Editor: Guanghua Zhao

Received: 6 April 2023
Revised: 19 May 2023
Accepted: 29 May 2023
Published: 1 June 2023

1. Introduction

Membrane filtration is a fractionation technology often utilised in dairy processing to obtain functional ingredients. In microfiltration, skimmed milk is passed over a membrane layer with pores large enough to allow for the transmission of whey proteins and small molecular weight constituents, which are typically 0.08–0.2 µm, to increase the casein micelle fraction. This process is often combined with diafiltration, which is the introduction of water or other media to enhance the transmission of permeable compounds while maintaining high efficiencies [1]. Microfiltration in combination with diafiltration is a common way to produce micellar casein concentrates (MCCs), i.e., milk protein concentrates with an increased ratio of casein to whey proteins [2,3].

For several years, the microfiltration of milk, with or without diafiltration, has been studied in detail from both the processing and product side. The permeate throughput per membrane area and time, referred to as flux, and whey protein permeability are the main target process parameters that need to be optimised by identifying the optimal conditions for, e.g., feed concentration, nominal pore size, and transmembrane pressure (TMP). The latter is the pressure gradient between the retentate and the permeate side. The permeate flux increases with TMP until it reaches a plateau, referred to as the limiting flux, where the formation of a deposit layer on the membrane with increasing TMP counteracts the permeation [4]. When the concentration factor of the feed increases, the limiting flux decreases and is reached at a lower TMP [5–7]. In addition, the composition of the permeate may change with TMP due to the formation of the deposit layer. For example, it has been shown that whey protein permeation decreases with increasing TMP once above the limiting flux [8–10], as the specific resistance of the deposit layer increases and its porosity decreases with increasing TMP [10,11]. The deposit layer was moreover shown to be composed of both casein micelles and the whey protein β-lactoglobulin [12], meaning a further decrease in whey protein permeation due to attachment to the deposit layer. Diafiltration with water was shown to increase flux due to a reduction in feed viscosity via the removal of lactose, minerals, and proteins [6,7]. However, the diafiltration medium has an effect. For example, lower fluxes were reached as the hardness of the diafiltration water was increased [13]. Continuous diafiltration while keeping the feed volume constant can achieve better flow than discontinuous batch diafiltration [14]. However, this may not correspond to efficiencies in whey protein permeation: while continuous diafiltration at a constant feed volume resulted in the highest flux, the highest whey protein permeation was achieved using continuous diafiltration with simultaneous volume reduction or intermittent diafiltration [14].

The microfiltration of milk is usually performed at either 10 or 50 °C. While at 50 °C it is possible to operate at high flow rates and whey proteins permeate efficiently through the membrane, the tendency towards fouling is lower and microbial quality better at a process temperature of 10 °C. However, as the feed viscosity is high, low temperatures result in lower flux values [8]. Additionally, β-casein is known to partly dissociate from the casein micelles at <10 °C, and its presence in the serum implies that it can be transmitted through the membrane and change the permeate composition.

Concentrating casein micelles by membrane filtration, especially in combination with diafiltration, affects the colloidal structure of casein micelles, which in turn, affects the techno-functional properties of the final concentrate [15]. In particular, the partial dissociation of casein molecules into the serum phase as a consequence of the milieu exchange and increased interactions between casein micelles have been reported [16–18]. The presence of non-sedimentable caseins in the serum phase can impart important differences onto the techno-functionality of the concentrates. For example, it can affect the rennet-induced aggregation of the casein micelles in the final product [19].

Thus far, little attention has been given to the impact of the deposit gel layer formed on the membrane on the physical and chemical characteristics of the resulting milk concentrate. It might be hypothesised that the presence of a deposit layer is related to a higher extent of casein dissociation from the micelles, as, during the filtration process, the shear forces are greatest close to the membrane surface and may drive the partial disruption of casein micelles from the deposit layer. This hypothesis was tested by processing a milk concentrate at two different TMPs. The aim of the present research was to study whether a difference in transmembrane pressure during the diafiltration of MCCs in cross-flow spiral-wound membranes can change the amount of non-sedimentable casein aggregates in the final retentate. This has never been investigated before, even though knowledge on the formation of non-sedimentable casein aggregates as affected by the processing conditions is crucial to predict the functionality of the resulting MCCs.

2. Materials and Methods

2.1. Sample Preparation

Pasteurised skimmed milk was obtained from Arla Foods amba (Viby, Denmark) and incubated in a water bath at ~50 °C for 30–60 min prior to the experiments. Microfiltration was performed using a modified pilot unit model R2, ID180334 (GEA Group AG, Düsseldorf, Germany), equipped with a ∅16 cm × 97 cm PVDF spiral-wound membrane module with a 0.1 μm cut-off and 17.1 m^2 membrane area (V0.1-3B-6438; Synder Filtration, Vacaville, CA, USA). The water in the system was displaced with 80 L of skimmed milk before the experiment was started. A total of ~420 L of skimmed milk was continuously added to the process while constantly removing the permeate in ~30–50 L intervals and recirculating the retentate to concentrate the casein micelle fraction in the feed tank. During the concentration step, the transmembrane pressure (TMP) was kept constant at 0.05 MPa. After removing ~360 L of permeate, continuous diafiltration with ~100 L of soft water was conducted at a TMP of either 0.05 or 0.1 MPa. The water was added automatically by the system to maintain the level in the feed tank. The experiments were conducted at 50 °C. The feed–retentate ratio was set to 1.2, with a targeted retentate flow of 100 L h^{-1}, and the circulation loop pump was run at 45% of the maximum capacity. Then, 0.2 g L^{-1} sodium azide was added to all withdrawn samples to prevent spoilage.

Skimmed milk and retentate samples were centrifuged at 100,000× *g* and 20 °C for 1 h (Optima L-80 XP; Beckman Coulter, Inc., Brea, CA, USA; rotor type Ti70) to obtain the serum phases for further investigations [16].

2.2. Protein Analysis

The protein contents of skimmed milk and retentates as well as their centrifugal supernatants were determined using a Gerhardt Dumatherm (C. Gerhardt GmbH & Co.KG, Königswinter, Germany) with a conversion factor of N × 6.38 [16].

Whey protein content of the permeates was analysed by reversed-phase high-performance liquid chromatography, as described previously in detail [16]. The concentrations of α-lactalbumin and β-lactoglobulin were calculated from the peak areas using calibration curves (0.1–40 mg L^{-1}) obtained from commercial standards (Sigma-Aldrich, Steinheim, Germany).

Proteins in the centrifugal supernatants were analysed in native conditions by size exclusion chromatography (20 mmol/L Bis-Tris propane buffer; Sephacryl S-500 high-resolution resin, GE Healthcare, Uppsala, Sweden), as described previously [20], to separate the whey proteins and non-sedimentable casein aggregates.

2.3. Retentate Viscosity

The viscosities of the skimmed milk and microfiltration retentates were measured according to Coşkun et al. [16] with slight modifications, using a stress-controlled AR-G2 rheometer (TA Instruments, New Castle, DE, USA) equipped with a DIN concentric cylinder geometry (d_i = 28 mm, d_o = 30 mm, h = 42 mm) and a Peltier element for temperature control. A logarithmic shear ramp from $\dot{\gamma}$ = 10 to 300 s^{-1} was run for 120 s, and 10 points per decade were recorded. The measurement temperature was 50 °C (equal to the filtration temperature), and the samples were heated in a water bath at 50 °C for 10–30 min before the measurements. The apparent viscosity is reported at $\dot{\gamma}$ = 200 s^{-1}.

2.4. Calcium and Phosphate Analysis

The calcium and phosphate contents of skimmed milk, microfiltration retentates and permeates, and centrifugal supernatants were determined using inductively coupled plasma mass spectrometry (ICP-MS 7900, Agilent Technologies, Inc., Santa Clara, CA, USA) as described previously [21]. Briefly, sample aliquots (20 μL for retentates, 100 μL for all other samples) were blended with 1 mL HNO_3 in glass vials and digested at 1400 W (max. temperature 145 °C) in a microwave oven (Multiwave 3000, Anton Paar GmbH, Graz, Austria) equipped with a rotor (model 64MG5, Anton Paar GmbH). The digested samples were diluted with ultrapure water before injection for ICP-MS. The general-purpose plasma mode was used to quantify ^{43}Ca and ^{31}P isotopes at a sample depth of 129 mm and a nebuliser flow rate of 0.5 rps. Helium was used to reduce polyatomic interference, and yttrium was used as the internal standard.

2.5. Statistical Analysis

All analyses were performed twice; the results are reported as mean values. Statistical differences between milk protein concentrates diafiltered at low and high TMPs were evaluated by Welch's t-test ($p < 0.05$).

3. Results

Figure 1 illustrates the flux (i.e., permeate flow per membrane area) and TMP during the filtration process. The flux decreased gradually over time with concentration. This is typically due to the gradual formation of a deposit layer on the membrane, causing a higher resistance to permeation [8]. Additionally, the continuous increase in feed viscosity with increasing concentration lowered the permeate flow, as previously demonstrated [6,7]. In the concentration step, the development of the flux with processing time was identical in both independent trials, demonstrating the reproducibility of the process. The continuous diafiltration with soft water (3.0 mg L^{-1} calcium and 0.3 mg L^{-1} phosphate) was initiated after removing ~360 L of permeate (Figure 1; dotted line), and the TMP was either kept at 0.05 MPa or increased to 0.1 MPa during this part of the process.

During diafiltration, the flux increased continuously because of a decrease in viscosity of the concentrates due to the milieu exchange [7]. At a higher TMP, there were no differences in the flux, indicating that 0.05 MPa was already close to the limiting flux for this process, which is in agreement with previous reports [5].

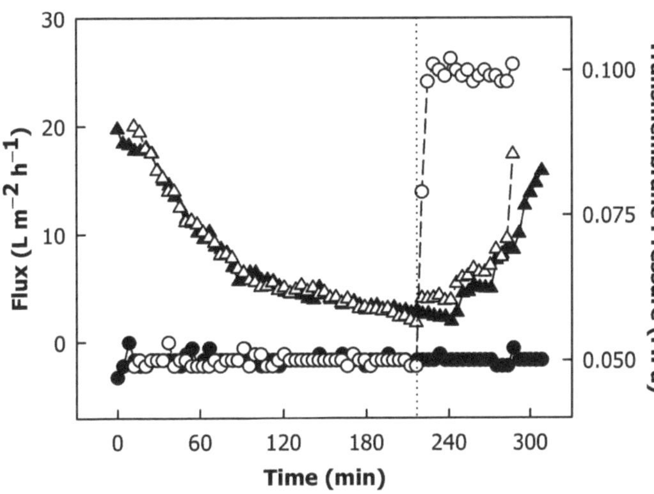

Figure 1. Flux (triangles) and transmembrane pressure (circles) during filtration of skimmed milk through a 0.1 μm spiral-wound PVDF membrane module in combination with diafiltration against soft water. The dotted line indicates the beginning of the continuous diafiltration step. Transmembrane pressure during diafiltration was 0.05 MPa (closed symbols) or 0.10 MPa (open symbols).

The total protein content increased from 34 g/L in skimmed milk to 176 g/L with concentration, corresponding to a concentration factor of ~5×. The concentration of non-sedimentable proteins (in the centrifugal supernatants) increased to a lesser extent due to the permeation of whey proteins (Figure 2A). Figure 2B shows the concentration of α-lactalbumin and β-lactoglobulin in the permeate samples collected at different time points of the process. The whey protein concentration in the permeates remained fairly constant during concentration. Due to the recirculation of the retentate, the whey proteins that did not pass the membrane were concentrated together with the casein micelles, leading to a higher concentration in the feed stream and thus a greater extent of permeation compared to a process where the feed concentration is kept constant [6]. Therefore, the whey protein concentration in the permeates decreased after the final concentration factor was reached and diafiltration started (Figure 2; dotted line). Differences between diafiltration at low and high TMPs were significant only for the first permeate sample (permeate volume of ~395 L), where a lower β-lactoglobulin content was found in the permeate obtained from diafiltration at a high TMP ($p < 0.05$). A lower extent of whey permeation at a higher TMP due to compression of the deposit layer was also reported by Hartinger et al. [8]. In contrast, France et al. [22] did not observe differences in whey protein permeation between filtration below and above the limiting flux. Diafiltration decreased the total protein content of both microfiltration retentates to ~120 g/L, which is a greater extent than expected from the amount of permeated whey proteins and concentration of serum proteins, indicating that casein was also transmitted through the membrane. In fact, it was found by RP-HPLC that casein was present in all permeate samples (data not shown). It was previously reported that a milieu exchange with soft water can result in the disintegration of casein micelles, especially at 50 °C [13]. This could explain the loss of some of the micellar caseins during the diafiltration step. Together with the protein content of the microfiltration retentates, their viscosity at 50 °C decreased by diafiltration from ~9.7 to ~3.3 mPa·s (Figure 3), which also explains the increase in the flux during diafiltration (Figure 1). The differences in protein content and viscosity between retentates obtained from diafiltration at low and high TMPs were not significant ($p > 0.05$).

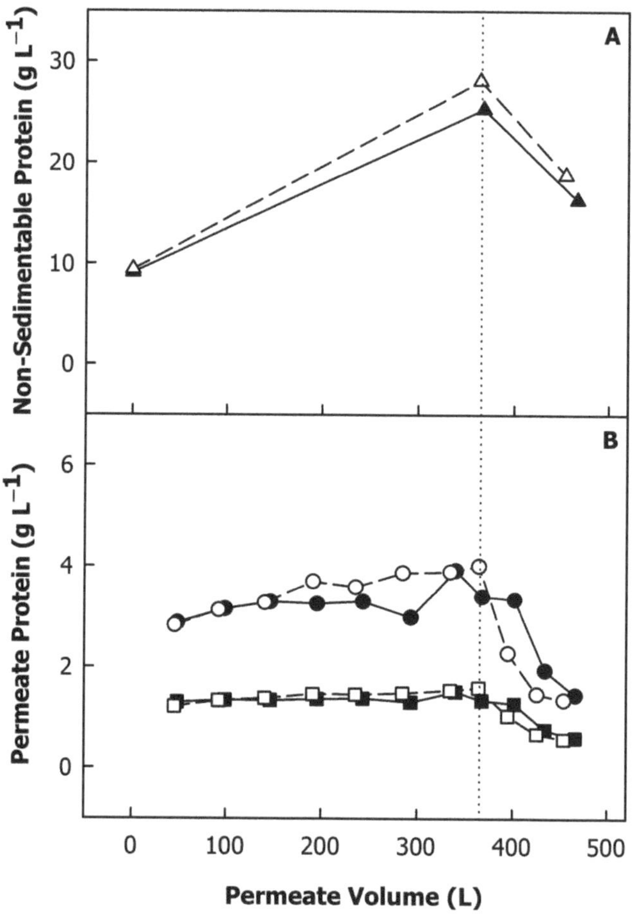

Figure 2. Non-sedimentable protein content of skimmed milk and milk protein concentrates during microfiltration and diafiltration (**A**), and α-lactalbumin (squares) and β-lactoglobulin concentration (circles) of the microfiltration permeates and different stages of the process (**B**). The dotted line indicates the beginning of the diafiltration step. Transmembrane pressure during diafiltration was 0.05 MPa (closed symbols) or 0.10 MPa (open symbols).

Figure 4 summarises the concentrations of calcium and phosphate of the original skimmed milk, the retentates, and the corresponding centrifugal supernatants (Figure 4A,C), as well as in all permeate samples (Figure 4B,D). The total calcium and phosphate (Figure 4; diamonds) of the retentates increased during the concentration step in the same way as the protein content, and the amount of micellar calcium and phosphate remained constant at ~32 and ~20 mg per g casein, respectively, which is in agreement with previous reports [16,18]. The micellar calcium and phosphate remained constant during diafiltration at a TMP of 0.05 MPa, whereas, in the case of the higher TMP, the concentrations decreased to ~29 and 17 mg per g casein. The differences in total calcium and phosphate between a low and high TMP were significant ($p < 0.05$). This indicates that a higher TMP can cause some dissociation of micellar salts.

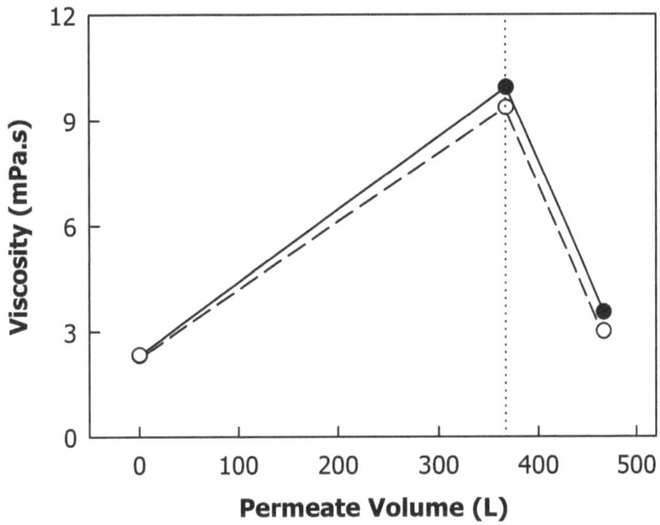

Figure 3. Viscosity of skimmed milk and milk protein concentrates during microfiltration and diafiltration. The dotted line indicates the beginning of the diafiltration step. Transmembrane pressure during diafiltration was 0.05 MPa (closed symbols) or 0.10 MPa (open symbols).

During the first half of the concentration step, the calcium and phosphate contents of the permeates were close to or higher than those of the serum phase of the skimmed milk (~0.48 g L^{-1} and ~0.53 g L^{-1}, respectively), whereas they were previously reported to be lower due to interactions with non-permeable proteins such as non-micellar caseins or aggregated whey proteins [16,18,23]. This underlines the permeation of some casein micelles through the membrane, which would increase the calcium and phosphate contents of the permeates due to the colloidal calcium phosphate associated with the casein micelles. The diafiltration step decreased both total and soluble calcium and phosphate due to the milieu exchange, i.e., the permeation of soluble salts including those dissociated from the casein micelles.

Figure 5 shows the size exclusion chromatograms of the centrifugal supernatants of skimmed milk and the microfiltration retentates, showing the typical pattern of non-protein material eluting at <60 min and >120 min, caseins at 75–90 min, and native whey proteins at 90–120 min [20]. There was a clear trend towards an increase in the casein peak with diafiltration, as was also reported previously [16]. The presence of caseins in the centrifugal supernatant is consistent with a change in the serum phase's ionic composition, causing dissociation of the casein micelles. A higher elution peak was found when the concentrates were prepared at a low TMP during diafiltration (0.05 MPa) compared to diafiltration at a TMP of 0.1 MPa. This result disputes the hypothesis that membrane filtration at a higher operating TMP causes more dissociation of casein due to the presence of high shear forces and a deposit layer on the membrane. It was concluded that a higher TMP, causing a greater compression of the deposit layer [8,11], will instead reduce the ratio of soluble to colloidal casein. This could be due to a higher amount of calcium present in the gel layer, as shown by the lower permeation of calcium in the permeates at a high TMP (Figure 4B).

Interestingly, Weinberger et al. [10] reported that the amount of calcium bridges in the deposit layer increases with increasing pH of the substrate, which typically occurs during diafiltration with water [16,18]. On the other hand, the proportion of calcium bridges was lower at a higher TMP; however, the sum of stabilising bonds in the deposit layer increased with increasing TMP [10], indicating also a stronger attraction of proteins in the bulk phase. Therefore, the non-sedimentable casein aggregates formed during DF might become embedded in the deposit layer to a greater extent in case of a higher TMP.

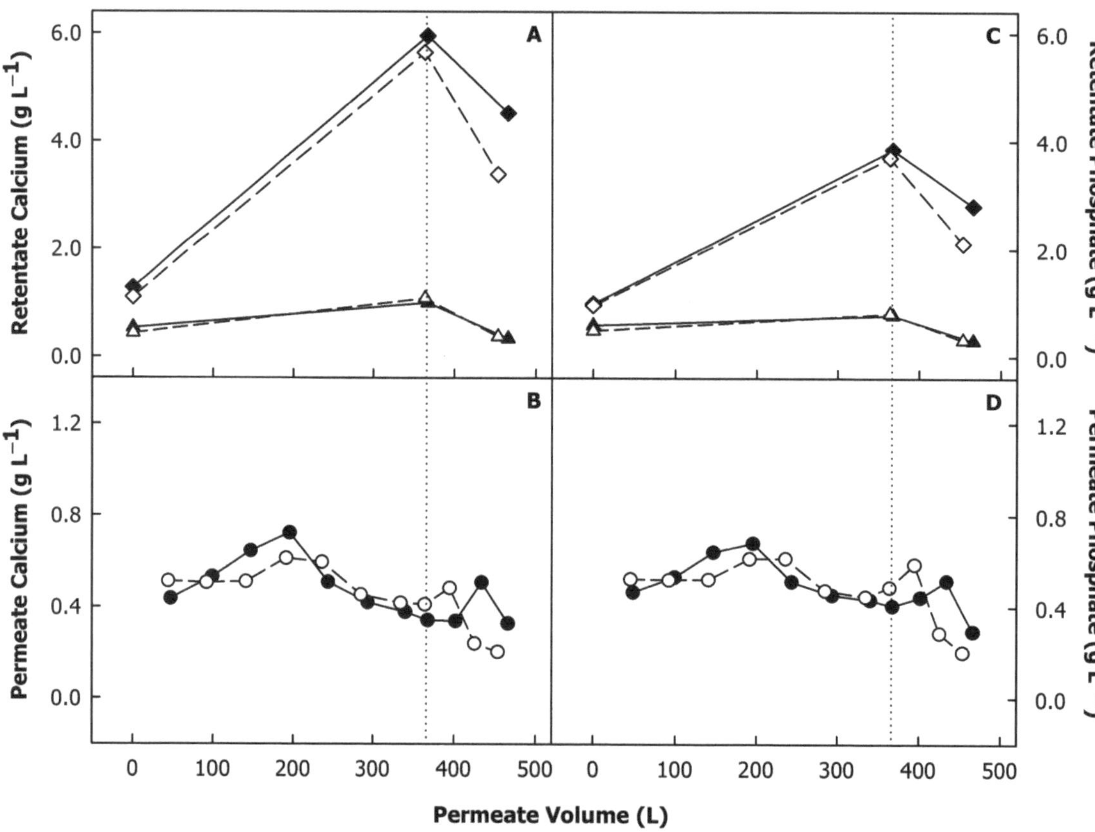

Figure 4. Total (diamonds) and soluble contents (triangles) of calcium (**A**) and phosphate (**C**) of skimmed milk and milk protein concentrates during microfiltration and diafiltration, and calcium (**B**) and phosphate (**D**) concentrations of the microfiltration permeates and different stages of the process. The dotted lines indicate the beginning of the diafiltration step. Transmembrane pressure during diafiltration was 0.05 MPa (closed symbols) or 0.10 MPa (open symbols).

It is also important to note that it has been previously shown that the concentration of milk proteins by membrane filtration causes a smaller release of caseins into the serum phase than using osmotic stressing, which is a gentle, osmotic pressure-driven concentration method carried out without shear [24]. In that case, it was hypothesised that the shear forces and transmembrane pressure during membrane filtration facilitates rearrangements of casein micelles, limiting their dissociation. Likewise, the dissociation of casein micelles might be limited when operating at a high TMP.

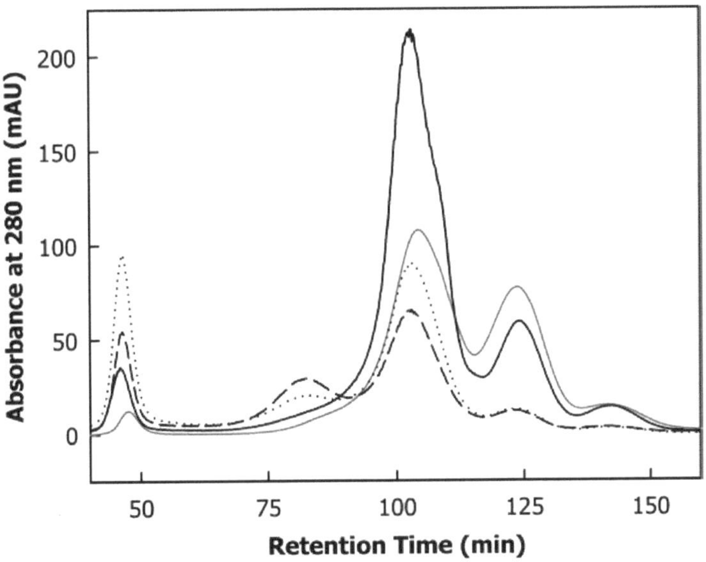

Figure 5. Size exclusion chromatograms of centrifugal supernatants of skimmed milk (grey line) and microfiltration retentate before diafiltration (black full line), after diafiltration at 0.05 MPa transmembrane pressure (black dashed line), and after diafiltration at 0.1 MPa transmembrane pressure (black dotted line). Curves are representative chromatograms.

4. Conclusions

Skimmed milk was concentrated by microfiltration and subsequently diafiltered against soft water using a tangential cross-flow membrane. The effect of TMP on protein permeation was studied in relation to the formation of non-sedimentable casein aggregates in the retentates. The differences in the operating TMP conditions during diafiltration showed little effect on the whey protein permeation. On the contrary, non-sedimentable casein aggregates were formed to a greater extent at a lower TMP. It was concluded that the greater compression of the deposit layer and/or more pronounced rearrangements of the casein micelles in the bulk phase at a higher TMP limited the dissociation of caseins into the serum phase. Both non-sedimentable casein aggregates and modifications in the casein micelle structure due to rearrangements during membrane filtration might have considerable effects on the techno-functional properties of MCCs. The present study therefore contributes important knowledge to the understanding of the processing–structure–function interrelations of milk protein concentrates, which is crucial to predict the functionality of MCCs depending on their processing history.

Author Contributions: Conceptualisation, N.R., Ö.C. and M.C.; methodology, N.R., Ö.C. and M.C.; validation, N.R. and Ö.C.; formal analysis, N.R. and Ö.C.; investigation, N.R. and Ö.C.; resources, M.C.; data curation, N.R. and Ö.C.; writing—original draft preparation, N.R.; writing—review and editing, Ö.C. and M.C.; visualisation, N.R.; supervision, M.C.; project administration, M.C.; funding acquisition, M.C. All authors have read and agreed to the published version of the manuscript.

Funding: This work was partly funded by the Danish Dairy Research Foundation. The support of the Center for Innovative Food (CiFOOD) at Aarhus University is also greatly acknowledged. Data were generated through accessing the research infrastructure at Aarhus University, including FOODHAY (Food and Health Open Innovation Laboratory, Danish Roadmap for Research Infrastructure).

Data Availability Statement: The data presented in this study are available on request from the corresponding author.

Acknowledgments: Special thanks goes to Bjarke Hammelsvang (KOLD College, Odense, Denmark) and Ole Lillevang (GEA, Skanderborg, Denmark) for their support in operating the pilot plant membrane filtration, Freja Mardal (Arla Food Ingredients, Videbæk, Denmark) for the stimulating discussions, and to Hanne Søndergaard Møller (Aarhus University, Aarhus, Denmark) for their assistance with the analytical work.

Conflicts of Interest: The authors declare no conflict of interest.

References

1. Raak, N.; Corredig, M. Manufacture of milk and whey products: Caseins, caseinates and micellar casein. In *Encyclopedia of Dairy Sciences*, 3rd ed.; McSweeney, P.L.H., McNamara, J.P., Eds.; Elsevier: Amsterdam, The Netherlands, 2022; Volume 5, pp. 8–17. [CrossRef]
2. Carter, B.G.; Cheng, N.; Kapoor, R.; Meletharayil, G.H.; Drake, M.A. Microfiltration-derived casein and whey proteins from milk. *J. Dairy Sci.* **2021**, *104*, 2465–2479. [CrossRef] [PubMed]
3. Hammam, A.R.A.; Martínez-Monteagudo, S.I.; Metzger, L.E. Progress in micellar casein concentrate: Production and applications. *Comp. Rev. Food Sci. Food Safety* **2021**, *20*, 4426–4449. [CrossRef] [PubMed]
4. Hausmann, A.; Duke, M.C.; Demmer, T. Principles of membrane filtration. In *Membrane Processing: Dairy and Beverage Applications*, 1st ed.; Tamime, A.Y., Ed.; Blackwell Publishing Ltd.: Hoboken, NJ, USA, 2013; pp. 18–51. [CrossRef]
5. Hartinger, M.; Kulozik, U. Milk protein fractionation by spiral-wound microfiltration membranes in diafiltration mode—Influence of feed protein concentration and composition on the filtration performance. *Int. Dairy J.* **2020**, *102*, 104606. [CrossRef]
6. Li, Z.; Liu, D.; Xu, S.; Zhang, W.; Zhou, P. Transmission of major and minor serum proteins during microfiltration of skim milk: Effects of pore diameters, concentration factors and processing stages. *Foods* **2021**, *10*, 888. [CrossRef]
7. Schäfer, J.; Mesch, I.; Atamer, Z.; Nöbel, S.; Kohlus, R.; Hinrichs, J. Calcium reduced skim milk retentates obtained by means of microfiltration. *J. Food Eng.* **2019**, *247*, 168–177. [CrossRef]
8. Hartinger, M.; Heidebrecht, H.-J.; Schiffer, S.; Dumpler, J.; Kulozik, U. Milk protein fractionation by means of spiral-wound microfiltration membranes: Effect of the pressure adjustment mode and temperature on flux and protein permeation. *Foods* **2019**, *8*, 180. [CrossRef]
9. Schiffer, S.; Hartinger, M.; Matyssek, A.; Kulozik, U. On the reversibility of deposit formation in low temperature milk microfiltration with ceramic membranes depending on mode of adjustment of transmembrane pressure and wall shear stress. *Sep. Purif. Technol.* **2020**, *247*, 116962. [CrossRef]
10. Weinberger, M.E.; Andlinger, D.J.; Kulozik, U. A novel approach for characterisation of stabilising bonds in milk protein deposit layers on microfiltration membranes. *Int. Dairy J.* **2021**, *118*, 105044. [CrossRef]
11. Steinhauer, T.; Kühnl, W.; Kulozik, U. Impact of protein interactions and transmembrane pressure on physical properties of filter cakes formed during filtration of skim milk. *Procedia Food Sci.* **2011**, *1*, 886–892. [CrossRef]
12. Steinhauer, T.; Kulozik, U.; Gebhardt, R. Structure of milk protein deposits formed by casein micelles and β-lactoglobulin during frontal microfiltration. *J. Membrane Sci.* **2014**, *468*, 126–132. [CrossRef]
13. Reitmaier, M.; Barbosa, B.; Sigler, S.; Heidebrecht, H.-J.; Kulozik, U. Impact of different aqueous phases on casein micelles: Kinetics of physicochemical changes under variation of water hardness and diafiltration conditions. *Int. Dairy J.* **2020**, *109*, 104776. [CrossRef]
14. Song, B.; Zhang, Y.; Lu, J.; Pang, X.; Wei, M.; Zheng, S.; Zhang, M.; Zhang, S.; Lv, J. Effect of different diafiltration process on the protein fractionation of skim milk by cross flow microfiltration. *LWT* **2021**, *152*, 112330. [CrossRef]
15. Corredig, M.; Krishnankutty Nair, P.; Li, Y.; Eshpari, H.; Zhao, Z. Understanding the behavior of caseins in milk concentrates. *J. Dairy Sci.* **2019**, *102*, 4772–4782. [CrossRef]
16. Coşkun, Ö.; Wiking, L.; Rahimi Yazdi, S.; Corredig, M. Molecular details of the formation of soluble aggregates during ultrafiltration or microfiltration combined with diafiltration of skim milk. *Food Hydrocoll.* **2022**, *124 Pt A*, 107244. [CrossRef]
17. Li, Y.; Corredig, M. Acid induced gelation behaviour of skim milk concentrated by membrane filtration. *J. Texture Stud.* **2020**, *51*, 101–110. [CrossRef]
18. Raak, N.; Pedersen, J.S.; Corredig, M. Structural and colloidal changes of casein micelles induced by membrane concentration and their dependence on the milieu exchange conditions. *Coll. Surf. A* **2023**, *670*, 131580. [CrossRef]
19. Krishnankutty Nair, P.; Corredig, M. Sodium caseinate hinders chymosin-induced aggregation of caseins in concentrated milk: The role of soluble caseins and calcium ions. *JDS Commun.* **2021**, *2*, 309–312. [CrossRef]
20. Li, Y.; Dalgleish, D.; Corredig, M. Influence of heating treatment and membrane concentration on the formation of soluble aggregates. *Food Res. Int.* **2015**, *76*, 309–316. [CrossRef]
21. Coşkun, Ö.; Raak, N.; Corredig, M. Heat induced interactions in whey protein depleted milk concentrates: Comparison of ultrafiltration and microfiltration. *Food Hydrocoll.* **2023**, *137*, 108354. [CrossRef]
22. France, T.C.; Kelly, A.L.; Crowley, S.V.; O'Mahony, J.A. The effects of temperature and transmembrane pressure on protein, calcium and plasmin partitioning during microfiltration of skim milk. *Int. Dairy J.* **2021**, *114*, 104930. [CrossRef]

23. Renhe, I.R.T.; Corredig, M. Effect of partial whey protein depletion during membrane filtration on thermal stability of milk concentrates. *J. Dairy Sci.* **2018**, *101*, 8757–8766. [CrossRef] [PubMed]
24. Krishnankutty Nair, P.; Corredig, M. Rennet-induced gelation of concentrated milk in the presence of sodium caseinate: Differences between milk concentration using ultrafiltration and osmotic stressing. *J. Dairy Sci.* **2015**, *98*, 27–36. [CrossRef] [PubMed]

foods

MDPI

Article

Ionic Strength Dependence of the Complex Coacervation between Lactoferrin and β-Lactoglobulin

Rima Soussi Hachfi, Pascaline Hamon, Florence Rousseau, Marie-Hélène Famelart and Saïd Bouhallab *

INRAE, Institut Agro, STLO, 65 Rue de Saint Brieuc, F-35042 Rennes, France
* Correspondence: said.bouhallab@inrae.fr

Abstract: Heteroprotein complex coacervation is an assembly formed by oppositely charged proteins in aqueous solution that leads to liquid–liquid phase separation. The ability of lactoferrin and β-lactoglobulin to form complex coacervates at pH 5.5 under optimal protein stoichiometry has been studied in a previous work. The goal of the current study is to determine the influence of ionic strength on the complex coacervation between these two proteins using direct mixing and desalting protocols. The initial interaction between lactoferrin and β-lactoglobulin and subsequent coacervation process were highly sensitive to the ionic strength. No microscopic phase separation was observed beyond a salt concentration of 20 mM. The coacervate yield decreased drastically with increasing added NaCl from 0 to 60 mM. The charge-screening effect induced by increasing the ionic strength is attributed to a decrease of interaction between the two oppositely charged proteins throughout a decrease in Debye length. Interestingly, as shown by isothermal titration calorimetry, a small concentration of NaCl around 2.5 mM promoted the binding energy between the two proteins. These results shed new light on the electrostatically driven mechanism governing the complex coacervation in heteroprotein systems.

Keywords: complex coacervation; ionic strength; proteins; desalting

Citation: Soussi Hachfi, R.; Hamon, P.; Rousseau, F.; Famelart, M.-H.; Bouhallab, S. Ionic Strength Dependence of the Complex Coacervation between Lactoferrin and β-Lactoglobulin. *Foods* **2023**, *12*, 1040. https://doi.org/10.3390/foods12051040

Academic Editors: Norbert Raak, Ruifen Li and Laura Roman

Received: 21 December 2022
Revised: 21 February 2023
Accepted: 21 February 2023
Published: 1 March 2023

1. Introduction

Oppositely charged polymers interacting mainly through electrostatic interactions can undergo a spontaneous liquid−liquid phase separation (LLPS) into polymer-rich dense phase called coacervates and a less concentrated phase called the dilute phase. Complex coacervation, known as LLPS, has been and still is a subject of intense experimental and theoretical interest since the pioneer research of Bungenberg De Jong and Kruyt one century ago [1]. Although several theoretical models have been provided to describe complex coacervation, none of them were able to perfectly explain it [2–6]. Several studies, based on those models, described complex coacervation phenomenon as a four-step process. The first step is the spontaneous formation of both symmetrical and random heterocomplexes by electrostatic attraction (building blocks). The building blocks, also called primary units, come together to form soluble complexes. The third step involves the rearrangement of soluble complexes into spherical micrometric droplets, characteristic of complex coacervation [7]. Finally, the droplets coalesce, thus forming the dense phase of the coacervates [8]. Interest in complex coacervation is remarkably increasing as it is an ambitious undertaking that will open enormous opportunities for numerous applications in the food, cosmetics, pharmaceutical and biological fields [9,10]. Moreover, complex coacervation can occur across a wide-variety of charged macromolecules, such as protein–polysaccharide, protein–synthetic polyelectrolyte, polyelectrolyte–polyelectrolyte and protein–proteins mixture. However, heteroprotein complex coacervation (HPCC), i.e., involving two or more proteins, is comparatively understudied [11–14]. Part of the published works on heteroprotein complex coacervation focused on the assembly between lactoferrin (LF) as basic protein and β-lactoglobulin (βLG) as acidic protein. In these studies, multiscale characterization of the

LF–βLG formed coacervates were conducted using biophysical tools such as Small Angle X ray Scattering (SAXS) measurements [15] or solid-state Nuclear Magnetic Resonance NMR [16]. The potential applications of LF–βLG coacervates in food and their effectiveness for encapsulation of bioactives were also reported [17]. In these studies, LF–βLG coacervates showed a high vitamin encapsulation yield, around 98% [17]. In addition to that, thanks to their high viscosity, the coacervates can be used as food texturizing agents that offer a good substitute for exogenous additives, such as polysaccharides, that will help in developing "clean label" functional food products [18]. For HPCC, the most subtle is to set the physico-chemical parameters (pH, stoichiometry, concentration, etc.) for optimal coacervation. Yan et al. [19] reported that LF–βLG coacervation is favored under a narrow range of pH 5.7 to 6.2, with lower ionic strength value and a total protein concentration between 10 and 40 g/L. For their part, Anema & de Kruif [20], noticed that LF–βLG coacervates were formed in the pH region of 5.0–7.3 and ionic strength lower than 100 mM. In a previous work, the pH range has been narrowed and optimal LF–βLG coacervation occurred at pH 5.5 [21]. Furthermore, the coacervates yield has been compared when mixing LF with the two isoforms A and B of βLG. Interestingly, the higher coacervates yield was recovered with the most acidic isoform, i.e., βLG isoform A (one more aspartic acid residue). These results underline the high sensitivity of LF–βLG coacervation to a small variation in the net protein charge [12,21]. Behind the pH, the ionic strength is another important parameter that controls the net protein charge and consequently the electrostatically driven heteroprotein complex coacervation. Therefore, the aim of this present work is to gain more insight on the effect of the presence of salt on the interaction and complex coacervation between LF and βLG.

2. Materials and Methods

LF with a purity of 90 g/100 g and iron saturation of 10–20 mol iron–mol protein according to technical specification was purchased from Fonterra Cooperative Group (Auckland, New Zealand) and used without further purification. Commercial bovine βLG, containing both A and B variants, was further purified before use; as βLG is prompt to self-aggregation during long storage, the non-native and aggregated species were regularly removed by dispersing in ultrapure water (30 g/L), adjusting to pH 5.2 with 1 M HCl and then storing at 30 °C for 10 min to precipitate aggregated and non-native forms. The dispersion was then centrifuged at 36,000× *g* for 10 min at 25 °C (Avanti, J-26S XP BioSafe Three-Phase Non-IVD Centrifuge, Beckman Coulter, Villepinte, France). The supernatant containing native βLG was adjusted to pH 7.0 with 1 M NaOH, freeze-dried and stored at −20 °C until later. The protein purity in obtained powder was around 95% as assessed by HPLC analysis. Sodium chloride (NaCl) was purchased from VWR Chemicals (Rosny-sous-Bois, France). (2-(N-morpholino) ethanesulfonic acid hydrate (MES) was purchased from Sigma-Aldrich (St. Louis, MO, USA) and all other chemicals were of analytical grade.

2.1. Preparation of Samples and Direct Mixing

MES buffer (10 mM) was used as solvent in all experiments and was prepared by solubilizing MES powder in ultra-pure water. Solid NaCl was added to reach targeted concentration and then adjusted to pH 5.5 using 1 mM NaOH solution.

LF and βLG Protein powders were solubilized in MES buffer with desired concentration of NaCl (0–400 mM), and the pH was readjusted to 5.5 when required. This pH value was found to be optimal for complex coacervation between the two whey proteins at the current stoichiometry and total protein concentration [17]. In fact, this pH value lies between the two isoelectric points of the proteins; 5.2 and 8.8 for βLG and LF, respectively [12]. The exact protein concentrations were determined in the two stock solutions by absorbance at 280 nm with a SAFAS UV MC2 spectrophotometer (SAFAS UV MC2, Safas, Monte-Carlo, Monaco) using 0.96 L/g.cm and 1.47 L/g.cm as extinction coefficients for βLG and LF, respectively.

For direct mixing experiments, conducted in duplicate, solutions of βLG and LF prepared in 10 mM MES buffer at various NaCl concentrations were mixed at room temperature to reach a final concentration of 0.5 mM and 0.05 mM for βLG and LF, respectively, which means a molar ratio of 10/1. The direct mixing experiments were conducted at least in duplicate, and the mean and the standard deviation are plotted

2.2. Complex Coacervates by Desalting

The formation of complex coacervates between the two proteins prepared at various salt concentrations (0, 100, 200 and 400 mM NaCl) was monitored during continuous desalting against 10 mM MES buffer, pH 5.5 using dialysis membranes. Ten mL of each mixture were put into a 6–8 KDa molecular weight cut-off (MWCO) dialysis membrane (diameter = 14.6 mm) (SpectraPor, Repligen Corporation, CA, USA). The dialysis membrane was then submerged into a dialysis bath containing 1 L of MES buffer under constant stirring at room temperature. Fourteen aliquots were taken from the dialysis bag at different times during the 24 h of dialysis for analyses. Dialysis experiments of the 10 mM MES buffer with studied NaCl concentrations and without proteins were also performed as dialysis control experiments. Successive low volume sampling does not affect subsequent measurements

During dialysis experiments, conductivity of the dialysis bath was measured using an electrical conductivity meter (HI98192, Hanna instrument, Strasbourg, France). The probe of the conductivity meter (HI763133, Hanna instrument, Strasbourg, France) was submerged into the bath throughout the dialysis time to make sure that the ion exchange between the dialysis bag and the bath is taking place. Different samples were collected from inside the bag, and the salt concentration was measured using an electrical conductivity meter (HI98192, Hanna instrument, Strasbourg, France). The dialysis experiments were conducted at least in duplicate.

2.3. Turbidity Measurements

The turbidity caused by the spontaneous formation of coacervates as spherical droplets in LF–βLG direct mixtures and during the dialysis experiments was measured at 600 nm using a microplate spectrophotometer (Multiskan™ GO, Fisher Scientific, Strasbourg, France). In fact, the spectrophotometer provides absorbance measurements that could be converted to turbidity using the following equation:

$$Turbidity\left(\mathrm{cm}^{-1}\right) = \frac{2.303 \times A}{L} \quad (1)$$

where A is absorbance at 600 nm and L (cm) is the light path length corresponding to the height of the liquid column into the microplate well [22].

2.4. Coacervate Yield

Protein partition was determined by protein quantification after phase separation. Dilute and coacervate phases were separated by centrifugation (Heraeus Biofuge Primo, Thermo Scientific, Waltham, MA, USA) at 28,000× g for 30 min. Protein content in the supernatant was quantified by liquid chromatography (UltiMate 3000 HPLC, Thermo Fisher Scientific, Waltham, MA, USA) as previously described [21]. A PLRP-S column (300 Å, 2.1 × 150 mm, 8 μm) with a flow rate of 0.2 mL/min (Agilent Technologies, Santa Clara, CA, USA) was used. Milli-Q water containing 1.06 ‰ (v/v) of trifluoroacetic acid and an 80/20 acetonitrile/milli-Q water (v/v) mixture containing 1.0 ‰ (v/v) of trifluoroacetic acid were used for elution. The absorbance at 280 nm was measured during the elution using a Waters 2487 detector.

The proteins content in the coacervate phase was calculated by subtracting the proteins content in the supernatant from the total initial protein concentration.

The coacervate yield (protein yield in the coacervate phase) was calculated using the following equation:

$$\text{The coacervate yield }(\%) = \frac{\text{final protein concentration in the coacervates}}{\text{Initial protein concentration}} \times 100 \quad (2)$$

2.5. Phase Contrast Microscopy

Optical microscopy was used to check that coacervates rather than amorphous aggregates were formed during direct mixing and dialysis experiments. Observations were conducted at room temperature using an Olympus phase contrast microscope (BX51TF, Rungis, France) set at the magnifications 100×.

2.6. Isothermal Titration Calorimetry (ITC)

The ITC experiment was performed at two temperatures, 25 °C and 35 °C, using a VP-ITC micro-calorimeter (MicroCal VP-ITC, Malvern Panalytical, Malvern, UK) by successive injections of LF solution (0.25 mM) into a βLG solution (0.1 mM) loaded in sample cell (1.425 mL). Titration experiments were performed at different NaCl concentrations between 0 and 20 mM. All solutions, prepared in 10 mM MES buffer pH 5.5, were degassed under vacuum before titration experiments. The reference cell was filled with NaCl solutions in the respective concentration for each measurement, and the sample cell was filled with βLG solution. The LF solution in the syringe was also set at the same NaCl concentration. βLG was titrated with 25 successive injections of 10 μL of LF solution. The initial delay was set at 60 s, and the stirring rate inside the sample cell was set at 300 rpm to ensure the homogeneity of the cell solution during titration. The interval between injections was 200 s to reach the thermodynamic equilibrium. For each ITC experiment, a reference titration was performed by titrating LF solutions directly into 10 mM MES buffer containing the studied NaCl concentration. A negligible signal was associated with this reference injection, which was subtracted from the corresponding experimental signal. The ITC data were fitted using graphical user interface (GUI) of PyTC (version 1.2.2, University of Oregon, OR, USA), an open-source python software [23]. The ITC experiments were performed at least in duplicate.

2.7. Statistical Analysis

The effect of salt concentration on coacervation yield and turbidity during direct mixing and dialysis experiments and on titration by ITC was tested using analyses of variance (ANOVA) with one factor. When the ANOVA was significant ($p < 0.05$), Tukey's multiple comparison test was used for paired comparisons of means with $p < 0.05$. Tests were performed using R Studio, a software package developed by a community (version 2021.09.1, Boston, MA, USA).

3. Results and Discussion

3.1. Interactions between βLG and LF

3.1.1. Direct Mixing Experiments

The effect of ionic strength on the interaction–complex coacervation was studied by preparing individual proteins in the required salt concentration before mixing. Figure 1 shows the turbidity and the coacervate yield at different ionic strengths from 0 mM to 100 mM. Turbidity and coacervate yield were maximal without added salt and decreased with increased ionic strengths to reach a value close to zero at 20 mM for both the turbidity and the coacervates yield. At this concentration, no LLPS was detected.

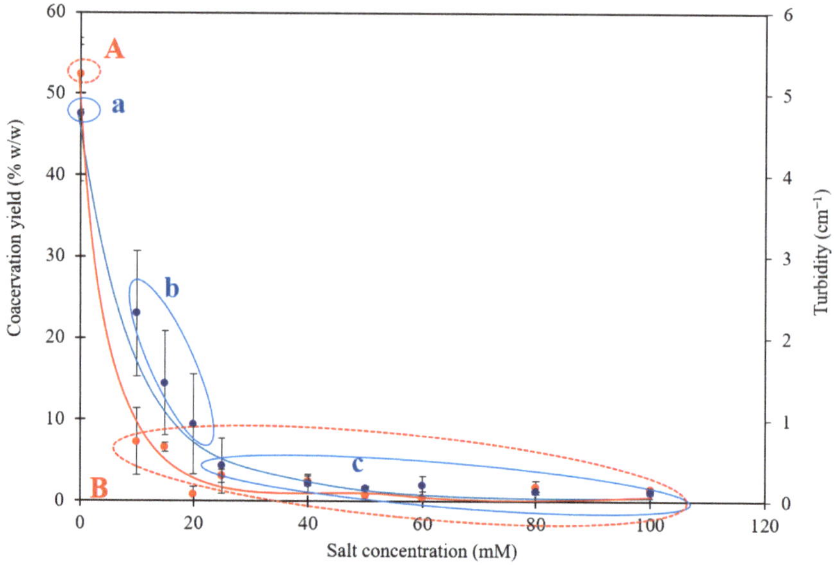

Figure 1. Evolution of lactoferrin–β-lactoglobulin coacervate yield (**red**) and turbidity (**blue**) as a function of salt concentration after direct mixing in 10 mM MES buffer, pH 5.5. Total protein concentration of 0.55 mM. The ANOVA on NaCl concentration was significant ($p < 4 \times 10^{-9}$ and $p < 3 \times 10^{-5}$ for the coacervates yield and turbidity, respectively). [a,b,c] and [A,B]: values not sharing the same letter are significantly different according to the Tukey's multiple comparison test ($p < 0.05$).

These results are in agreement with those reported by Yan et al. [19] for the same coacervates system at a close pH value, say pH 5.9. In their study, the yield in βLG and LF has been monitored as a function of NaCl concentrations (0−100 mM) using size exclusion chromatography. Both proteins showed a decrease in their yield with the increase of the ionic strength. A concentration of 20 mM of NaCl was also found to be a critical salt concentration for coacervation. Moreover, in the same paper, the authors noticed that at this salt concentration, and after centrifugation, a white precipitate of βLG aggregates was observed instead of LF–βLG coacervates. The effect of NaCl on the heteroprotein coacervates was also conducted on another protein system, β-casein–LF [19]. Similar ionic strength dependency of complex coacervation between these two proteins at pH value of 6.5 was reported. However, for β-casein–LF, complex coacervation was still observed even for ionic strength higher than 140 mM, a value 7 times higher than that found for LF–βLG. Hence, the concentration of NaCl tolerated by LF–βLG system is lower than that tolerated by β-casein–LF system. This difference might be explained by the random coil structure of β-casein as intrinsically disordered protein with high hydrophobicity compared to βLG [24]. Consequently, the subtle sensitivity to salt is completely protein structure dependent.

The coacervate yield decreased drastically with increasing ionic strength (Figure 1). The yield of whey protein and gum arabic complex coacervates at pH 4 decreased when increasing NaCl concentration and was completely suppressed at ionic strength higher than 60 mM [25]. Hence, the value found for protein–polysaccharide was higher than that found in the present work for heteroprotein LF–βLG system. In addition to that, the same authors reported that the kinetics of the phase separation slows down upon addition of salt because the coalescence of the coacervate droplets takes more time and water being slowly released from the coacervate phase.

Even though the decrease in the coacervate yield is a great proof of the effect of ionic strength on the formation of coacervates, turbidity measurement is also a powerful indicator of the coacervation. A slight increase of NaCl concentration induces a rapid decrease of turbidity (Figure 1). As checked by microscopic observations, addition of salt decreased both the size and the number of formed droplets; although further and systematic image analyzes are required, it seems that the droplet diameter decreased significantly after addition of NaCl. This behavior is not specific to HPCC since it was generally observed for several systems, such as polysaccharide–polysaccharide and polysaccharide–protein [26–29]. Such evolution of the turbidity as a function of the increase in the ionic strength highlights the predominant role of attractive electrostatic forces in the complex coacervation process. Hence, stronger attractive interactions between biopolymers lead to a more turbid solution [30]. The major difference lies in the salt sensitivity threshold, which is strongly dependent on the structures of the mixed macromolecules.

This huge dependency of LF–βLG complex coacervation on ionic strength obeys the mechanism explained a long time ago by Bungenberg de Jong [1]: the presence of microions screens the charges of the polymers, which weakened attractive forces between them and disrupted the intermolecular electrostatic interactions. As complex coacervation is an electrostatically driven process, the weakening of the electrostatic interaction hampers the complex formation and hinders the occurrence of LLPS. As a matter of fact, the theory of Overbeek and Voorn [2] elucidates that, before reaching a critical salt concentration, the polymer mixture was able to get separated into a polymer-rich and a polymers-poor phase (the LLPS). However, with the addition of microions, the composition of the two phases became very close until reaching a critical point, beyond which LLPS is abolished [2]. Overall, A high sensitivity of LF–βLG complex coacervation to salt with a critical NaCl concentration of 20 mM above which the coacervation can no longer occur was reported. Addition of increasing salt concentration (from 0 to 20 mM) decreased the overall net charge of the two proteins, as the measured zeta potential decreased from −14 to −12 mV for βLG and from +12 to +4 mM for LF (results not shown). Consequently, increasing NaCl concentration affected the protein surface net charge, shifted the protein's isoelectric points and reduced the long-range electrostatic attractions.

3.1.2. LF–βLG Interaction Energy

ITC experiments were conducted to provide a detailed thermodynamic description on how low ionic strength affects the interaction and association between βLG and LF. The heat flow versus time (raw data) profiles associated with the titrations of LF into βLG at different NaCl concentration and at 25 °C are shown in the top panels of Figure 2. The actual heat associated with the interactions (Figure 2 bottom panels) was obtained by integrating the peaks of the top panels and subsequently subtracting the heat produced from the titration of LF into the buffer. Based on the resulting titration profiles, the heat flow was negative ($\Delta H < 0$), meaning that the sum of interactions and other phenomena taking place is an exothermic process. The presence of NaCl affects signal intensity but did not change the negative signature, suggesting that the nature of non-Coulombic interactions (such as hydrophobic interactions and hydrogen bonding) was not altered by the salt-shielding effect. Exothermic processes during the interaction between two oppositely charged macromolecules have been reported for various other systems [21,31–35]. The exothermic nature of an enthalpically driven complexation process is generally attributed to the predominance of electrostatic interactions. During the titration at very low added salt, a higher response in heat change was observed during the first injections, which gradually decreased over the titration and tended to saturation with increased protein molar ratio.

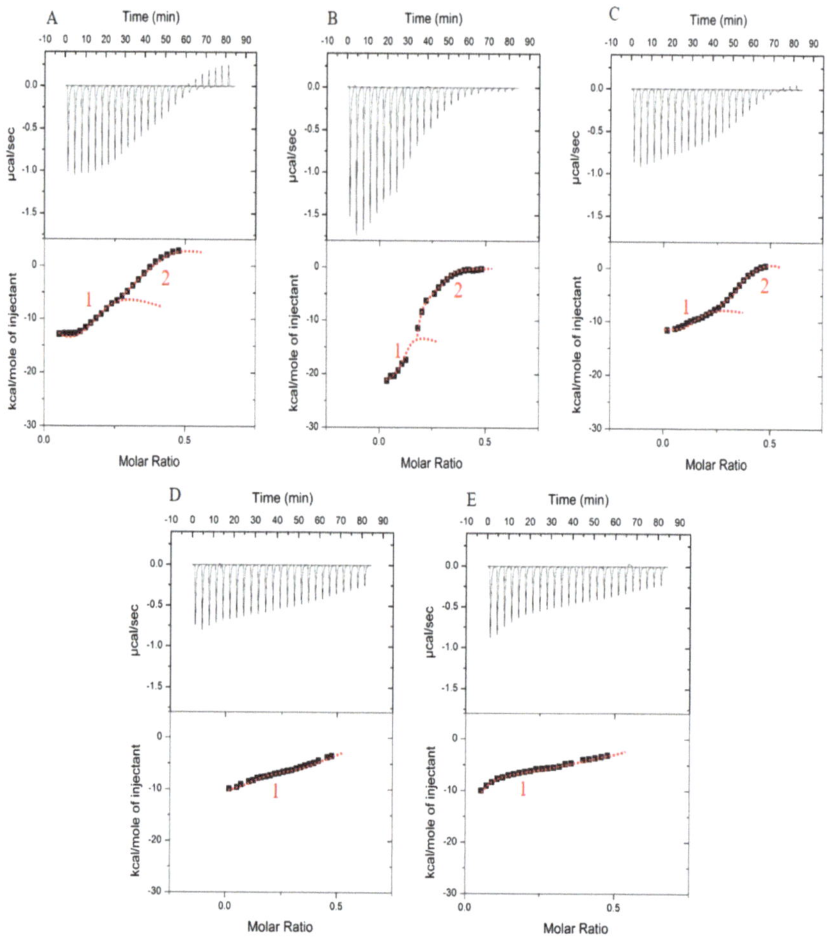

Figure 2. Heat flow thermograms as a function of time (**upper panel**) obtained during the titration of βLG (0.1 mM) by LF (0.25 mM) at different salt concentrations in 10 mM MES buffer pH 5.5 and at 25 °C. **Bottom panel**: graphical representation of the integrated data of enthalpy versus the molar ratio of LF: βLG. (**A**): without added salt; (**B**): with 2.5 mM added NaCl; (**C**): with 5 mM added NaCl; (**D**): with 15 mM added NaCl; (**E**): with 20 mM added NaCl. The red line is just to guide the eyes to distinguish when applicable the two inflection points. Inflection points are indicated by numbers 1 and 2.

The thermograms at very low salt concentration (Figure 2A–C) show two inflexion points that could describe two successive steps, which means that two possible events came into play during the complexation of βLG with LF. The second inflexion point (above molar ration of 0.2) disappeared with higher ionic strength values (Figure 2D,E). Since increasing the ionic strength can enhance the hydrophobic interaction [36], the second event might be hampered by the reinforcement of the hydrophobic effects. To confirm this hypothesis, the same ITC experiment was conducted without added salt, but at 35 °C as a temperature increment promotes hydrophobic effects. As illustrated in Figure 3, when temperature increased from 25 to 35 °C without added salt, the second signal was lost. Hence, hydrophobic effects might prevent the second mechanism that normally took place at a high molecular ratio. Given that an increase in temperature or ionic strength enhance the hydrophobic effects causing protein aggregation [37], a possible explanation could be that the interaction–coacervation between βLG and LF is a question of predominance between electrostatic interactions and hydrophobic effects. A decrease in electrostatic attractive forces being predominant over the effect of an increase in hydrophobic effects.

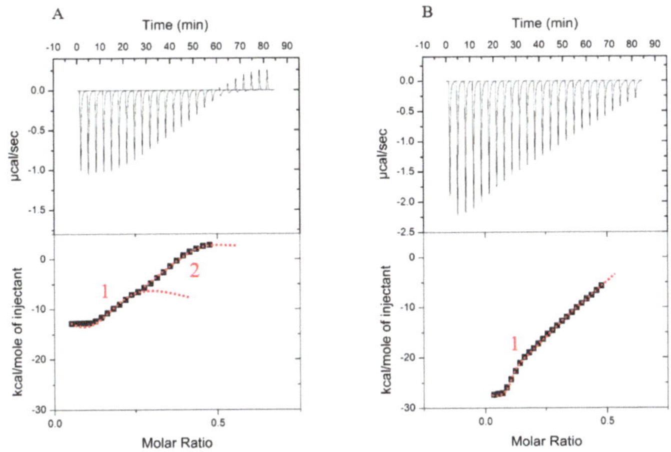

Figure 3. Heat flow thermogram as a function of the time (**upper panel**) obtained during the titration of βLG (0.1 mM) by LF (0.25 mM) in 10 mM MES buffer pH 5.5 at 25 °C (**A**) and 35 °C (**B**). **Bottom panel**: corresponding graphical representation of the integrated data of enthalpy versus the molar ratio of LF:βLG with no added salt. The red line is just to guide the eyes to distinguish when applicable the two inflection points. Inflection points are indicated by numbers 1 and 2.

The formation of coacervates, according to a two-step process as detected by ITC, has been described for other macromolecular systems [32,38,39]. The first enthalpy-driven step is attributed to electrostatic interactions or ion pairing and leads to the formation of soluble complexes. The second step, rather entropy-driven, is attributed to the self-aggregation of these complexes into coacervates. This explanation fits well for the studied LF–βLG system since the second inflexion point was suppressed with increasing ionic strength, concomitantly to the disappearance of LLPS as shown by turbidity measurements presented above. The second event can thus be assigned to the complex coacervation step between the two proteins.

The explanation of what happens during the two steps is not easy since, for macromolecular interactions, each thermodynamic signal is a result of the contribution of several phenomena: interaction, protein conformational change, release of water, protons and other ions, complexation, reorganizations, aggregation, etc. The overall measured signal therefore comes from endothermic and exothermic reactions whose final absolute value is the result of the dominant energy. To go further in the exploration of the thermodynamic changes occurring during titration, the binding isotherms were fitted using PyTC to determine the enthalpy and associated binding constant (Table 1). Without added salt, the interaction exhibited high enthalpy change and high affinity constant K_a in the micromolar range. Interestingly, the addition of 2.5 mM NaCl further promotes the interaction with a +25% gain in enthalpy value and twofold increase of the affinity constant. Hence, a small amount of added salt favors the interactions between the two oppositely charged proteins. For higher added NaCl concentrations, the screening effect of salt on the interaction and association between the two proteins started to be observed as reflected in the ITC signals and in the significant decrease of K_a and ΔH values (Table 1).

Table 1. The binding constant (K_a) and the enthalpy (ΔH) as a function of the salt concentration measured [NaCl] by the GUI of PyTC.

[NaCl] (mM)	K_a (M^{-1}) $\times 10^5$	ΔH (Kcal/mol)
0	4.21 ± 0.06 [ab]	-18.8 ± 0.04 [a]
2.5	9.72 ± 0.05 [ac]	-24.7 ± 0.02 [c]
5	2.75 ± 0.05 [ab]	-14.6 ± 0.04 [ae]
15	0.847 ± 0.02 [b]	-12.09 ± 0.01 [bde]
20	0.332 ± 0.015 [b]	-8.944 ± 0.03 [d]

a,b,c,d,e: values in a column not sharing the same superscript are significantly different according to the Tukey's multiple comparison test ($p < 0.05$). The ANOVA on NaCl concentration was significant ($p < 4 \times 10^{-5}$ and $p < 0.02$ for K_a and ΔH, respectively).

Burgess [40] reported that the general trend of gelatin–acacia coacervate yield increased with an increase in ionic strength up to a maximum and then decreased with a further increase in ionic strength. This "salting-in like" trend was explained as a consequence of the effect of added salt on the extent of coiling and charge densities of the involved macromolecules. The results found here for LF and βLG are consistent with such reported data except that the concentration of NaCl tolerated by LF–βLG (HPCC system) is much less than that tolerated by gelatin–acacia coacervates. In general and compared to HPCC, higher concentration of salt is needed to screen the interaction–complexation in protein–polysaccharide systems, and a relatively small amount can indeed promote the complexation as confirmed recently for pea protein–chitosan and ovalbumin–carboxymethylcellulose [41,42].

The enthalpy change reflects the contribution of hydrogen bonds, electrostatic and van der Waals interactions [34], and its decrease is hence expected with screening effect at increasing salt concentration. In fact, the presence of Na^+ ions and Cl^- ions reduced the strength of the electric field around charged groups in the proteins causing the saturation of their binding sites, and lowering their interactions. The decrease of enthalpy and binding affinity with increasing NaCl (5 to 60 Mm) were also reported for β-conglycinin–lysozyme system [35].

The ITC experiments revealed that βLG and LF complexation was enthalpically favorable and that a small amount of salt ions promoted the interactions between the two proteins. Those results provided one more proof of the major role played by the ionic strength on the complex coacervation of βLG and LF. Understanding the kinetics of the formation of coacervates via a desalting technique is the next step of this study.

3.2. *LF–βLG Complex Coacervation via Desalting*

Unlike the direct mixing presented in Section 3.1.1, desalting is an alternative protocol to better determining the ionic strength sensitivity of a complex coacervation process. Desalting was proposed as a gentle method to build and better control assemblies of nanoparticles and complex biological assemblies [43]. The desalting protocol involves, first, the mixing of the two macromolecules at a sufficiently high ionic strength in which the interactions are inhibited (charge screening). This mixed "inactive" solution is then dialyzed to decrease continuously the ionic strength around the mixed macromolecules [44,45].

In the current study, the behavior during continuous dialysis against MES buffer of mixtures of LF and βLG prepared at three high salt concentrations prior to mixing them has been monitored. The decrease of salt concentration inside the dialysis tubes monitored by conductivity measurements followed an expected exponential behavior (Figure 4). The higher the initial concentration, the longer the dialysis time to reach the final equilibrium concentration. Whatever the initial salt concentration in the range 100 to 400 mM, almost total elimination of salt is achieved after 3.5 h of dialysis in our experimental conditions. Costalat et al. [44] reported the similar kinetics of chloride elimination where 100% of salt was removed after 5 h and 6 h for NaCl starting concentrations of 2 M and 6 M, respectively.

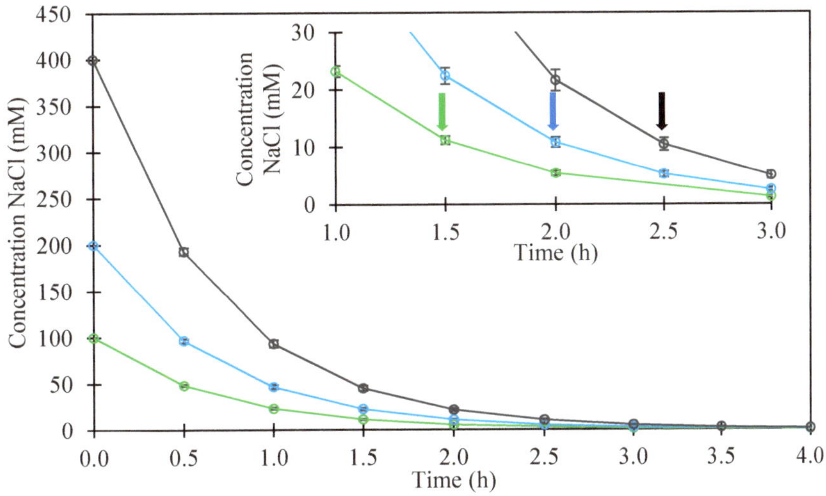

Figure 4. Evolution of ionic strength inside the dialysis tube as monitored by conductivity during 4 h of dialysis of lactoferrin–β-lactoglobulin mixed at various initial ionic strength: 100 mM (**green**), 200 mM (**blue**) and 400 mM (**black**). Insert: zoom on the three first hours. Arrows indicate the time of the maximum of turbidity for each experiment. Dialysis experiments were conducted against 10 mM MES buffer at pH 5.5.

The visual aspect and the corresponding microscopic images of the mixture in the dialysis tube at 100 mM NaCl during 24 h of dialysis are shown in Figure 5. At the beginning of the dialysis, the solution in the tube was almost transparent and no interaction was detected between the two proteins at this ionic strength. Once the dialysis began, the solution in the dialysis tube started getting turbid until reaching a maximum. At the end of the 24 h of dialysis, the solution was once again transparent as the LLPS took place (Figure 5C). This move toward a transparent solution is explained by the progressive formation of highly turbid micro-droplets (microphase separation) that progressively coalesce leading to the observed LLPS.

Turbidity measurements over time during dialysis confirmed the visual and microscopic observations as shown in Figure 6. First, it is worth mentioning that, without added salt, the protein solution reached spontaneously and quickly the maximum value of turbidity as shown for direct mixing experiments (Figure 1). The turbidity decreased continuously in parallel with the occurrence of LLPS inside the dialysis tube. On the other hand, for LF–βLG mixture exposed to high ionic strengths, the turbidity during the dialysis first increased progressively until reaching a maximum and then decreased to almost zero. The areas under the curves are substantially identical for the three-dialysis experiments. The dialysis time needed for maximum turbidity depended on the initial salt concentration, which is to be correlated to the desalting kinetics shown in Figure 4. The ANOVA of the effect of initial NaCl concentration on the maximum of turbidity was significant ($p < 0.04$). The maximum turbidity values at 0 mM and 400 mM were significantly different according to the Tukey's multiple comparison tests ($p < 0.05$). As illustrated in Figure 4, the highest turbidity value was reached when the salt concentration inside the tube was around 10 mM for the three desalting experiments. The transition from unassociated state to complexation–coacervation between LF and βLG occurred at an optimal and constant ionic strength of around 10 mM, confirming the high salt sensitivity of the studied heteroprotein system. By superimposing the results of Figures 4 and 6, a significant level of turbidity (complex coacervation) was reached around 10 mM NaCl inside the dialysis tubes.

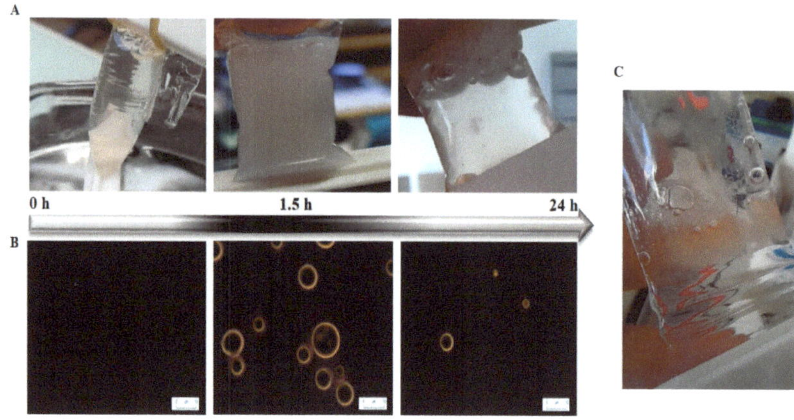

Figure 5. Appearance of lactoferrin–β-lactoglobulin coacervates during desalting protocol as a function of dialysis time for the initial salt concentration of 100 mM. (**A**): visual aspect inside the dialysis tube; (**B**): corresponding microscopic images showing the formation of coacervate droplets; (**C**): an image of the coacervates at the bottom of the dialysis tube at the end of the dialysis experiment. Dialysis was performed against 10 mM MES buffer at pH 5.5. Scale bar: 10 μm.

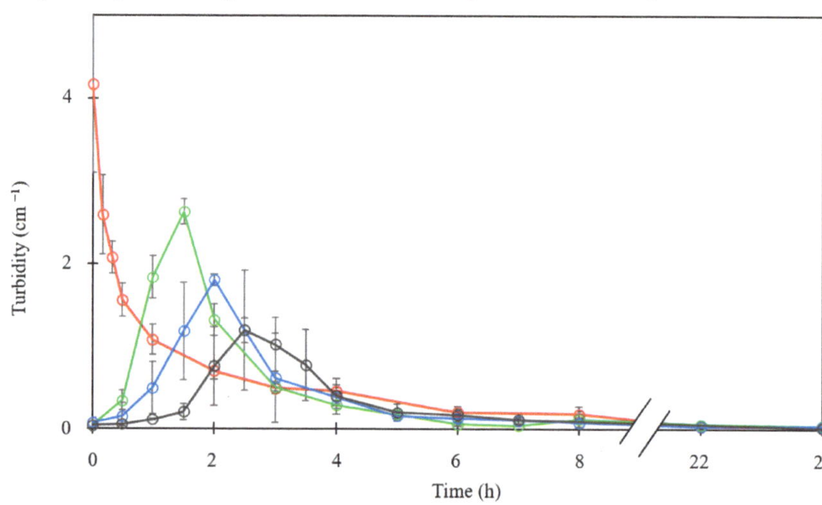

Figure 6. Evolution of turbidity inside the dialysis tube during 24 h desalting experiment of lactoferrin–β-lactoglobulin mixed at total protein concentration of 0.55 mM in 10 mM MES buffer, pH 5.5 at various initial salt concentrations. No added salt (**red**); with added NaCl at concentrations of 100 mM (**green**), 200 mM (**blue**) and 400 mM (**black**).

The overall size of the coacervates droplets formed during dialysis experiments varied from 1 to 10 μm as assessed by microscopic observations. However, their number was significantly lower for the sample with an initial NaCl concentration of 400 mM; at the maximum of turbidity, the number of coacervates was equal to 11 ± 3 and 6 ± 2 for 100 mM and 400 mM, respectively (measurements were conducted for images covering a field of 88 × 66 μm). Fresnais et al. [43] reported that dialysis leads to the formation of nanoparticle–polymer clusters with a size three- to fivefold larger than that obtained with direct mixing protocol at the same ionic strength. The same research group concluded that the decrease of the desalting rate led to an increase in the hydrodynamic diameter of copolymers complexes [45]. This explication can fit with the results of the LF–βLG system, as the decrease of the desalting rate is equivalent to conducting dialysis with a higher initial concentration. A higher initial salt concentration delays the on-set of the complexation–coacervation process during dialysis experiments (i.e., turbidity change, Figure 6).

The evolution of the coacervate yield was also monitored during desalting experiments (Figure 7). During the initial dialysis time, like turbidity, the increase of the coacervate yield was slowed down by a higher salt concentration. It is obvious that, without added salt, the coacervate yield sharply increased to a plateau value after 3 h of dialysis. However, in the presence of high salt concentrations, the coacervate yield increased slowly during time owing to the strengthening of the electrostatic interactions and the release of the binding sites that caused the coacervates to form progressively.

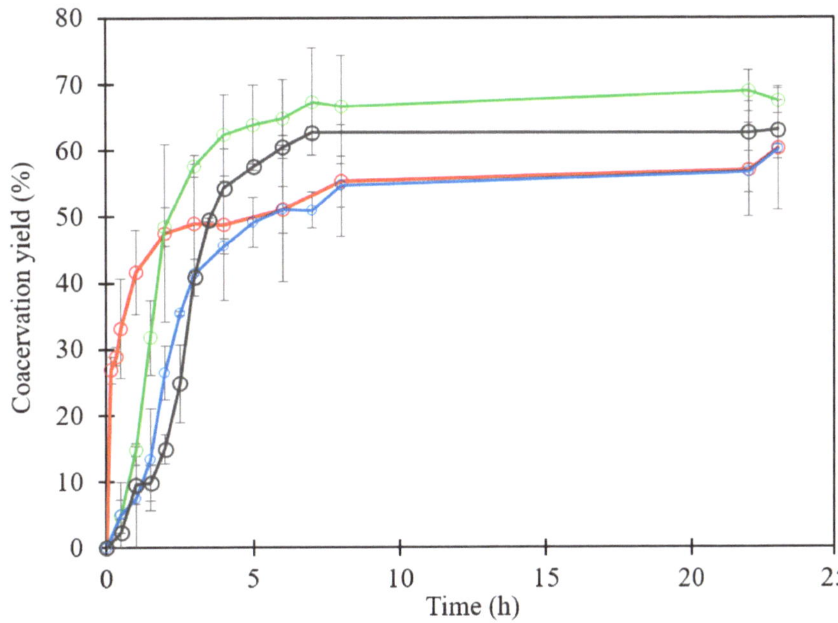

Figure 7. Evolution of the coacervate yield inside the dialysis tube during desalting experiment of lactoferrin–β-lactoglobulin mixed at total protein concentration of 0.55 mM in 10 mM MES buffer, pH 5.5 at various initial salt concentrations. No added salt (**red**); with added NaCl at concentrations of 100 mM (**green**), 200 mM (**blue**) and 400 mM (**black**).

Initial high salt concentrations did not affect the final coacervate yield recovered after desalting, which varied from 55% to 65% as statistically assessed by a non-significant ANOVA test ($p = 0.6$) (Figure 7).

4. Conclusions

In a previous work, the ability of lactoferrin and β-lactoglobulin to form heteroprotein complex coacervates under optimal conditions of pH and molar stoichiometry of 5.5 and 10, respectively, has been reported in a previous work. Here, the influence of ionic strength on such complex coacervation process was demonstrated using direct mixing and desalting protocols. We showed that, whatever the protocol used, the interaction and subsequent assembly of LF and βLG are highly sensitive to ionic strength of the medium. Compared to the effect of salt on other published systems, the heroprotein complex coacervation between βLG and LF seems to be one of the most sensitive to ionic strength; the molecular interaction between the two proteins can be tuned at a very narrow window of added salt concentration (i.e., 0–5 mM). A concentration of about 20 mM of added NaCl was enough to abolish LLPS between the two proteins, while molecular interactions are still detected at this ionic strength, as proved by ITC experiments. Therefore, the first steps of the coacervation, i.e., the interactions leading to the formation of primary units and soluble complexes, were more resistant to ionic strength than the final steps, i.e., the formation of turbid micrometric droplets. Further research using more sensitive techniques should be conducted in order to identify the salt effect on the soluble complexes and its influence on the structure of the coacervates network. These results are useful for targeting potential applications of LF–βLG complex coacervates as efficient microcapsules for bioactives and texturing agents for food products.

Author Contributions: Conceptualization, S.B.; methodology, R.S.H., P.H. and F.R.; software, R.S.H. and M.-H.F.; validation, M.-H.F. and S.B.; formal analysis, R.S.H., M.-H.F. and S.B.; investigation, R.S.H., P.H. and F.R.; writing—original draft preparation, R.S.H.; writing—review and editing, M.-H.F. and S.B.; supervision, S.B. and M.-H.F.; project administration, S.B.; funding acquisition, S.B. All authors have read and agreed to the published version of the manuscript.

Funding: This research was funded by public institutions: Regional council of Brittany and INRAE.

Data Availability Statement: Data are contained within the article.

Acknowledgments: We gratefully acknowledge the financial support provided by the Regional council of Brittany and INRAE and Julien Jardin for statistical analyses.

Conflicts of Interest: The authors declare no conflict of interest.

References

1. Bungenberg De Jong, H.G.; Kruyt, H.R. Coacervation (Partial Miscibility in Colloid Systems). In *Proceedings of the Royal Academy of Science at Amsterdam*; Koninklijke Nederlandse Akademie Wetenschappen: Amsterdam, The Netherlands, 1929; Volume 32, pp. 849–856.
2. Overbeek, J.T.G.; Voorn, M.J. Phase Separation in Polyelectrolyte Solutions. Theory of Complex Coacervation. *J. Cell. Comp. Physiol.* **1957**, *49*, 7–26. [CrossRef]
3. Veis, A.; Bodor, E.; Mussell, S. Molecular Weight Fractionation and the Self-Suppression of Complex Coacervation. *Biopolymers* **1967**, *5*, 37–59. [CrossRef]
4. Tainaka, K. Study of Complex Coacervation in Low Concentration by Virial Expansion Method. I. Salt Free Systems. *Biopolymers* **1967**, *46*, 37–59. [CrossRef]
5. Sato, H.; Nakajima, A. Complex Coacervation in Sulfated Polyvinyl Alcohol-Aminoaeetalyzed Polyvinyl Alcohol System If. Formationof Coacervatedroplets. *Colloid Polym. Sci.* **1974**, *252*, 944–948. [CrossRef]
6. Veis, A. A Review of the Early Development of the Thermodynamics of the Complex Coacervation Phase Separation. *Adv. Colloid Interface Sci.* **2011**, *167*, 2–11. [CrossRef]
7. Wang, X.; Zhang, P.; Tian, L. Spatiotemporal Organization of Coacervate Microdroplets. *Curr. Opin. Colloid Interface Sci.* **2021**, *52*, 101420. [CrossRef]
8. Jho, Y.; Yoo, H.Y.; Lin, Y.; Han, S.; Hwang, D.S. Molecular and Structural Basis of Low Interfacial Energy of Complex Coacervates in Water. *Adv. Colloid Interface Sci.* **2017**, *239*, 61–73. [CrossRef]
9. Xiao, Z.; Liu, W.; Zhu, G.; Zhou, R.; Niu, Y. A Review of the Preparation and Application of Flavour and Essential Oils Microcapsules Based on Complex Coacervation Technology: Microcapsules Based on Coacervation Technology. *J. Sci. Food Agric.* **2014**, *94*, 1482–1494. [CrossRef]
10. Costalat, M.; Alcouffe, P.; David, L.; Delair, T. Controlling the Complexation of Polysaccharides into Multi-Functional Colloidal Assemblies for Nanomedicine. *J. Colloid Interface Sci.* **2014**, *430*, 147–156. [CrossRef]
11. Adal, E.; Sadeghpour, A.; Connell, S.; Rappolt, M.; Ibanoglu, E.; Sarkar, A. Heteroprotein Complex Formation of Bovine Lactoferrin and Pea Protein Isolate: A Multiscale Structural Analysis. *Biomacromolecules* **2017**, *18*, 625–635. [CrossRef]
12. Croguennec, T.; Tavares, G.M.; Bouhallab, S. Heteroprotein Complex Coacervation: A Generic Process. *Adv. Colloid Interface Sci.* **2017**, *239*, 115–126. [CrossRef]
13. Desfougères, Y.; Croguennec, T.; Lechevalier, V.; Bouhallab, S.; Nau, F. Charge and Size Drive Spontaneous Self-Assembly of Oppositely Charged Globular Proteins into Microspheres. *J. Phys. Chem. B* **2010**, *114*, 4138–4144. [CrossRef]
14. Zheng, J.; Gao, Q.; Ge, G.; Wu, J.; Tang, C.; Zhao, M.; Sun, W. Heteroprotein Complex Coacervate Based on β-Conglycinin and Lysozyme: Dynamic Protein Exchange, Thermodynamic Mechanism, and Lysozyme Activity. *J. Agric. Food Chem.* **2021**, *69*, 7948–7959. [CrossRef]
15. Kizilay, E.; Seeman, D.; Yan, Y.; Du, X.; Dubin, P.L.; Donato-Capel, L.; Bovetto, L.; Schmitt, C. Structure of Bovine β-Lactoglobulin–Lactoferrin Coacervates. *Soft Matter* **2014**, *10*, 7262. [CrossRef]

16. Peixoto, P.D.S.; Tavares, G.M.; Croguennec, T.; Nicolas, A.; Hamon, P.; Roiland, C.; Bouhallab, S. Structure and Dynamics of Heteroprotein Coacervates. *Langmuir* **2016**, *32*, 7821–7828. [CrossRef]
17. Chapeau, A.-L.; Tavares, G.M.; Hamon, P.; Croguennec, T.; Poncelet, D.; Bouhallab, S. Spontaneous Co-Assembly of Lactoferrin and β-Lactoglobulin as a Promising Biocarrier for Vitamin B9. *Food Hydrocoll.* **2016**, *57*, 280–290. [CrossRef]
18. Soussi Hachfi, R.; Famelart, M.-H.; Rousseau, F.; Hamon, P.; Bouhallab, S. Rheological Characterization of β-Lactoglobulin/Lactoferrin Complex Coacervates. *LWT* **2022**, *163*, 113577. [CrossRef]
19. Yan, Y.; Kizilay, E.; Seeman, D.; Flanagan, S.; Dubin, P.L.; Bovetto, L.; Donato, L.; Schmitt, C. Heteroprotein Complex Coacervation: Bovine β-Lactoglobulin and Lactoferrin. *Langmuir* **2013**, *29*, 15614–15623. [CrossRef]
20. Anema, S.G.; de Kruif, C.G.K. Complex Coacervates of Lactotransferrin and β-Lactoglobulin. *J. Colloid Interface Sci.* **2014**, *430*, 214–220. [CrossRef]
21. Tavares, G.M.; Croguennec, T.; Hamon, P.; Carvalho, A.F.; Bouhallab, S. Selective Coacervation between Lactoferrin and the Two Isoforms of β-Lactoglobulin. *Food Hydrocoll.* **2015**, *48*, 238–247. [CrossRef]
22. Rabek, J. Absorption of Radiation. *Photochem. React. Polym. Mech. Photophysical Process.* **1987**, *26*, 5–9. [CrossRef]
23. Duvvuri, H.; Wheeler, L.C.; Harms, M.J. *Pytc*: Open-Source Python Software for Global Analyses of Isothermal Titration Calorimetry Data. *Biochemistry* **2018**, *57*, 2578–2583. [CrossRef]
24. Anema, S.G.; de Kruif, C.K. Co-Acervates of Lactoferrin and Caseins. *Soft Matter* **2012**, *8*, 4471. [CrossRef]
25. Weinbreck, F.; Tromp, R.H.; de Kruif, C.G. Composition and Structure of Whey Protein/Gum Arabic Coacervates. *Biomacromolecules* **2004**, *5*, 1437–1445. [CrossRef] [PubMed]
26. Boral, S.; Bohidar, H.B. Effect of Ionic Strength on Surface-Selective Patch Binding-Induced Phase Separation and Coacervation in Similarly Charged Gelatin−Agar Molecular Systems. *J. Phys. Chem. B* **2010**, *114*, 12027–12035. [CrossRef]
27. Xiong, W.; Ren, C.; Jin, W.; Tian, J.; Wang, Y.; Shah, B.R.; Li, J.; Li, B. Ovalbumin-Chitosan Complex Coacervation: Phase Behavior, Thermodynamic and Rheological Properties. *Food Hydrocoll.* **2016**, *61*, 895–902. [CrossRef]
28. Zhang, J.; Du, H.; Ma, N.; Zhong, L.; Ma, G.; Pei, F.; Chen, H.; Hu, Q. Effect of Ionic Strength and Mixing Ratio on Complex Coacervation of Soy Protein Isolate/Flammulina Velutipes Polysaccharide. *Food Sci. Hum. Wellness* **2023**, *12*, 183–191. [CrossRef]
29. Han, J.; Yan, J.; Du, Y.; Wu, H.; Zhu, B. Formation and Stability of Electrostatic Complexes Formed between Scallop Female Gonad Protein Isolates and Sodium Alginate: Influence of PH, Total Concentration, Blend Ratio, and Ionic Strength. *J. Food Sci.* **2022**, *87*, 2504–2514. [CrossRef]
30. Mjahed, H.; Voegel, J.-C.; Chassepot, A.; Senger, B.; Schaaf, P.; Boulmedais, F.; Ball, V. Turbidity Diagrams of Polyanion/Polycation Complexes in Solution as a Potential Tool to Predict the Occurrence of Polyelectrolyte Multilayer Deposition. *J. Colloid Interface Sci.* **2010**, *346*, 163–171. [CrossRef]
31. Dong, D.; Li, X.; Hua, Y.; Chen, Y.; Kong, X.; Zhang, C.; Wang, Q. Mutual Titration of Soy Proteins and Gum Arabic and the Complexing Behavior Studied by Isothermal Titration Calorimetry, Turbidity and Ternary Phase Boundaries. *Food Hydrocoll.* **2015**, *46*, 28–36. [CrossRef]
32. Kayitmazer, A.B. Thermodynamics of Complex Coacervation. *Adv. Colloid Interface Sci.* **2017**, *239*, 169–177. [CrossRef]
33. Kutscher, M.; Cheow, W.S.; Werner, V.; Lorenz, U.; Ohlsen, K.; Meinel, L.; Hadinoto, K.; Germershaus, O. Influence of Salt Type and Ionic Strength on Self-Assembly of Dextran Sulfate-Ciprofloxacin Nanoplexes. *Int. J. Pharm.* **2015**, *486*, 21–29. [CrossRef]
34. Santos, M.B.; da Costa, A.R.; Garcia-Rojas, E.E. Heteroprotein Complex Coacervates of Ovalbumin and Lysozyme: Formation and Thermodynamic Characterization. *Int. J. Biol. Macromol.* **2018**, *106*, 1323–1329. [CrossRef]
35. Zheng, J.; Gao, Q.; Ge, G.; Wu, J.; Tang, C.; Zhao, M.; Sun, W. Sodium Chloride-Programmed Phase Transition of β-Conglycinin/Lysozyme Electrostatic Complexes from Amorphous Precipitates to Complex Coacervates. *Food Hydrocoll.* **2022**, *124*, 107247. [CrossRef]
36. Dubin, P.; Bock, J.; Davis, R.; Schulz, D.N.; Thies, C. (Eds.) *Macromolecular Complexes in Chemistry and Biology*; Springer: Berlin/Heidelberg, Germany, 1994; ISBN 978-3-642-78471-2.
37. Lin, T.; Dadmohammadi, Y.; Davachi, S.M.; Torabi, H.; Li, P.; Pomon, B.; Meletharayil, G.; Kapoor, R.; Abbaspourrad, A. Improvement of Lactoferrin Thermal Stability by Complex Coacervation Using Soy Soluble Polysaccharides. *Food Hydrocoll.* **2022**, *131*, 107736. [CrossRef]
38. Priftis, D.; Laugel, N.; Tirrell, M. Thermodynamic Characterization of Polypeptide Complex Coacervation. *Langmuir* **2012**, *28*, 15947–15957. [CrossRef]
39. Vitorazi, L.; Ould-Moussa, N.; Sekar, S.; Fresnais, J.; Loh, W.; Chapel, J.-P.; Berret, J.-F. Evidence of a Two-Step Process and Pathway Dependency in the Thermodynamics of Poly(Diallyldimethylammonium Chloride)/Poly(Sodium Acrylate) Complexation. *Soft Matter* **2014**, *10*, 9496–9505. [CrossRef]
40. Burgess, D.J. Practical Analysis of Complex Coacervate Systems. *J. Colloid Interface Sci.* **1990**, *140*, 227–238. [CrossRef]
41. Zhang, Q.; Jeganathan, B.; Dong, H.; Chen, L.; Vasanthan, T. Effect of Sodium Chloride on the Thermodynamic, Rheological, and Microstructural Properties of Field Pea Protein Isolate/Chitosan Complex Coacervates. *Food Chem.* **2021**, *344*, 128569. [CrossRef]
42. Xiong, W. Complex Coacervation of Ovalbumin-Carboxymethylcellulose Assessed by Isothermal Titration Calorimeter and Rheology: Effect of Ionic Strength and Charge Density of Polysaccharide. *Food Hydrocoll.* **2017**, *73*, 41–50. [CrossRef]
43. Fresnais, J.; Lavelle, C.; Berret, J.-F. Nanoparticle Aggregation Controlled by Desalting Kinetics. *J. Phys. Chem. C* **2009**, *113*, 16371–16379. [CrossRef]

44. Costalat, M.; David, L.; Delair, T. Reversible Controlled Assembly of Chitosan and Dextran Sulfate: A New Method for Nanoparticle Elaboration. *Carbohydr. Polym.* **2014**, *102*, 717–726. [CrossRef] [PubMed]
45. Berret, J.-F. Controlling Electrostatic Co-Assembly Using Ion-Containing Copolymers: From Surfactants to Nanoparticles. *Adv. Colloid Interface Sci.* **2011**, *167*, 38–48. [CrossRef] [PubMed]

foods

MDPI

Article

Effect of Oxidative Modification by Peroxyl Radical on the Characterization and Identification of Oxidative Aggregates and In Vitro Digestion Products of Walnut (*Juglans regia* L.) Protein Isolates

Jinjin Zhao, Miaomiao Han, Qingzhi Wu, Xiaoying Mao *, Jian Zhang and Zhenkang Lu

School of Food Science and Technology, Shihezi University, Shihezi 832003, China
* Correspondence: maoxiaoying99@163.com

Abstract: Walnut protein is a key plant protein resource due to its high nutritional value, but walnuts are prone to oxidation during storage and processing. This article explored the oxidative modification and digestion mechanism of walnut protein isolates by peroxyl radical and obtained new findings. SDS-PAGE and spectral analysis were used to identify structural changes in the protein after oxidative modification, and LC-MS/MS was used to identify the digestion products. The findings demonstrated that as the AAPH concentration increased, protein carbonyl content increased from 2.36 to 5.12 nmol/mg, while free sulfhydryl content, free amino content, and surface hydrophobicity decreased from 4.30 nmol/mg, 1.47 μmol/mg, and 167.92 to 1.72 nmol/mg, 1.13 μmol/mg, and 40.93 nmol/mg, respectively. Furthermore, the result of Tricine-SDS-PAGE in vitro digestion revealed that protein oxidation could cause gastric digestion resistance and a tendency for intestinal digestion promotion. Carbonyl content increased dramatically during the early stages of gastric digestion and again after 90 min of intestine digestion, and LC-MS/MS identified the last digestive products of the stomach and intestine as essential seed storage proteins. Oxidation causes walnut proteins to form aggregates, which are then re-oxidized during digestion, and proper oxidative modification may benefit intestinal digestion.

Keywords: oxidative modification mechanism; walnut protein isolates; LC-MS/MS identification; aggregates; digestion products

Citation: Zhao, J.; Han, M.; Wu, Q.; Mao, X.; Zhang, J.; Lu, Z. Effect of Oxidative Modification by Peroxyl Radical on the Characterization and Identification of Oxidative Aggregates and In Vitro Digestion Products of Walnut (*Juglans regia* L.) Protein Isolates. *Foods* **2022**, *11*, 4104. https://doi.org/10.3390/foods11244104

Academic Editors: Ruifen Li, Laura Roman and Norbert Raak

Received: 17 November 2022
Accepted: 16 December 2022
Published: 19 December 2022

1. Introduction

Walnuts (*Juglans regia* L.) are becoming popular as a high-quality plant-based protein with a good nutritional and health profile [1,2]. However, because walnut kernels contain high levels of unsaturated fatty acids (62–68%) [3], they are often processed into walnut oil, and the process of processing into walnut oil produces a large number of defatted walnut meals. Since defatted walnut meal contains 50–60% protein [4], it is an important source of walnut protein. In the pretreatment process of walnut oil, with the destruction of the walnut cell structure, the polyunsaturated fatty acids in walnut are prone to lipid peroxidation under the catalysis of lipoxygenase (LOX), resulting in lipid free radicals and active lipid oxidation products [5,6], which indirectly lead to protein oxidation. Moreover, walnut protein can affect the quality of the composite product as a raw material or ingredient due to its early oxidation. Oxidation first leads to changes in the structure of walnut protein, which then affects its nutritional properties. At present, the relationship between digestion and health is a hot focus of research, and walnut is also widely used in the food processing industry; therefore, research into the structure and digestive properties of walnut protein will provide a theoretical foundation for quality monitoring during the product's processing and storage.

Protein is the material basis of life, and the absorption and utilization of nutrients by the human body are mainly carried out in the stomach and small intestine of the digestive tract. Protein source, protein modification, digestive enzyme secretion, and activity lead to differences in protein digestion and absorption [7]. Protein oxidation can lead to breakage and cross-linking of protein molecules, producing soluble and insoluble aggregates that may have effects on protein digestion, and the oxidation of amino acid side chains may cause changes in the antioxidant capacity of the oxidation products. However, the effect of oxidation on protein digestion, on the other hand, still remains debatable. Previous studies found that oxidized rice bran protein incubated with malondialdehyde has a lower digestibility index [8], whereas moderate oxidation of SPI increases its digestibility via linoleic acid modification [9]. Digestion is a complex physiological process influenced by the digestive phase's pH, distinct oxidation systems, and degree of oxidation [10].

In the study of protein oxidative damage, the carbonyl content is often used as a key indicator of protein oxidative damage. Proteins in food are digested into oligopeptides or amino acids before being absorbed and utilized [11], which also allows the possibility of the re-oxidation of proteins during digestion [12]. Much of the current research on protein oxidation has focused on the effects of oxidation on protein structure, functional properties, and digestibility. There are few studies on the oxidative aggregation of walnut proteins and its specific effects on digestion products.

In recent years, there has been a surge of interest in the relationship between dietary protein oxidation and LC-MS/MS. Fang et al. [13] explored the effect of oxidation on protein amino acids and used LC-MS/MS to evaluate the site and mechanism of oxidation of myosin at different ozone treatment durations. Zhang et al. [14] used HPLC-HCD-MS/MS to explore the mechanism of lipid oxidation on the IgG/IgE binding capability of ovalbumin (OVA) and discovered nine oxidation sites on the α-helix of the protein. The explanation of these studies mostly starts with amino acids. However, mass spectrometry, meanwhile, has also been increasingly used for the identification of digestive products. Wang et al. [15] studied the digestibility of various proteins in the elderly gastrointestinal tract and concluded that the digestive profile of soy proteins is more susceptible to altered digestive conditions than meat proteins. Thus, amino acid analysis techniques and mass spectrometry identification techniques are important in the characterization of protein oxidation and digestion.

2,2-azobis (2-methylpropionamidine) dihydrochloride (AAPH) is frequently used as a representative of the typical free radicals of lipid peroxidation in oxidation models [16]. The formation of aggregates and their impact on digestion were investigated in this experiment by simulating in vitro AAPH oxidation and gastrointestinal digestion. To account for the digestive processes of walnut proteins, we measured the carbonyl content of proteins in different stages of gastric and intestinal digestion; we used the Tricine-SDS-PAGE profiling to characterize changes in digestion, and the digestion results were identified by LC-MS/MS. This study provides the basis for the quality control of walnut protein products. This study provides a theoretical foundation for the quality control of walnut protein products and walnut-based composite products during storage and processing.

2. Materials and Methods

2.1. Materials

Walnuts were purchased from the farmers' market (Shihezi, China). 2,2-azobis (2-methylpropionamidine) dihydrochloride (AAPH), 2,4-dinitrophenylhydrazine (DNPH), 5,5′-dithiobis (2-nitrobenzoic acid) (DTNB), and 8-anilino-1-naphthalenesulfonic acid (ANS) were obtained from Yuanye (Shanghai, China). Pepsin (3000 units/mg) and trypsin (250 units/mg) were purchased from Boao (Shanghai, China). The Tris-Tricine-SDS-PAGE and BCA kits were provided by Solarbio (Beijing, China). All the other chemicals were of analytical reagent grade and obtained in China.

2.2. Preparation of Walnut Protein Isolates

The defatted walnut flour (DWF) was made using the method described by Mao and Hua et al. [3]. In brief, walnuts were shelled, peeled, and crushed, and walnut powder was soaked with n-hexane at 1:5 (*w*/*v*) for 1 h, and the precipitation was obtained after filtration. The precipitation and releaching were repeated three times. Finally, the filtered degreased walnut powder was placed in the fume hood to evaporate the organic reagent, and the dried walnut powder was sifted through an 80-mesh screen The protein content of the Kjeldahl procedure was 61.04 ± 0.13%. The walnut protein isolate was prepared according to the method described by Zhu et al. [17]. In brief, the DWF was mixed with water in a 1:26 (*w*/*v*) ratio, the pH was adjusted to 9, the mixture was stirred at 53 °C for 1.5 h, and the precipitate was centrifuged to remove the precipitate. The supernatant was adjusted to pH 4.5, stirred for 1 h, and then centrifuged to collect the precipitate. Finally, the precipitate was dissolved in deionized water, the pH was neutralized, and the solution was lyophilized at 4 °C. The Kjeldahl procedure yielded a protein content of 90.39 ± 0.30% (in the above Kjeldahl method, the conversion coefficient of protein is 6.25).

2.3. Modification of Walnut Proteins with AAPH

Samples of oxidized walnut protein were prepared according to and slightly modified by the method of Wang et al. [18]. Walnut protein isolates were dispersed at a concentration of 20 mg/mL in 10 mmol/L phosphate buffer (pH 8.0) (containing 0.5 mg/mL NaN_3). They were then combined with a 1:1 mixture of walnut protein suspension and AAPH solution. AAPH concentrations in the protein solution were 0, 0.04, 0.2, 1, 5, and 25 mmol/L at the end. The resulting walnut protein solution was shaken and incubated at 37 °C under light protection for 24 h. After stopping the reaction in an ice bath, it was dialyzed for 72 h at 4 °C to remove residual AAPH before being lyophilized and stored at −20 °C.

2.4. Determination of Oxidation and Aggregation Formation in Walnut Proteins

2.4.1. Carbonyl Group

The content of protein-bound carbonyls was determined according to the method of Wu et al. [19]. The 2,4-dinitrophenylhydrazine colorimetric method was used to measure the absorbance at 370 nm, and the molarity was calculated as 22,000/M·cm extinction coefficient per mg of protein carbonyl derivative.

$$\text{Carbonyl content (n mol/mg)} = \text{A} \times 45.45 \times \frac{\text{D}}{\text{P}} \tag{1}$$

where D is the dilution factor, P is the concentration of the protein sample in milligrams per milliliter, and A is the absorbance at 370 nm.

2.4.2. Free Sulfhydryl Group

The DNTB colorimetric method was used, with slight modifications based on Fu et al. [20]. Walnut protein was dissolved in 0.1 mol/L PBS containing l mmol/L EDTA and 1% SDS pH 8.0 to fully dissolve and adjust the protein concentration. Take 3 mL protein solution, add 3 mL PBS (containing 1 mmol/L EDTA and 1% SDS), add 0.1 mL DNTB, 25 °C water bath for 1 h. After the reaction, centrifuge at 8000 r/min for 30 min. Calculate the mercapto group content by measuring absorbance at 412 nm without DNTB as control. The results were expressed as nmole of -SH per milligram of soluble protein with a molar extinction coefficient of 13,600/M·cm.

$$\text{Free sulfhydryl content (n mol/mg)} = \text{B} \times 70.67 \times \frac{\text{D}}{\text{P}} \tag{2}$$

where D is the dilution factor, P is the concentration of the protein sample in milligrams per milliliter, and B is the absorbance at 412 nm.

2.4.3. Free Amino Group

The free amino content was determined by its derivatization with OPA, according to Chen et al. [21]. Dissolve 40 mg phthalaldehyde in 1 mL methanol, 2.5 mL 20% SDS, 25 mL 0.1 mol/L borax, 100 µL β-mercaptoethanol, mix well and keep volume constant to

50 mL. Then mix 200 μL of each protein solution with 4 mL of OPA reagent, react at 35 °C for 2 min, and measure the absorbance at 340 nm. The content of the free amino group was calculated from a standard curve made from tryptophan with linearity (R^2 = 0.9976; $p < 0.05$).

2.4.4. Secondary Structure (FTIR)

The protein's secondary structure was determined using FTIR spectrometry according to the method of Zhu et al. [22]. Add 100 mg of potassium bromide powder to the sample, mix, and tablet. The blank control was potassium bromide, and the infrared spectrometer was used to perform full-band scanning (4000–400 cm^{-1}) with a resolution of 4 cm^{-1} and a total of 32 scans.

2.4.5. Surface Hydrophobicity

The ANS fluorescence probe method was slightly modified from Wu et al. [23]. The protein concentrations in the samples ranged from 0.005–0.5 mg/mL. A total of 4 mL of the diluted sample was added to 50 μL of ANS solution (0.01 mol/L pH 7.0 phosphate buffer at 8 mmol/L), and the excitation and emission wavelengths were set at 395 nm and 473 nm, respectively.

2.4.6. UV Spectrum Analysis

The ultraviolet (UV) spectrum was measured using the method previously described by Hellwig et al. [24] with minor modifications. The protein sample concentration was 0.5 mg/mL, and the instrument scan range was 190 nm–500 nm.

2.4.7. Fluorescence Determination of Tryptophan, NFk, and Schiff Bases

Walnut protein was dissolved in phosphate buffer (10 mmol/L, pH 7.0), magnetically stirred for 2 h, and then centrifuged at 8000 rpm for 30 min at 4 °C. The protein concentration in the sample was 0.5 mg/mL. The fluorescence of tryptophan was measured using a slightly modified method developed by Wu et al. [23]. The emission spectra were scanned at 300–400 nm, using 285 nm as the excitation wavelength. The Schiff bases and N′-formyl-L-kynurenine (NFk) were determined using a slightly modified method of Fu et al. [20]. The width of the slit was 5 nm, and the voltage was 500 mV. The excitation wavelength was set to 350 nm, and the fluorescence value at 460 nm was used to calculate the Schiff base intensity, while the fluorescence value at 440 nm was used to calculate the NFk intensity.

2.5. Digestion In Vitro

2.5.1. Simulation of the Digestive Process

By referring to Brodkorb et al. [25], the digestion protocol was slightly modified. Berify used various concentrations of KCl, KH_2PO_4, NaCl, $MgCl_2$ $(H_2O)_6$, $(NH_4)_2CO_3$, and $CaCl_2(H_2O)_2$ to simulate 1.2X stomach gastric electrolyte solution (SGF) and intestinal electrolyte solution (SIF). The pH of the SGF was adjusted to 3.0 with hydrochloric acid before use, and the pH of the SIF was adjusted to 7.0 with NaOH before being mixed with the enzyme solution (4 parts electrolyte solution plus 1 part enzymes). The electrolyte solutions of SGF and SIF were preheated to 37 °C. To avoid precipitation during storage, $CaCl_2(H_2O)_2$ should be added immediately before the digestion experiment. Gastric digestion was performed at 37 °C for 2 h with stirring at 100 rpm, with samples taken at 0, 30, 60, 90, and 120 min, labeled G-30, G-60, G-90, and G-120. The extracted samples were boiled for 5 min to inactivate the enzyme, and digestion was continued for 2 h with a 1:1 addition of intestinal digestion solution, with samples taken at 0, 30, 60, 90, and 120 min, labeled I-30, I-60, I-90, and I-120. The supernatant was collected by centrifugation at 6000 rpm for 20 min.

2.5.2. Gel Electrophoresis Analysis

Follow the method of Zhu et al. [22], with minor modifications, after adding 10 μL of the same volume of 2X SDS sample buffer with β-mercaptoethanol (reducing), the samples were boiled for 5 min. Concentrates and separates were run at 80 and 120 volts, respectively, stained with R-250 Coomassie Brilliant Blue for 30 min, and then decolorized with 7.5% glacial acetic acid and 25% methanol. Following the method of Xu et al. [26] with minor modifications, a Tricine-SDS-PAGE kit was used to determine the molecular weight composition of walnut peptides from the gastric and subsequent intestinal digestion. The samples were boiled for 10 min after the 5X loading buffer with β-mercaptoethanol (reducing) was added at a volume ratio of 1:5. The gel concentrations were as follows: 16% of separation gel, 10% of interlayer gel, and 4% of concentration gel. Cathodic electrophoresis buffer was used in the inner bath, while anodic electrophoresis buffer was used in the outer bath. The voltage was set at 30 volts at the start of the experiments and increased to 90 volts once the bands were in the interlayer gel. The gel was first fixed for 1 h with fixative, then stained for 30 min with R-250 Coomassie Brilliant Blue, and finally decolored with a 7.5% acetic acid and 25% methanol solution.

2.5.3. Determination of the Protein Carbonyl Indexes in Gastrointestinal Digestion

According to 2.4.1, the carbonyl values of walnut protein samples after oxidation of AAPH with different concentrations were measured, and samples were taken at 30, 60, 90, and 120 min during gastric digestion and labeled G-30, G-60, G-90, and G-120, while the samples were taken at 30, 60, 90, and 120 min during subsequent intestinal digestion and labeled I-30, I-60, I-90, and I-120.

2.5.4. Amino acid Decay Rate before and after Digestion of Oxidized Protein

An automatic amino acid analyzer was used to measure 17 different amino acids (Asp, Thr, Ser, Glu, Gly, Ala, Cys, Val, Met, Ile, Leu, Tyr, Phe, Lys, His, Arg, and Pro) using the method of Kuipers and Gruppen et al. [27]. The amino acid results are expressed in grams of amino acids per 100 g of liquid supernatant.

2.5.5. Determination of Antioxidant Activity of Hydrolysates

Adopting the method of Georgiou et al. [28], the ABTS free radical scavenging ability, DPPH free radical scavenging ability, and hydroxyl free radical scavenging ability of the hydrolysate were determined.

2.6. Identification of Oxidized Protein Digestive Products

2.6.1. LC-MS/MS

In-Gel Digestion

Referring to the method of Fang et al. [13], with minor modifications, the gel was carefully placed into a 1.5 mL EP tube and washed twice with ultrapure water for 5 min each time before being decolorized with 400 μL (25 mmol/L NH_4HCO_3, 50% acetonitrile), rinsed until transparent, and freeze-dried. Lyophilized samples were hydrolyzed in 10 μL of a 20 ng/μL trypsin solution for 30 min at 4 °C. This is a pre-trypsin digestion step that completely absorbs the gel block, improving and fully hydrolyzing the gel during subsequent digestion. After being treated with 25 mmol/L of NH_4HCO_3 buffer, the gel was hydrolyzed at 37 °C for 16 h (without trypsin). After that, the hydrolyzed solution was extracted and transferred to a new EP tube. The original tube contained 0.1% trifluoroacetic acid (20 mL). The solution was siphoned after 15 min of incubation, added to the previous solution for 15 min of ultrasound, and then freeze-dried.

2.6.2. Easy-n LC 1200 Combined with a Q-Exactive Mass Spectrometer

Reversed-phase nano-liquid chromatography–tandem mass spectrometry (LC-MS/MS) was used to analyze each sample. The peptide amino acid sequences were obtained from the Uniprot database at https://www.uniport.org/ (accessed on 11 December 2022).

2.7. Statistical Analysis

The data were analyzed using SPSS 26.0 statistical software, and the graphs were created using Origin 2022. To assess significant differences ($p < 0.05$), Duncan's multiple range test was used.

3. Results

3.1. Effect of Oxidation on Protein Carbonyl Groups, Free Sulfhydryl Groups, and Free Amino Groups

Protein carbonyl groups and free sulfhydryl groups are often used to characterize the extent of protein oxidation. According to Soudani et al. [29], protein side chains (especially Pro, Arg, Lys, and Thr) readily generate carbonyl (CO) groups (aldehydes and ketones) during oxidation. Under constant rate thermal decomposition conditions, AAPH is a water-soluble azo compound that generates peroxyl radicals [30]. The amount of peroxyl radicals produced by AAPH is proportional to the concentration of AAPH [31]. As shown in Table 1, the carbonyl content increased significantly with increasing AAPH concentration ($p < 0.05$), and a similar trend was reported by Mao et al. [5]. In this study, the initial carbonyl content of the walnut sample was 2.36 nmol/mg, and the carbonyl content increased 0.35-fold and 1.17-fold at AAPH concentrations of 1 mmol/L and 25 mmol/L, respectively, compared to the control.

Table 1. Effects of oxidation on the carbonyl group, free sulfhydryl group, and free amino group of walnut samples with increasing concentrations of AAPH [a].

AAPH (mmol/L)	Carbonyl (nmol/mg)	Free SH (nmol/mg)	Free Amino (μmol/mg)
0	2.36 ± 0.09 f	4.30 ± 0.03 a	1.28 ± 0.03 c
0.04	2.42 ± 0.11 e	3.80 ± 0.05 b	1.39 ± 0.01 b
0.2	2.68 ± 0.05 d	3.66 ± 0.02 c	1.47 ± 0.02 a
1	3.18 ± 0.02 c	2.22 ± 0.01 d	1.17 ± 0.01 d
5	4.23 ± 0.07 b	2.02 ± 0.01 e	1.15 ± 0.02 d
25	5.12 ± 0.10 a	1.72 ± 0.08 f	1.13 ± 0.02 d

[a] Each index of the sample is the average of three determinations, and the data are expressed as mean ± standard error. Different letters in the same column indicate significant differences at the $p < 0.05$ level.

The free sulfhydryl content of walnut blank treated samples was 4.30 nmol/mmg, and it was already reduced by 48.37% at an AAPH concentration of 1 mmol/L. This may be because oxidation causes the unfolding of the protein structure, turning it into intramolecular and intermolecular disulfide bonds [32]. This suggests that a significant loss of free sulfhydryl groups had already occurred in walnut samples when the concentration of AAPH was 1 mmol/L and that this loss was linked to the development of protein aggregates. He et al. [33] studied the structural effect of heat treatment on quinoa and found that heat treatment increased the content of disulfide bonds and decreased the digestibility in vitro.

Another important marker of protein oxidation is amino acid composition [34]. The free amino acid content increased from 1.28 μmol/mg to 1.47 μmol/mg with increasing oxidation concentration but notably dropped at 1 mmol/L, with a fall of 23.1% from 0.2 mmol/L. The function of amino acids containing NH or NH_2 in the side chain in carbonyl group synthesis was supported by the decline in free amino acid content, which is likewise consistent with the rise in carbonyl content [35]. The results suggested that the reduction in the amount of free amino acids in walnut samples might be caused by their interaction with an oxidizing agent to produce more adducts.

3.2. Changes in Protein Structure after Oxidative Modifications

Protein tertiary structure alterations are frequently studied using spectroscopy. The UV spectrum of the protein is mainly derived from the absorption of the side chains of tryptophan and tyrosine residues at the UV wavelength of 280 nm [36]. Figure 1A shows that when the AAPH concentration was 1 mmol/L, the UV absorption intensity of protein at 280 nm increased significantly, indicating that the protein structure had been disrupted.

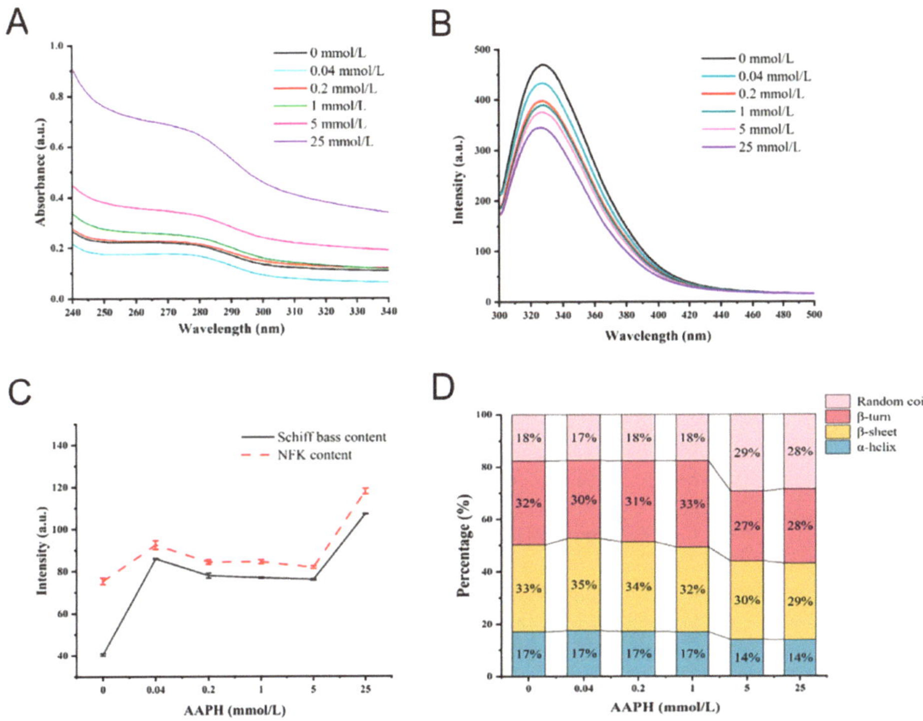

Figure 1. (**A**) UV absorption; (**B**) endogenous fluorescence; (**C**) Schiff base and NFk content; (**D**) secondary structure. Values of 0, 0.04, 0.2, 1, 5, and 25 represent various AAPH concentrations (unit: mmol/L), and different letters (a–d) indicate significant differences at $p < 0.05$.

Tryptophan is environmentally sensitive and highly susceptible to oxidation, and the endogenous fluorescence intensity typically demonstrates changes in tryptophan residues [37]. As shown in Figure 1B, the control sample had the highest fluorescence intensity of 431.55, and as the concentration of AAPH increased, the endogenous fluorescence intensity decreased to 344.88. This could be due to the low single-electron oxidation potential of tryptophan residues in the protein, which are susceptible to free radical capture and degradation to kynurenine, resulting in a decrease in the walnut protein's endogenous fluorescence intensity [38]. In addition, tryptophan's maximum absorption wavelength shifted blue, and the blue shift illustrates that as the concentration of AAPH increases, the non-polarity of the tryptophan microenvironment also increases, demonstrating that oxidative aggregation of the walnut protein leads to the return of unoxidized tryptophan in the protein molecule to the non-polar environment.

N′-formyl-L-kynurenine (NFk) is a tryptophan metabolism intermediate product [39]. The NFk content appeared at a higher level when the AAPH concentration was 0.04 mmol/L, as shown in Figure 1C, which may be the result of mild oxidative modification. However, the trends in NFk content and tryptophan intensity were not fully complementary, explaining that NFk is not the only tryptophan degradation product. This result is consistent with Li et al. [40] Schiff bases are organic compounds that have imine or methylimine character groups (-RC=N-) and are formed by the condensation of amines with reactive carbonyl groups [41]. In Figure 1C, Schiff bases show a similar trend to NFk, which may be due to the similar detection wavelengths of NFk and Schiff (440 nm for Schiff bases and 460 nm for

NFk), and it is consistent with the findings of Xi and Chen et al. [42]. Furthermore, protein carbonyl group formation can be covalently attached to the adjacent free amino group in the lysine residue, condensing the Schiff base and causing protein covalent cross-linking, confirming that protein carbonyl group formation is a manifestation of oxidative damage and aggregation.

The effect of oxidation on the secondary structure of walnut proteins was assessed using the band of amide I (1600–1700 cm^{-1}) in the protein's FTIR spectrum. It has been demonstrated that the α-helix content in the secondary structure is closely related to protein in vitro digestibility, and that it is positively correlated with in vitro digestibility [43]. Figure 1D shows that when the concentration of AAPH was 5 mmol/L, the proteins of the α-helix were visibly reduced, while the random coil was increased in the secondary structure. This demonstrates that as oxidation levels increase, the structure of proteins changes from orderly to disorderly, which may also have some effect on protein digestion.

3.3. *Effects of Oxidation on Protein Aggregation*

SDS-PAGE has the advantages of being fast, having a high resolution, and having high sensitivity, making it especially suitable for the determination of the molecular weight of proteins [44]. Figure 2A shows the profile of walnut protein incubated with different concentrations of AAPH, where the intensity bands of 10–15 kDa (Decomposition 1) and 17–20 kDa (Decomposition 2) deepened with increasing concentrations of AAPH (1 mmol/L), with apparent cross-linking of bands of 55–70 kDa (Cross-linking 2) and 30–35 kDa (Cross-linking 1) occurring when the concentration of AAPH was 1 mmol/L. Feng et al. [45] used reactive oxygen species-modified whey protein concentrate to conclude that mild oxidation can produce soluble aggregates, while increasing the oxidizing agents causes insoluble cross-linking. In this experiment, at the oxidizing agents of 0.04 and 0.2 mmol/L, the bands of 55–70 kDa were blurred more than in the control group, which may be the reason for having produced the soluble aggregates. On the other hand, when the oxidizing agent increased, the 55–70 kDa subunits were enhanced and the insoluble aggregates formed. The aforementioned phenomenon is consistent with the change in the free sulfhydryl group, as shown in Table 1. When AAPH was 0.5 mmol/L, free sulfhydryl decreased by 14.88%, and when AAPH was 1 mmol/L, it decreased by 48.37%. This implies that the decomposition of oxidized disulfide bonds is greater than the synthesis and decreases as the oxidant concentration increases. The formation of soluble aggregates may be caused by the Cys-S-S-linking formed by the reversible reaction of free sulfhydryl groups and cysteine.

Protein aggregates have previously been reported to form as a result of surface hydrophobic interactions [45]. As demonstrated in Figure 2B, the surface hydrophobicity falls with an AAPH concentration of 0.04 mmol/L and experiences a dramatic reduction at an AAPH concentration of 1 mmol/L. Peroxy radicals can cause the conversion of hydrophobic side-chain amino acid residues into hydrophilic groups, which can then be incorporated into the protein, resulting in the oxidation of amino acid residues in the protein and the unfolding and exposure of hydrophobic groups [37]. In Figure 1C, NFk content and Schiff base content were significantly increased when AAPH concentrations were below 1 mmol/L, indicating that tryptophan metabolism might be related to the formation of soluble aggregates. As a result of hydrophobic interactions, the exposed hydrophobic groups form aggregates, resulting in a decrease in surface hydrophobicity.

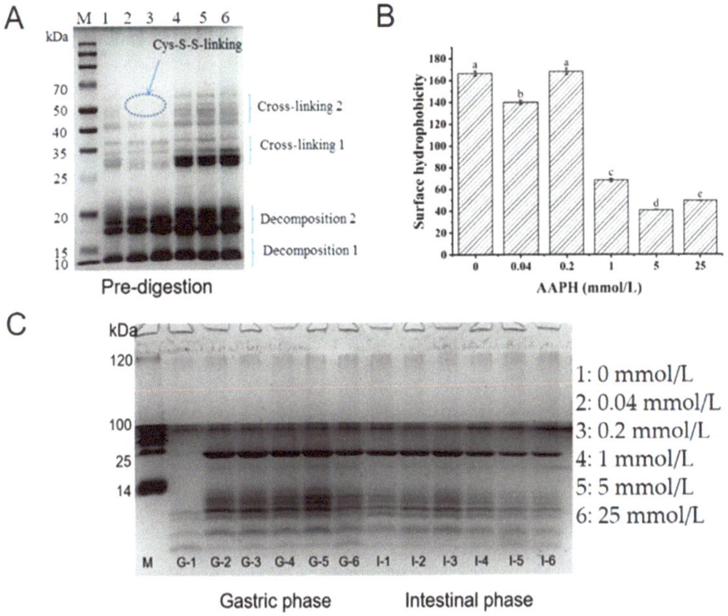

Figure 2. (**A**) Electrophoresis profiles of walnut isolated proteins after 24 h of incubation with different concentrations of AAPH, the circle shows the cross-linking of two sulfhydryl groups of cysteine due to oxidation; (**B**) surface hydrophobicity; and (**C**) small-molecule electrophoresis profiles of walnut isolated proteins after AAPH oxidation in gastric and subsequent intestinal digestion for 2 h, respectively, G: gastric phase, I: intestinal phase, and different letters (a–e) indicate significant differences at $p < 0.05$.

3.4. Protein Oxidation and Digestibility during Digestion In Vitro Simulation

The digestibility of walnut samples with varying degrees of oxidation in the process of gastric and subsequent intestinal digestion can be observed from the Tricine-SDS-PAGE profiles. Figure 2C shows that there were significant differences in the gastric phase between the control and oxidized proteins (G-1 with other channels), with the control sample (G-1) only having blurred bands with small molecules (14 kDa or less), and the band distribution of oxidized walnut proteins was extremely similar throughout the digestion time, but with an increasing intensity of bands (G-2 to G-5) with increasing oxidation degree in the gastric phase. The G-5 band is clearly deepened relative to the G-2 band as seen on the gel profile, which suggests that oxidation would have a resistant effect on the stomach due to the greater concentration of G-5 oxidation. The shallow band of the G-6 channel may be responsible for the excessive oxidation. Sante-L et al. [46] believe that oxidative aggregation of proteins reduces the recognition site of pepsin and thus reduces protein digestibility. In the intestinal phase, some interesting facts are that the intensity of the control band (I-1) is stronger than the gastric (G-1), and as the degree of oxidation increases, the bands (I-2 to I-6) become shallower. These phenomena are possibly due to the re-oxidation of the digestion phase, which can also explain the control band of I-1 with G-1, the hydrolysis of walnut in the gastric phase, and the change in sample structure with oxidation, which may result in more trypsin recognition sites and make the oxidation protein more digestible.

3.5. Dynamic Changes in Carbonyl Content during the Digestion of Walnut Protein

The amount of protein carbonyls may be an indicator of oxidative protein damage in the body [47]. The presence of re-oxidation during the digestion of walnut protein and the impact of different levels of oxidation on digestion were tested using the carbonyl content during digestion. When compared to the sample's carbonyl content without the digesting treatment (Table 1), we can see that Figure 3A–F with G-30 has a greater carbonyl content, and thus the pH is thought to be an easy trigger for the rise in oxidation-induced protein carbonyl concentration. It can also be noticed that the carbonyl content of late gastric digestion is reduced to a certain extent, which may be because the digestion concentration has changed. Additionally, after 90 min of intestinal digestion, carbonyl levels at nearly all oxidation concentrations increased. This finding may be related to the fact that the protein that is hydrolyzed into the active peptide is more likely to be oxidized. According to Van-H et al. [48], the digestion of red meat considerably raised the levels of protein carbonylation and lipid oxidation. Therefore, during digestion, proteins from both plants and meat can undergo further oxidation. Therefore, it is crucial to take into account the protein products' active oxidation protection.

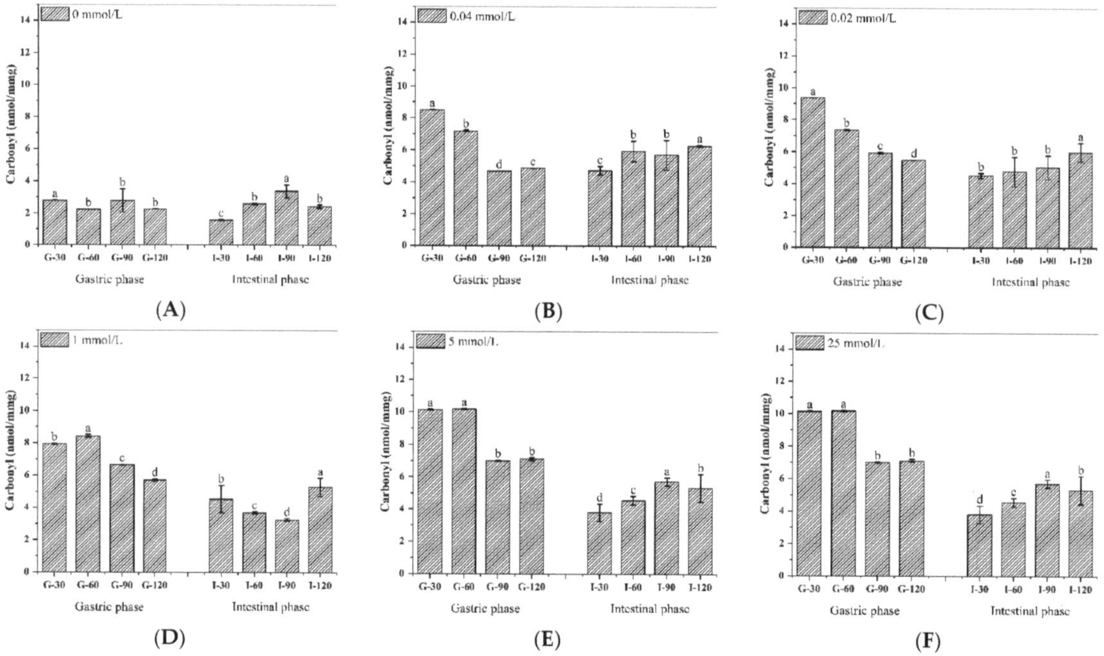

Figure 3. Carbonyl content of walnut isolated proteins after AAPH oxidation in the process of gastric and subsequent intestinal digestion at 30, 60, 90, and 120 min, respectively. (**A–F**) represent different incubation concentrations of AAPH, and different letters (a–d) indicate significant differences at $p < 0.05$.

3.6. *Antioxidant Capacity of Digestion Products*

Figure 4 depicts the antioxidant capacity of digestion products. The gastric digestion products of walnut proteins showed reduced scavenging activity against ABTS, DPPH, and hydroxyl radicals after oxidation by peroxyl radicals. However, at mild oxidation doses, hydroxyl radicals in the intestinal phase were slightly elevated, most likely due to changes in peptide fragments. Certain amino acids and their derivatives, such as Cys, His, Try, Met, Tyr, and Pro, for example, have antioxidant properties and can effectively scavenge free radicals [49]. Furthermore, hydroxyl radical antioxidant activity was lower in ABTS after intestinal digestion than in the gastric phase, most likely because the stomach digestion comprised peptides comprising the hydrophobic amino acids alanine, valine, and leucine, which act as antioxidants by binding to oxygen.

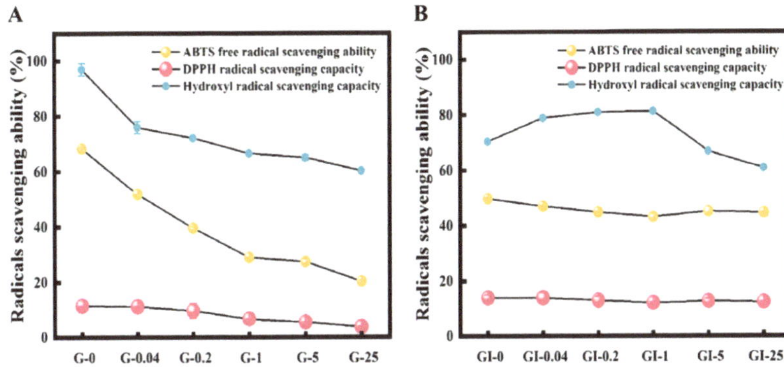

Figure 4. The antioxidant capacity of walnut isolated proteins after AAPH oxidation in gastric digestion (**A**) and subsequent intestinal digestion (**B**) for 2 h, respectively. G: gastric phase; GI: intestinal phase; 0, 0.04, 0.2, 1, 5, and 25 represent various AAPH concentrations (unit: mmol/L).

3.7. Mechanism of Selective Oxidation with Amino Acids

The quantity and composition of amino acids can affect the degree of protein absorption and utilization. Here, we have chosen a representative concentration that is typical (AAPH = 5 mmol/L). As shown in Figure 5, almost all amino acid levels were decreased to varying degrees during in vitro oxidation and digestion. The decay rates of amino acids in walnut samples before and after oxidation are ranked from high to low as: Cys, Met, Tyr, Lys, His, Val, and Pro, and the decay rates of amino acids in walnut samples before and after digestion are ordered from high to low as: Lys, Met, Pro, Gln, Ile, Ala, and Val. Fang et al. [13] showed that Cys, Met, Tyr, and Lys are vulnerable. The most frequent sites of oxidation were determined to be His, Lys, Pro, Arg, Met, and Cys by Stadtman et al. [50]. From the results of the present study, the oxidation of amino acids is selective, and cysteine is the most vulnerable amino acid and is typically the first to be oxidized, despite the fact that theoretically all amino acid side chains are vulnerable to free radical attack. The main byproduct of the oxidation of sulfur-containing amino acids such as Met is methionine sulfoxide, which accounts for the reduced Met level. Since residues such as Cys, Try, Met, Arg, and Lys are frequently the targets of free radicals, amino acids with active side chains, such as those found in sulfhydryl, thioether, amino, mitochondrial, and indole rings, are similarly vulnerable to oxidation.

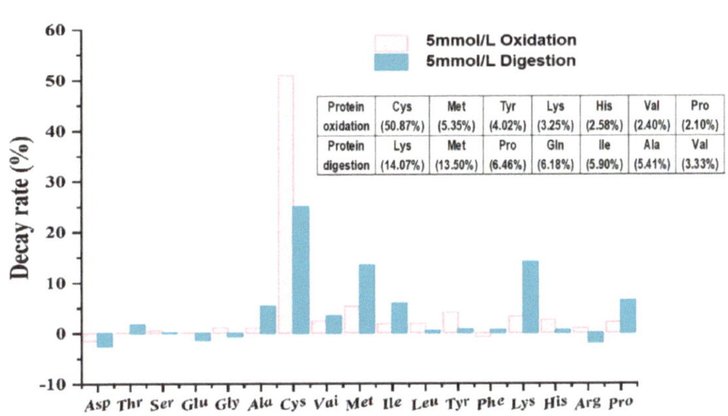

Figure 5. The decay rate of 17 amino acids for walnut isolated proteins after AAPH oxidation and digestion for 2 h at the concentration of AAPH is 5 mmol/L, respectively.

3.8. Antidigestive Protein Identification and Peptide Sequence Alignment

We later characterized the representative samples of G-0, G-5, and I-5 by using the LC-MS/MS technology. The identification results are shown in Table 2. In three samples, it obtained 91, 534, and 450 peptide sequences and identified 22, 31, and 25 proteins, respectively. The major proteins identified as resistant to gastric and intestinal digestion from the detected peptides were the 11S globulin seed storage protein, legumin, vicilin, 2S sulfur-rich seed storage protein, and other important seed storage proteins, such as oleosin, actin, glutaredoxin, salicylate carboxymethyltransferase-like, and ubiquitin-40S ribosomal protein S27a-like. It found that aquaporin, cell wall structural proteins, histone, ankyrin repeat-containing proteins, and some enzymes were resistant to pepsin after oxidative modification, but no corresponding peptides were detected at G-0 and GI-5. It is possible that the oxidative modifications are resistant to digestion in the stomach and may be digested further in the intestine, or that the oxidatively modified proteins promote trypsin hydrolysis, which corresponds to the results shown by Tricine-SDS-PAGE in the intestinal phase. As shown in the total ion chromatography in Figure 6, it can also be clearly seen that oxidation affects the digestibility of proteins, among which the sample of G-0 has the lowest detected abundance, indicating that the unoxidized sample has a higher degree of digestion, and G-5 and I-5 were less digested than the unoxidized sample based on the higher detected abundance. Subsequently, we performed a peptide sequence summary of the top 10 proteins identified, the results of which are available in the Supplementary File (Document S1). The gray-labeled sequences are the same peptide sequences as G-0, G-5, and I-5. From the sequence comparison of G-0 and G-5, it can be found that a large number of new peptides are produced by oxidation during gastric digestion, and some peptides are increased in the intestine, which indicates that oxidation may cause cross-linked aggregation, mask the pepsin recognition site, and thus produce resistance to gastric digestion.

Table 2. The identification results of walnut protein digestion products (G-0, G-5, and I-5) by LC-MS/MS G-0: gastric digestion that is not oxidized; G-5: gastric digestion at the AAPH concentration of 5 mmol/L; I-5: subsequent intestinal digestion at the AAPH concentration of 5 mmol/L; UP: unique peptides.

	ACCESSION	DESCRIPTION	MW	PI	G−0		G−5		I−5	
					UP	SCORE	UP	SCORE	UP	SCORE
1	XP_018818401.1	11S globulin seed storage protein 2-like	54.3	6.81	11	40.43	62	9.91	63	965.24
2	XP_018827137.1	11S globulin seed storage protein 2-like	53.4	7.8	2	8.39	20	5.51	18	186.72
3	XP_018827329.1	11S globulin seed storage protein Jug r 4	58.1	7.25	5	14.63	46	38.36	52	757.22
4	XP_018827328.1	11S globulin-like	58.3	7.77	13	45.10	66	119.56	59	908.08
5	XP_018842296.1	legumin B-like	55.8	7.5	12	74.54	71	12.29	64	1504.61
6	XP_018812171.1	vicilin Car i 2.0101	94.3	6.83	9	24.86	23	1048.8	13	73.32
7	XP_018814692.1	Vicilin Jug r 6.0101	57.4	7.3	5	19.99	22	30.3	15	73.04
8	XP_035546314.1	vicilin-like seed storage protein At2g28490	59.6	6.92	2	5.38	6	199.66	5	17.19
9	XP_018824007.1	2S sulfur-rich seed storage protein 2	17.1	6.52	5	27.03	3	2.7	3	65.45
10	XP_018837699.1	oleosin 18.2 kDa-like	16.6	9.91	2	7.39	4	2.63	3	5.78
11	XP_018842295.1	legumin B-like	55.8	8.18	1	28.82	2	21	2	399.8
12	XP_018856265.1	oleosin 1-like	14.7	10.14	4	8.72	1	2.2	1	2.92
13	XP_018842385.1	oleosin 5-like	16.3	10.08	2	5.3	5	0	2	2.46
14	XP_018814488.1	actin-97	41.7	5.49	3	6.52	5	7.3	3	8.54
15	XP_018821252.1	glutaredoxin-like	13.7	6.09	1	2.05	2	2.15	1	1.73
16	XP_035549483.1	salicylate carboxymethyltransferase-like	40.7	6.54	1	2.39	1	4.65	1	2.35
17	XP_018805239.1	ubiquitin-40S ribosomal protein S27a-like	17.7	9.79	1	2.09	2	3.09	1	2.87
18	XP_018810059.1	peroxygenase-like	26.9	5.87			2	2.6	2	2.17
19	XP_018831031.1	oleosin 1	14.8	9.63			1	0	1	2.62
20	XP_018809740.1	oleosin 18.2 kDa-like	15.9	9.72			3	0	1	0
21	XP_018818883.1	haloacid dehalogenase-like hydrolase domain-containing protein Sgpp	33.7	5.63	1	2.23			1	1.96
22	XP_018841715.1	histone H4-like	14.2	10.84	2	6.31	1	2.17		
23	XP_018812582.1	40S ribosomal protein S8-like	25.4	10.43	1	1.66				
24	XP_018845470.1	oleosin	15.6	9.54	1	1.94				
25	XP_018811862.1	aspartic proteinase-like isoform X1	57.9	5.39	1	2.04				
26	XP_018851532.1	probable aquaporin TIP3-2	27.4	7.71			1	5.54		
27	XP_018849277.1	probable aquaporin TIP3-2	27.4	7.66			1	3.21		
28	XP_018822092.1	protein FAM135B-like	91.8	8			1	874.88		
29	XP_035550528.1	putative glycine-rich cell wall structural protein 1 isoform X1	21.3	9.11			1	2.56		
30	XP_018841706.1	histone H2A	15.9	10.67			1	1.74		
31	XP_035543376.1	glyceraldehyde-3-phosphate dehydrogenase GAPCP2, chloroplastic-like	45.4	9.01			1	5.1		
32	XP_018826340.2	vicilin-like seed storage protein At2g18540	79	5.44			1	498.72		

Table 2. *Cont.*

	ACCESSION	DESCRIPTION	MW	PI	G−0		G−5		I−5	
					UP	SCORE	UP	SCORE	UP	SCORE
33	XP_018811454.1	11-beta-hydroxysteroid ehydrogenase B-like	40.9	6.35			2	2.1		
34	XP_018809799.1	ankyrin repeat-containing protein ITN1-like	25.6	7.46			1	2.79		
35	XP_018821986.2	malate synthase, glyoxysomal	64.2	7.55			1	128.13		
36	XP_018835846.1	polygalacturonase QRT3	52.5	6.54					1	0
37	XP_018847613.1	histone H2AX-like	14.9	10.36					2	4.05
38	XP_018826287.1	barwin-like	20.4	7.62					1	0
39	XP_018828748.1	elongation factor 1-alpha	49.5	9.07					1	2.35

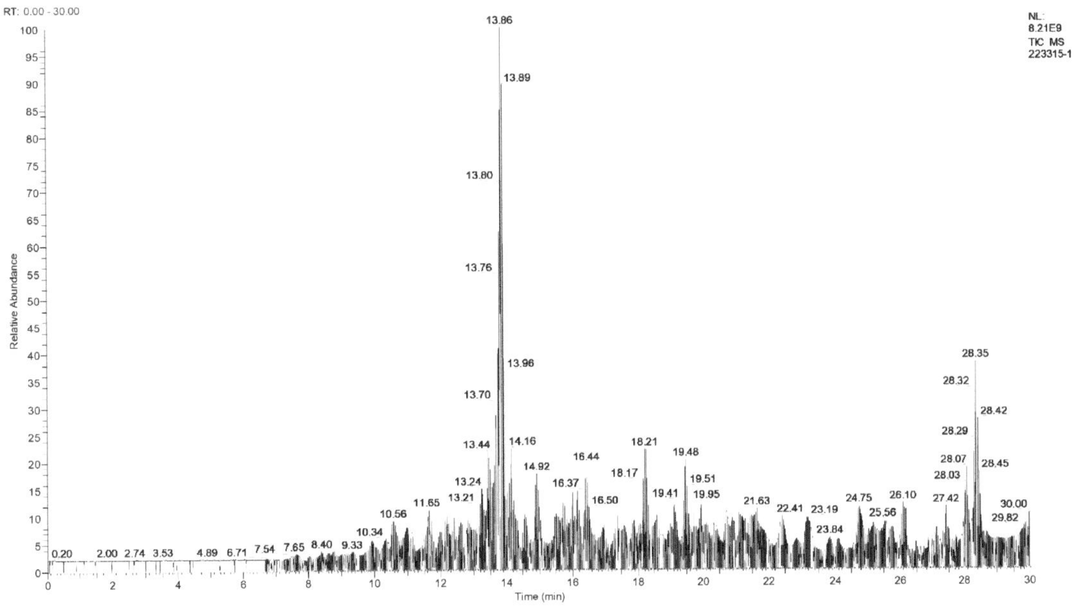

(**A**) Gastric–0 mmol/L

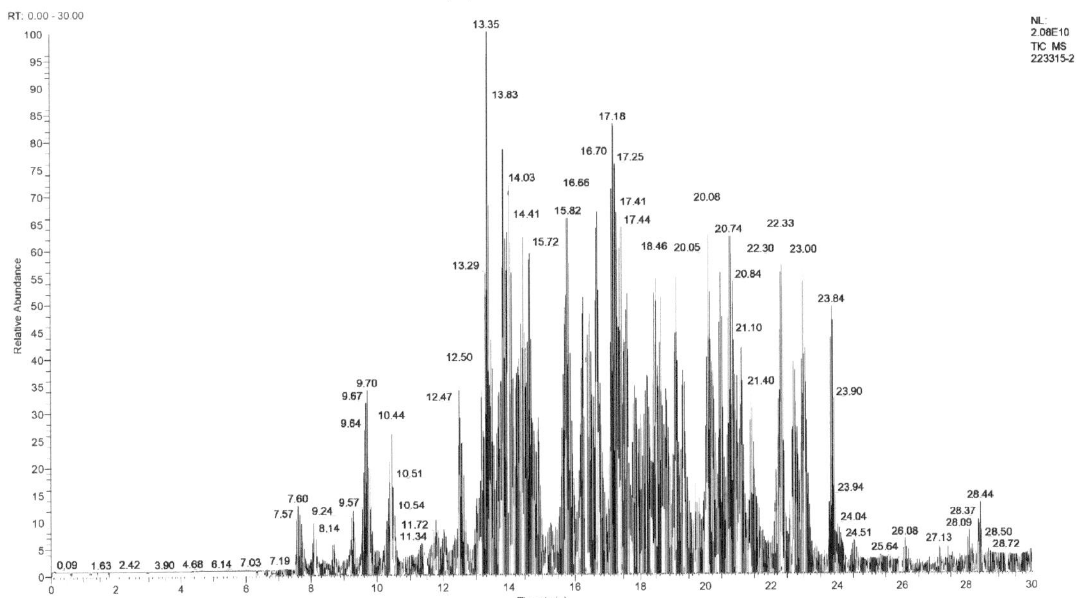

(**B**) Gastric–5 mmol/L

Figure 6. *Cont.*

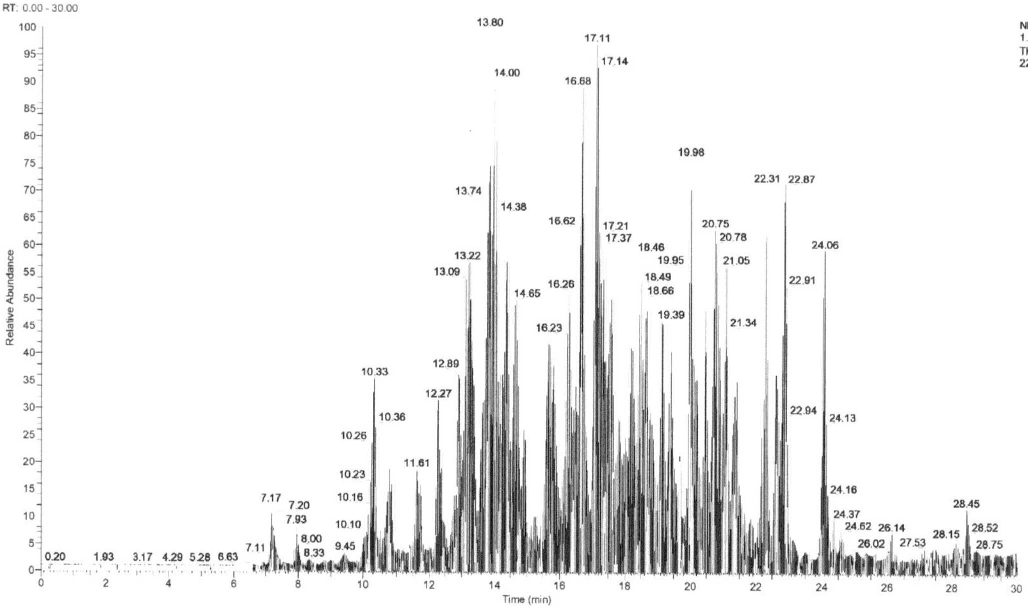

(C) Intestinal–5 mmol/L

Figure 6. Total ions (X: time, Y: intensity) of walnut protein samples. (**A**) Sample of gastric digestion that is not oxidized; (**B**) sample of gastric digestion at the AAPH concentration of 5 mmol/L; and (**C**) sample of subsequent intestinal digestion at the AAPH concentration of 5 mmol/L.

4. Conclusions

Alkoxy radical oxidation induced by AAPH causes a series of structural changes in proteins, such as a decrease in free amino groups, a loss of sulfhydryl groups, decreased surface hydrophobicity, and increased tryptophan metabolite NFk content and Schiff base content. In amino acid analysis, before and after oxidation and digestion, cysteine is usually the first amino acid to be oxidized, which also implies the importance of the disulfide bond in protein aggregates. Protein oxidation unfolding and aggregation were found to occur simultaneously in SDS-PAGE profiles, and the Cys-S-S-linking generated from the cysteine and free sulfhydryl group reactions and the increased content of NFk and Schiff bases may all be responsible for the soluble aggregation.

Protein oxidation leads to an increase in the carbonyl content, and the generation of the carbonyl groups is a complex process often associated with protein cross-linking and forming aggregates, which has an impact on digestion. The major proteins identified as resistant to gastric and intestinal digestion from the detected peptides were mostly seed storage protein. As a result, it is critical to investigate the properties of the oxidation and digestion products of walnut protein structure. The mechanism of walnut protein oxidation and digestion is beneficial to the quality of the protein processing process because it provides the foundation and direction for research into the digestion characteristics of walnut-based complex proteins.

Supplementary Materials: The following supporting information can be downloaded at: https://www.mdpi.com/article/10.3390/foods11244104/s1, Document S1: A peptide sequence summary of the top 10 proteins identified.

Author Contributions: Conceptualization, J.Z. (Jinjin Zhao); methodology, J.Z. (Jinjin Zhao) and M.H.; software, J.Z. (Jinjin Zhao) and M.H.; validation, J.Z. (Jinjin Zhao) and X.M.; formal analysis, J.Z. (Jinjin Zhao) and Q.W.; investigation, J.Z. (Jinjin Zhao), M.H. and Z.L.; resources, X.M. and J.Z.

(Jian Zhang); data curation, J.Z. (Jinjin Zhao); writing—original draft preparation, J.Z. (Jinjin Zhao) and X.M.; writing—review and editing, J.Z. (Jinjin Zhao) and X.M.; visualization, J.Z. (Jinjin Zhao); supervision, X.M.; project administration, X.M.; funding acquisition, X.M. and Q.W. All authors have read and agreed to the published version of the manuscript.

Funding: This research was funded by the National Natural Science Foundation of China (No. 31560433); Scientific and Technological Research Projects in the Key Areas of the Xinjiang Production and Construction Corps of China (No. 2018AB015), Key Laboratory for Nutrition and Safety of Xinjiang Specialty Foods, Shihezi City of Eighth Division (grant number 2022PT02), and Innovative team-building plan for key field projects in Xinjiang Production and Construction Corps (grant number 2019CB007).

Data Availability Statement: Data is contained within the article or Supplementary Material.

Conflicts of Interest: The authors declare no conflict of interest.

References

1. Sze-Tao, K.W.C.; Sathe, S.K. Walnuts (*Juglans regia* L): Proximate composition, protein solubility, protein amino acid composition and protein in vitro digestibility. *J. Sci. Food Agric.* **2000**, *80*, 1393–1401. [CrossRef]
2. Oke, O.; Oyaniyi, T.; Adewumi, O.; Bamigboye, O.; Lawah, M.; Jatto, K.; Adara, C.; Marizu, J.; Ogunbela, A. Economic, nutritional and medicinal values of African walnut (*Tetracarpidium conophorum*) in Nigeria (Hutch. & Dalziel): A review. *J. Res. For. Wildl. Environ.* **2020**, *12*.
3. Mao, X.-Y.; Hua, Y.-F. Chemical composition, molecular weight distribution, secondary structure and effect of NaCl on functional properties of walnut (*Juglans regia* L) protein isolates and concentrates. *J. Food Sci. Technol.-Mysore* **2014**, *51*, 1473–1482. [CrossRef] [PubMed]
4. Chen, Y.; Li, J.; Dong, N.; Zhang, Y.; Lu, X.; Hao, Y.; Qi, J. Separation and identification of ACE inhibitory peptides from defatted walnut meal. *Eur. Food Res. Technol.* **2020**, *246*, 2029–2038. [CrossRef]
5. Mao, X.; Wang, D.; Sun, L.; Zhang, J.; Wu, Q. Effect of peroxyl-radicals-induced oxidative modification in the physicochemical and emulsifying properties of walnut protein. *J. Am. Oil Chem. Soc.* **2021**, *98*, 903–910. [CrossRef]
6. Sun, L.; Wu, Q.; Mao, X. Effects of oxidation modification by malondialdehyde on the structure and functional properties of walnut protein. *Foods* **2022**, *11*, 2432. [CrossRef]
7. Xi, J. Research progress of nutritional diet for the disabled elderly. *Int. J. Front. Med.* **2021**, *3*. [CrossRef]
8. Li, F.; Wu, X.; Wu, W. Effects of oxidative modification by malondialdehyde on the in vitro digestion properties of rice bran protein. *J. Cereal Sci.* **2021**, *97*, 103158. [CrossRef]
9. Zhao, J.; Su, G.; Zhao, M.; Sun, W. Physicochemical changes and in vitro gastric digestion of modified soybean protein induced by lipoxygenase catalyzed linoleic acid oxidation. *J. Agric. Food Chem.* **2019**, *67*, 13978–13985. [CrossRef]
10. van Lieshout, G.A.A.; Lambers, T.T.; Bragt, M.C.E.; Hettinga, K.A. How processing may affect milk protein digestion and overall physiological outcomes: A systematic review. *Crit. Rev. Food Sci. Nutr.* **2020**, *60*, 2422–2445. [CrossRef]
11. Goodman, B.E. Insights into digestion and absorption of major nutrients in humans. *Adv. Physiol. Educ.* **2010**, *34*, 44–53. [CrossRef] [PubMed]
12. Hu, L.; Ren, S.; Shen, Q.; Ye, X.; Chen, J.; Ling, J. Protein oxidation and proteolysis during roasting and in vitro digestion of fish (Acipenser gueldenstaedtii). *J. Sci. Food Agric.* **2018**, *98*, 5344–5351. [CrossRef] [PubMed]
13. Fang, M.; Xiong, S.; Yin, T.; Hu, Y.; Liu, R.; Du, H.; Liu, Y.; You, J. Proteomic profiling and oxidation site analysis of gaseous ozone oxidized myosin from silver carp (*Hypophthalmichthys molitrix*) with different oxidation degrees. *Food Chem.* **2021**, *363*, 130307. [CrossRef] [PubMed]
14. Zhang, J.J.; Tu, Z.C.; Wang, H.; Hu, Y.M.; Du, P.C.; Yang, Y.P. Mechanism of the effect of 2,2′-azobis (2-amidinopropane) dihydrochloride simulated lipid oxidation on the IgG/IgE binding ability of ovalbumin. *Food Chem.* **2020**, *327*, 127037. [CrossRef]
15. Wang, C.; Zhao, F.; Bai, Y.; Li, C.; Xu, X.; Kristiansen, K.; Zhou, G. Effect of gastrointestinal alterations mimicking elderly conditions on in vitro digestion of meat and soy proteins. *Food Chem.* **2022**, *383*, 132465. [CrossRef]
16. Yin, H.Y.; Xu, L.B.; Porter, N.A. Free radical lipid peroxidation: Mechanisms and analysis. *Chem. Rev.* **2011**, *111*, 5944–5972. [CrossRef]
17. Zhu, Z.B.; Zhu, W.D.; Yi, J.H.; Liu, N.; Cao, Y.G.; Lu, J.L.; Decker, E.A.; McClements, D.J. Effects of sonication on the physicochemical and functional properties of walnut protein isolate. *Food Res. Int.* **2018**, *106*, 853–861. [CrossRef]
18. Wang, Z.M.; He, Z.F.; Gan, X.; Li, H.J. Effect of peroxyl radicals on the structure and gel properties of isolated rabbit meat myofibrillar proteins. *Int. J. Food Sci. Technol.* **2018**, *53*, 2687–2696. [CrossRef]
19. Wu, X.J.; Li, F.; Wu, W. Effects of rice bran rancidity on the oxidation and structural characteristics of rice bran protein. *Lwt-Food Sci. Technol.* **2020**, *120*, 108943. [CrossRef]
20. Fu, Q.; Liu, R.; Wang, H.; Hua, C.; Song, S.; Zhou, G.; Zhang, W. Effects of oxidation in vitro on structures and functions of myofibrillar protein from beef muscles. *J. Agric. Food Chem.* **2019**, *67*, 5866–5873. [CrossRef]

21. Chen, N.N.; Zhao, Q.Z.; Sun, W.Z.; Zhao, M.M. Effects of malondialdehyde modification on the in vitro digestibility of soy protein isolate. *J. Agric. Food Chem.* **2013**, *61*, 12139–12145. [CrossRef] [PubMed]
22. Zhu, Z.F.; Mao, X.Y.; Wu, Q.Z.; Zhang, J.; Deng, X.R. Effects of oxidative modification of peroxyl radicals on the structure and foamability of chickpea protein isolates. *J. Food Sci.* **2021**, *86*, 824–833. [CrossRef] [PubMed]
23. Wu, W.; Li, F.; Wu, X. Effects of rice bran rancidity on oxidation, structural characteristics and interfacial properties of rice bran globulin. *Food Hydrocoll.* **2021**, *110*, 106123. [CrossRef]
24. Hellwig, M. Analysis of Protein oxidation in food and feed products. *J. Agric. Food Chem.* **2020**, *68*, 12870–12885. [CrossRef] [PubMed]
25. Brodkorb, A.; Egger, L.; Alminger, M.; Alvito, P.; Assuncao, R.; Ballance, S.; Bohn, T.; Bourlieu-Lacanal, C.; Boutrou, R.; Carriere, F.; et al. INFOGEST static in vitro simulation of gastrointestinal food digestion. *Nat. Protoc.* **2019**, *14*, 991–1014. [CrossRef] [PubMed]
26. Xu, Z.C.; Chen, Y.M.; Zhang, C.M.; Kong, X.Z.; Hua, Y.F. The heat-induced protein aggregate correlated with trypsin inhibitor inactivation in soymilk processing. *J. Agric. Food Chem.* **2012**, *60*, 8012–8019. [CrossRef]
27. Kuipers, B.J.H.; Gruppen, H. Prediction of molar extinction coefficients of proteins and peptides using UV absorption of the constituent amino acids at 214 nm to enable quantitative reverse phase high-performance liquid chromatography-mass spectrometry analysis. *J. Agric. Food Chem.* **2007**, *55*, 5445–5451. [CrossRef]
28. Georgiou, C.D.; Zisimopoulos, D.; Argyropoulou, V.; Kalaitzopoulou, E.; Salachas, G.; Grune, T. Protein and cell wall polysaccharide carbonyl determination by a neutral pH 2,4-dinitrophenylhydrazine-based photometric assay. *Redox Biol.* **2018**, *17*, 128–142. [CrossRef]
29. Soudani, N.; Ben Amara, I.; Troudi, A.; Hakim, A.; Bouaziz, H.; Makni, F.A.; Zeghal, K.M.; Zeghal, N. Oxidative damage induced by chromium (VI) in rat erythrocytes: Protective effect of selenium. *J. Physiol. Biochem.* **2011**, *67*, 577–588. [CrossRef]
30. Werber, J.; Wang, Y.J.; Milligan, M.; Li, X.H.; Ji, J.A. Analysis of 2,2′-azobis (2-amidinopropane) dihydrochloride degradation and hydrolysis in aqueous solutions. *J. Pharm. Sci.* **2011**, *100*, 3307–3315. [CrossRef]
31. Gomez-Pastor, R.; Perez-Torrado, R.; Cabiscol, E.; Ros, J.; Matallana, E. Engineered Trx2p industrial yeast strain protects glycolysis and fermentation proteins from oxidative carbonylation during biomass propagation. *Microb. Cell Fact.* **2012**, *11*, 4. [CrossRef] [PubMed]
32. Kunath, S.; Schindeldecker, M.; De Giacomo, A.; Meyer, T.; Sohre, S.; Hajieva, P.; von Schacky, C.; Urban, J.; Moosmann, B. Prooxidative chain transfer activity by thiol groups in biological systems. *Redox Biol.* **2020**, *36*, 101628. [CrossRef] [PubMed]
33. He, X.; Wang, B.; Zhao, B.; Yang, F. Ultrasonic Assisted extraction of quinoa (Chenopodium quinoa Willd.) Protein and effect of heat treatment on its in vitro digestion characteristics. *Foods* **2022**, *11*, 771. [CrossRef] [PubMed]
34. Moldogazieva, N.T.; Mokhosoev, I.M.; Mel'nikova, T.I.; Porozov, Y.B.; Terentiev, A.A. Oxidative Stress and advanced lipoxidation and glycation end products (ALEs and AGEs) in Aging and age-related diseases. *Oxidative Med. Cell. Longev.* **2019**, *2019*, 3085756. [CrossRef]
35. Pan, J.; Lian, H.; Jia, H.; Hao, R.; Wang, Y.; Ju, H.; Li, S.; Dong, X. Dose affected the role of gallic acid on mediating gelling properties of oxidatively stressed Japanese seerfish myofibrillar protein. *LWT-Food Sci. Technol.* **2020**, *118*, 108849. [CrossRef]
36. Wang, P.-Y.; Yang, C.-T.; Chu, L.-K. Differentiating the protein dynamics using fluorescence evolution of tryptophan residue(s): A comparative study of bovine and human serum albumins upon temperature jump. *Chem. Phys. Lett.* **2021**, *781*, 138998. [CrossRef]
37. Wu, W.; Zhang, C.M.; Kong, X.Z.; Hua, Y.F. Oxidative modification of soy protein by peroxyl radicals. *Food Chem.* **2009**, *116*, 295–301. [CrossRef]
38. Wang, Z.; He, Z.; Gan, X.; Li, H. Interrelationship among ferrous myoglobin, lipid and protein oxidations in rabbit meat during refrigerated and superchilled storage. *Meat Sci.* **2018**, *146*, 131–139. [CrossRef]
39. Wang, Q.X.; Liu, D.X.; Song, P.; Zou, M.H. Tryptophan-kynurenine pathway is dysregulated in inflammation, and immune activation. *Front. Biosci.-Landmark* **2015**, *20*, 1116–1143. [CrossRef]
40. Li, Q.; Zheng, J.; Ge, G.; Zhao, M.; Sun, W. Impact of heating treatments on physical stability and lipid-protein co-oxidation in oil-in-water emulsion prepared with soy protein isolates. *Food Hydrocoll.* **2020**, *100*, 105167. [CrossRef]
41. Scheidegger, D.; Pecora, R.P.; Radici, P.M.; Kivatinitz, S.C. Protein oxidative changes in whole and skim milk after ultraviolet or fluorescent light exposure. *J. Dairy Sci.* **2010**, *93*, 5101–5109. [CrossRef] [PubMed]
42. Xi, J.; Chen, Y. Analysis of the relationship between heterocyclic amines and the oxidation and thermal decomposition of protein using the dry heated soy protein isolate system. *LWT* **2021**, *148*, 111738. [CrossRef]
43. Yang, Y.; Wang, Z.; Wang, R.; Sui, X.; Qi, B.; Han, F.; Li, Y.; Jiang, L. Secondary structure and subunit composition of soy protein in vitro digested by pepsin and its relation with digestibility. *BioMed. Res. Int.* **2016**, *2016*, 5498639. [CrossRef]
44. Sharma, N.; Sharma, R.; Rajput, Y.S.; Mann, B.; Singh, R.; Gandhi, K. Separation methods for milk proteins on polyacrylamide gel electrophoresis: Critical analysis and options for better resolution. *Int. Dairy J.* **2021**, *114*, 104920. [CrossRef]
45. Feng, Y.L.; Wang, J.M.; Hu, H.Y.; Yang, C. Effect of oxidative modification by reactive oxygen species (ROS) on the aggregation of whey protein concentrate (WPC). *Food Hydrocoll.* **2022**, *123*, 107189. [CrossRef]
46. Sante-Lhoutellier, V.; Aubry, L.; Gatellier, P. Effect of oxidation on in vitro digestibility of skeletal muscle myofibrillar proteins. *J. Agric. Food Chem.* **2007**, *55*, 5343–5348. [CrossRef] [PubMed]
47. Dalle-Donne, I.; Rossi, R.; Giustarini, D.; Milzani, A.; Colombo, R. Protein carbonyl groups as biomarkers of oxidative stress. *Clin. Chim. Acta* **2003**, *329*, 23–38. [CrossRef]

48. Van Hecke, T.; Vossen, E.; Bussche, J.V.; Raes, K.; Vanhaecke, L.; De Smet, S. Fat Content and nitrite-curing influence the formation of oxidation products and NOC-specific DNA adducts during in vitro digestion of meat. *PLoS ONE* **2014**, *9*, e0101122. [CrossRef]
49. Karami, Z.; Akbari-adergani, B. Bioactive food derived peptides: A review on correlation between structure of bioactive peptides and their functional properties. *J. Food Sci. Technol.* **2019**, *56*, 535–547. [CrossRef]
50. Stadtman, E.R. Oxidation of free amino acids and amino acid residues in proteins by radiolysis and by metal-catalyzed reactions. *Annu. Rev. Biochem.* **1993**, *62*, 797–821. [CrossRef]

foods

Article

Ultrasound-Assisted Extraction of Protein from Pumpkin Seed Press Cake: Impact on Protein Yield and Techno-Functionality

Deniz Sert, Harald Rohm * and Susanne Struck

Chair of Food Engineering, Institute of Natural Materials Technology, Technische Universität Dresden, 01062 Dresden, Germany
* Correspondence: harald.rohm@tu-dresden.de

Abstract: Conventional solvent-based methods widely used for isolating plant proteins may deliver an unsatisfactory protein yield and/or result in protein degradation. The present study started with the optimization of pumpkin seed protein from press cake by alkaline extraction and subsequent isoelectric precipitation. Subsequently, extraction was supported by ultrasound under three conditions: ultrasonic treatment followed by alkaline extraction (US+AE), concomitant ultrasonic treatment and alkaline extraction (UAE), and alkaline extraction followed by ultrasonic treatment (AE+US). Compared to the control group, an increase in protein yield was achieved after ultrasonic treatment, while the highest protein yield was obtained with AE+US (57.8 ± 2.0%). Isolates with a protein content of 94.04 ± 0.77 g/100 g and a yield of 43.6 ± 0.97% were obtained under optimized conditions. Following ultrasonic treatment applied during extraction, solubility, foaming capacity, foam stability, and denaturation enthalpy of the isolated protein increased, and water binding capacity decreased as compared to non-sonicated samples. The d_{90} particle size percentile of the extracted suspensions was 376.68 ± 38.32 μm for the control experiments, and particle size was significantly reduced in ultrasound-assisted treatments down to d_{90} = 179.93 ± 13.24 μm for the AE+US treatment). Generally, ultrasonication resulted in a significant increase in protein yield and improved techno-functional properties of the isolates.

Keywords: ultrasonic treatment; alkaline extraction; solubility; foaming stability; foaming capacity

Citation: Sert, D.; Rohm, H.; Struck, S. Ultrasound-Assisted Extraction of Protein from Pumpkin Seed Press Cake: Impact on Protein Yield and Techno-Functionality. *Foods* **2022**, *11*, 4029. https://doi.org/10.3390/foods11244029

Academic Editor: David Bongiorno

Received: 21 November 2022
Accepted: 10 December 2022
Published: 13 December 2022

1. Introduction

The demand for food protein has increased in recent years as consumers become more aware of the need for a sufficient intake. As a consequence, the search for plant-based protein has become more intense [1], also because of critical aspects concerning animal proteins, such as animal welfare and high greenhouse gas emissions in industrial meat production [2,3]. Although residues of plant food production occur irregularly in the annual production cycle and require further processing, they can be considered prospective sources for protein extraction [4]. Using agro-industrial production losses for protein recovery might be a long-term solution that improves the economic potential of these by-products [5].

The oilseed press cake is a by-product of vegetable oil processing and contains a considerable amount of protein. Canola and sunflower seeds are the most popular and best-studied oilseeds [6], but pumpkin seeds can also be considered a valuable source of protein [7]. Pumpkin is a plant within the *Cucurbitaceae* family, mainly cultivated in Asia and several European regions. According to the Food and Agriculture Organization of the United Nations, worldwide pumpkin production reached over 27 million tons in 2020—63% from Asia, with China (approx. 7.5 megatons) being the largest producer [8]. Oil pumpkin is a special mutant variety with a tasteless and almost unpalatable flesh, exclusively cultivated for gaining the seeds, which are only covered by a thin membrane. A total annual seed harvest of approx. 75,000 tons obtained from ~100,000 hectares in Middle and Eastern Europe is used for the production of cold-pressed pumpkin oil with a yield of ~0.4 L/kg. The oil is sold at a price from 20 €/L upwards and, for a region in the southeast of Austria, registered as a "Protected Geographical Indication" (PGI) product ("Steirisches Kürbiskernöl") by the European Union [9]. The popularity of pumpkin seeds for direct consumption is also progressively growing [10] since they are an excellent source of protein (content: 24.5–36 g/100 g), unsaturated fatty acids and secondary plant metabolites [11]. It is, for instance, recommended by the American Heart Association to consume approximately 30 g of pumpkin seeds per day due to its content of several nutrients showing positive effects on heart and bone health [12]. After removing residual oil from the pumpkin press cake, the protein content of the de-oiled fraction may increase to up to 65 g/100 g [13].

Extraction and precipitation techniques exhibit a significant impact on the structure and techno-functionality of the isolated protein and hence, on the applicability in processed foods. Water absorption capacity, amino acid composition, molecular shape and mass, net charge, size, solubility, isoelectric point, heat stability, hydrophobicity, and emulsification properties are just several of the important parameters that define the processing properties of plant proteins [3].

Ultrasound is a technology frequently used in protein extraction from plant-based sources [1,14,15]. Ultrasound-assisted extraction has several advantages, including improved yield, lower processing time, reduced solvent consumption, and minimal environmental influence [15–17]. Ultrasonic sound waves with a frequency of approx. 20 kHz usually result in severe cavitation effects. The energy released by the collapse of bubbles encourages the solvent to penetrate deeper into the suspended cell material, facilitating the mass transfer to and from the interface [18]. Studies using ultrasound support for protein extraction were, for instance, performed on barley [19], pea [20], peanut [1,21], canola [22], sunflower meal and seeds [16,23,24], rice bran [25,26], and walnut [27]. Several recent studies also showed that ultrasound application improves protein yield and functionality [19,21,28–30]. For instance, it was shown that ultrasound-assisted alkali extraction improves foaming and emulsifying capacity, solubility, and gel formation capacity [20,24].

Although there are many studies dealing with protein isolation from different plant sources, only little research is available on pumpkin seeds. The first aim of the current study was to improve the extraction of pumpkin seed protein from press cake by determining optimum parameters, with special emphasis on the protein content of the isolates and protein yield. Based on these preliminary findings, the effects of ultrasonic support during extraction on the physicochemical and techno-functional properties of the protein isolates were examined. Unlike previous research, the intention of our study was to assess the impact of ultrasound when introduced at different stages of the extraction procedure.

2. Materials and Methods

2.1. Materials

Milled press cake from pumpkin (*Cucurbita pepo* L. subsp. pepo var. styriaca GREB.) seed oil production was kindly provided by Estyria Naturprodukte GmbH (St. Ruprecht an der Raab, Austria). Prior to protein isolation, the fraction with a particle size of 300–600 μm was suspended in hexane at a ratio of 1:3 (w/v) for 40 min at room temperature while stirring at 500 rpm (MS-MP8, Witeg Labortechnik GmbH, Germany) [31]. After filtering the slurry through a Whatman #1 filter paper and repeating the hexane extraction twice, the de-oiled material was dried for 16 h at room temperature and stored at 6 °C until use.

2.2. Gross Composition Analysis

Protein and fat content, dietary fiber, moisture, and ash content of untreated and de-oiled pumpkin seed press cake powder were determined in duplicate using standard analytical techniques [32]. Protein content was determined by the Kjeldahl method (nitrogen conversion factor: 6.25), and fat content was determined by Soxhlet extraction with petroleum ether. Based on AOAC 991.43, dietary fiber was analyzed using an enzyme kit

(Megazyme Ltd., Bray, Ireland). Moisture content was assessed by drying at 105 °C, and ash content was determined gravimetrically after incineration at 550 °C.

2.3. Protein Isolation

2.3.1. Control Procedure

To identify the most appropriate conditions for the extraction of protein from milled press cake, dispersions in deionized water were adjusted to pH 8.0, 9.5, or 11.0 by using 0.5 N NaOH at a solvent-to-solid ratio of 10:1, 20:1, or 30:1, and then stirred at room temperature for 1 h, 2 h, or 3 h at 500 rpm on a magnetic stirrer. After centrifugation in a Biofuge Stratos (Thermo Scientific, Dreieich, Germany) at 8,000 rpm for 20 min, the supernatant was precipitated at pH 4.5, using HCl, and kept at room temperature for 1 h. After a second centrifugation step, the pellet was dissolved in a small amount of deionized water, pH was set to 7.0, followed by freeze drying (Beta 1–8 LDplus, Martin Christ Gefriertrocknungsanlagen GmbH, Osterode, Germany). The protein content of the isolate was determined by the Kjeldahl method, and protein yield calculated by equation [1] served as the second target variable:

$$\text{Protein yield (\%)} = \frac{m_i \times c_{pi}}{m_s \times c_{ps}} \times 100\ \% \tag{1}$$

m_i and m_s refer to the mass of the isolate and the initial sample, respectively. The protein content of the isolate is given by c_{pi}, while the protein content of the initial sample is indicated by c_{ps}.

To minimize the potential of unexpected deviation, the experimental Box–Behnken design contained 17 runs with five repetitions at the center point, and the response data of protein yield and protein content were analyzed using a quadratic model.

2.3.2. Protein Extraction with Ultrasonic Support

Ultrasonic support during protein extraction from press cake was realized by using three approaches:

- ultrasonic treatment followed by alkaline extraction (US+AE),
- concomitant ultrasonic treatment and alkaline extraction (UAE), and
- alkaline extraction followed by ultrasonic treatment (AE+US).

Ultrasonic treatments were carried out in four independent trials per condition, performed on two days (Figure 1), using a 25 kHz UDS 751/UP 200 S sonication unit with a 40 mm diameter sonotrode (Topas GmbH, Dresden, Germany) operated at a power density of 600 W/cm^2 and 50% amplitude.

A solvent-to-solid ratio of 29, which was determined to be optimum in the preliminary control experiments, was applied in all treatments. Prior to ultrasonic support, dispersions were stirred on a magnetic stirrer at room temperature for 10 min at 500 rpm. In the US+AE treatment, pH was adjusted to the pre-determined optimum (pH 11) after 10 min of continuous ultrasonic application, and then the alkaline extraction was carried out for 60 min. In the UAE treatment, ultrasound was continuously applied for 10 min at optimum pH and solvent-to-solid ratio. In AE+US treatment, the alkaline extraction was carried out under the optimum conditions, followed by 10 min of continuous ultrasonic treatment. In every case, an ice bath was used to keep the temperature at 20 °C. All the treatments were followed by isoelectric precipitation and lyophilization of the isolate. The protein content of each isolate was determined in duplicate. The protein isolates obtained from the two runs performed on a single day were then pooled before further analysis.

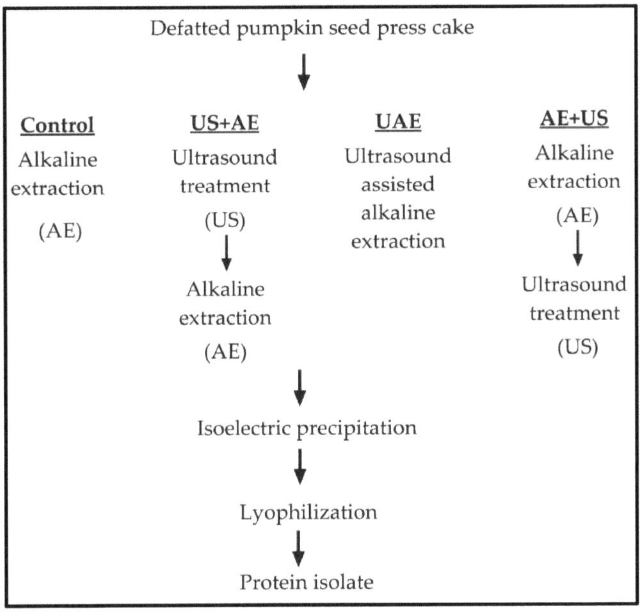

Figure 1. Outline of the protein isolation processes.

2.4. Particle Size Distribution in Alkaline Extracts

Following one day of storage at 6 °C, the size distributions of the particles in the dispersions after alkaline extraction were measured by laser diffraction (Helos KR with Sucell, Sympatec GmbH, Clausthal–Zellerfeld, Germany) at pH 7.0. Measurements were carried out in the range of 10–15% optical density. After 2 min equilibration under stirring, size distributions were recorded, and the diameters at 10% (d_{10}), 50% (d_{50}), and 90% (d_{90}) volume fractions were taken as particle size indicators. Analysis of each dispersion (four independent extractions per condition) was carried out in triplicate.

2.5. Determination of Techno-Functional Properties

2.5.1. Protein Solubility

Protein solubility was determined using the method of Gonzalez–Perez [33] with modification. Each protein isolate (pooled samples) was dissolved in deionized water at a final concentration of 5 mg/mL, and solution pH was adjusted to 3, 5, 7, or 9 using 0.5 N NaOH or 0.5 N HCl. Prior to centrifugation at 6,000 rpm for 15 min, the respective solutions were agitated for 2 h at room temperature and 300 rpm. The Kjeldahl procedure was used to determine protein solubility, defined as the amount of protein in the supernatant (p_2) related to the amount of protein in the initial sample (p_1):

$$\text{Solubility } (\%) = \frac{p_2}{p_1} \times 100\% \tag{2}$$

2.5.2. Water Binding Capacity

The method developed by Chen et al. [34] was used to determine the water binding capacity (WBC) of each pooled protein isolate in duplicate. Thirty mL of deionized water was added to 0.5 ± 0.01 g protein isolate and mixed for 30 s. After shaking in a horizontal position for 30 min at 20 °C, the mixture was centrifuged at $4000\times g$ for 10 min. The dry matter of the supernatant was analyzed using an MA30 moisture analyzer (Sartorius AG, Göttingen, Germany) at 95 °C to determine the amount of soluble protein. WBC refers to the quantity of water bound per gram of dry protein that remains in the centrifuged sediment.

2.5.3. Foaming Properties

To induce foam formation, deionized water was added to 0.5 ± 0.01 g of each protein isolate in duplicate, bringing the total volume to 50 mL (V_2). The dispersion was homogenized for 2 min at 20,000 rpm using a T25 dispersion unit (IKA–Werke GmbH & CO. KG, Staufen, Germany). The samples were then immediately transferred into a measuring cylinder, where the total volume (V_1) was measured. The foaming capacity was calculated according to Moure et al. [35]:

$$\text{Foaming capacity } (\%) = \frac{V_1 - V_2}{V_2} \times 100\% \tag{3}$$

The foam volume was determined at t = 0 min (V_0) and after 10, 30, 60, 90, and 120 min (V_t) to calculate foaming stability:

$$\text{Foaming stability } (\%) = \frac{V_t}{V_0} \times 100\% \tag{4}$$

2.6. *Different Scanning Calorimetry (DSC)*

A Discovery DSC25 differential scanning calorimeter connected to an RCS90 cooling unit (TA Instruments, Eschborn, Germany) was used to record thermograms. Gallium and indium were used as standards for instrument calibration, and nitrogen at a flow rate of 50 mL/min served as purge gas.

An empty standard aluminum pan served as control, and approx. 5 mg sample of each pooled protein isolate was weighed into the sample pan in duplicate. Following equilibration, the temperature was adjusted to 0 °C and then increased to 200 °C at a rate of 10 K/min. Measurements were made in duplicate. Peak denaturation temperature (T_{den}) and denaturation enthalpy (ΔH) were determined using the TRIOS 5.1.1 software.

2.7. *Sodium Dodecyl Sulphate Polyacrylamide Gel Electrophoresis*

The isolates were subjected to SDS-PAGE using the Laemmli method [36]. A 2 mg/mL protein solution was prepared with deionized water and mixed with 2x Laemmli buffer at a 1:1 (v/v) ratio. The mixtures were then heated to 95 °C for 5 min. Subsequently, 10 µL of the mixture was loaded into the electrophoresis equipment (C.B.S. Scientific Company Inc., Del Mar, CA, USA) and run for 1 h at 100 V. The running gel was then stained with Coomassie brilliant blue.

2.8. *Color Profile*

The color of each pooled protein isolate was measured in triplicate using a Luci 100 spectral colorimeter equipped with a D65 xenon lamp using the 10° observer (Hach Lange GmbH, Düsseldorf, Germany). Equation [5] was used to calculate color difference ΔE between the control and protein extracted using ultrasonic support:

$$\Delta E = \sqrt{(\Delta L^*)^2 + (\Delta a^*)^2 + (\Delta b^*)^2} \tag{5}$$

Apart from lightness L^*, chroma C^* and the hue angle h_{ab} were calculated for interpretation [37].

2.9. *Statistical Analysis*

The statistical significance of the effects of the extraction procedures on the properties of the protein isolate was evaluated using analysis of variance (ANOVA) and subsequent Duncan multiple comparison tests (SPSS Inc., Chicago, IL, USA). To set up the preliminary optimization procedure, Design Expert version 7.0.0 (State-Ease, Inc., Minneapolis, MN, USA) was used.

3. Results and Discussion

3.1. Composition of Pumpkin Press Cake

Apart from 4.73 g/100 g moisture and approx. 13 g/100 g residual oil, the native pumpkin seed press cake contained 60.24 g/100 g protein and significant amounts of dietary fiber and salts (Table 1). De-oiling with hexane reduced residual oil content to below 1 g/100 g. The moisture content of the de-oiled sample was similar to that of the untreated powder, and the fraction of all other components increased accordingly. The protein content of the de-oiled press cake was 68.68 g/100 g.

Table 1. Composition of untreated and de-oiled pumpkin seed press cake.

Component (g/100 g)	Untreated Press Cake	De-Oiled Press Cake
Moisture	4.73 ± 0.32 [a]	4.84 ± 0.19 [a]
Fat	13.38 ± 0.12 [a]	0.77 ± 0.14 [b]
Protein	60.24 ± 0.05 [a]	68.68 ± 0.13 [b]
Total dietary fibre	13.13 ± 0.71 [a]	17.39 ± 0.54 [b]
Ash	8.02 ± 0.06 [a]	9.11 ± 0.07 [b]

Mean values ± half deviation range ($n = 2$) in a row labeled with different letters differ significantly ($p < 0.05$).

As indicated by the data, pumpkin press cake is rich in protein and may serve as an excellent protein source when compared to, e.g., chia, sesame, rape, flax, sunflower, or hemp [38,39]. Bučko et al. [38] also de-oiled pumpkin press cake with hexane and found a residual protein content of 63.5 g/100 g. In their study on pumpkin protein isolation, Vinayashree and Vasu [7] specified the protein content of seed flour as 35.18 g/100 g, which increased to 51.85 g/100 g after de-oiling. The fat content of untreated and de-oiled press cake is similar to that reported in the literature [40,41] and appears typical for pumpkin. The same is true for the amount of salts and dietary fiber [42–44].

3.2. Optimization of the Protein Extraction Procedure

During optimizing the control extraction procedure, it turned out that pH had the highest impact on protein yield, with increased yield obtained after extraction at higher pH. Higher alkalinity generally improves extraction efficiency as protein solubility and the negative charge of the side amino groups of basic amino acids is significantly affected [45]. It was also shown for sunflower protein that, in the range of 2–10, extractability increased with increasing pH [46]. ANOVA provided the conditions for maximizing protein yield and protein content of the isolate. These conditions were a pH of 11, a solvent-to-solid ratio of 29, and a soaking time of 60 min, which was not different from previous research on other substrates [47,48], and resulted in a protein yield of 43.6 ± 1.0% and a protein content of the isolate of 94.04 ± 0.77 g/100 g. All subsequent isolation procedures comprising ultrasonic support were carried out under these conditions. Details of the optimization routine can be accessed from Supplementary Materials.

3.3. Effect of Ultrasonic Treatment on Protein Yield and Content

The effects of ultrasonic treatment on protein yield and protein content of the isolate are depicted in Figure 2. Compared to the protein yield of the control, the application of ultrasound under the given conditions (10 min, 50% amplitude) generally increased protein yield, which was 50.7 ± 1.5% with US+AE and 55.2 ± 1.8% with UAE. The highest protein yield was obtained with AE+US (57.8 ± 2.0%). However, enhanced acoustic cavitation might cause the denaturation of soluble proteins and hence reduce protein extraction efficiency [43]. In preliminary experiments with higher energy input (100% amplitude), we observed difficulties concerning temperature control but no significant increase in protein yield.

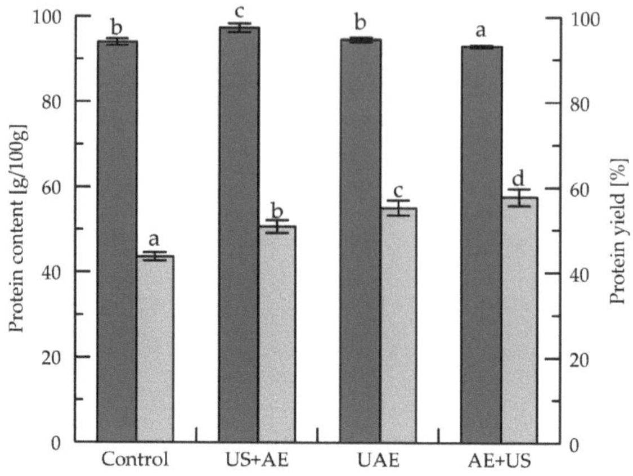

Figure 2. Protein content (dark bars) of dried pumpkin extract and protein yield obtained during extraction (light bars). Mean values ± standard deviations (n = 8) within the same bar type, labeled with different letters differ significantly ($p < 0.05$).

The highest protein recovery was achieved when the ultrasonic treatment followed alkaline extraction. Presumably, the underlying mechanical vibration intensifies the contact between the alkaline solution and the press cake powder, promotes cell destruction and opening of pores, and results in an improved mass transfer and, hence, protein release [16]. In a study on ultrasound-assisted protein extraction from cauliflower, protein content was 77.62 g/100 g, while extraction yield was 53.1% [49]. Another study used pulsed ultrasonic treatment at different power densities following alkaline extraction at pH 8.0 and room temperature for 1 h to extract protein from sunflower meal, and the protein yield was, depending on the respective conditions, 28.0–54.3% [16]. Tu et al. [43], who investigated ultrasonic extraction of albumin from de-oiled pumpkin seed using neutral extraction media, reported an extraction yield of 17.0%.

3.4. Particle Size Distribution of Alkaline Extracts

The extract produced using the control treatment showed the largest particles, and their size was reduced when ultrasound was applied (Table 2). This effect is caused by shear forces and cavitation; ultrasound affects electrostatic and hydrophobic interactions as well as hydrogen bonds and, as a result, decreases particle size [22,50]. In addition, the decrease in particle size was significantly ($p < 0.05$) more pronounced when alkaline extraction was followed by ultrasonic treatment (AE+US). Particle size reduction following ultrasonic treatment was also observed for canola protein isolate [22], pea protein isolate [20], sunflower protein isolate [16], and barley protein isolate [19]. As outlined in a study conducted by Malik et al. [24], the median particle size of sunflower protein isolates decreased from 114.6 µm to 94.3 µm after 10 min of ultrasonic treatment.

Table 2. The particle size of dispersions after alkaline extraction.

Treatment	d_{10} [µm]	d_{50} [µm]	d_{90} [µm]
Control	4.82 ± 0.44^{b}	72.12 ± 14.58^{b}	376.68 ± 38.32^{c}
US+AE	3.49 ± 0.20^{a}	31.50 ± 2.51^{a}	252.96 ± 24.85^{b}
UAE	3.38 ± 0.25^{a}	27.59 ± 4.02^{a}	245.83 ± 34.60^{b}
AE+US	8.05 ± 0.72^{c}	36.25 ± 5.71^{a}	179.93 ± 13.24^{a}

US+AE, ultrasonic treatment followed by alkaline extraction; UAE, ultrasonic treatment combined with alkaline extraction; AE+US, alkaline extraction followed by ultrasonic treatment. d_{10}, d_{50} and d_{90} refer to particle diameters at 10%, 50% and 90% volume fraction. Mean values ± standard deviations (n = 12) in a column labeled with different letters differ significantly ($p < 0.05$).

3.5. Protein Solubility

Solubility, counteracted by protein aggregation and denaturation, can be considered an indicator of protein functionality [14]. Table 3 shows that the solubility of the protein isolated from pumpkin press cake was largely affected by pH, showing a minimum (2.42 ± 0.06%) at pH 5. Generally, solubility is low near the isoelectric point but increases considerably at increased (pH ≤ 3.0) or reduced (pH ≥ 7.0) acidity because of the absence of electrostatic repulsive forces at the isoelectric point, resulting in protein molecules with neutral charge [51].

Table 3. Solubility and water binding capacity (WBC in g per g dry matter) of extracted pumpkin seed protein.

Treatment	Solubility [%]				WBC [g/g dm]
	pH 3	pH 5	pH 7	pH 9	
Control	35.73 ± 1.42 [a]	2.42 ± 0.06 [a]	15.17 ± 0.20 [a]	37.82 ± 1.07 [a]	3.95 ± 0.04 [b]
US+AE	37.33 ± 0.41 [b]	3.83 ± 0.42 [b]	18.72 ± 0.13 [b]	39.88 ± 0.30 [b]	3.81 ± 0.08 [b]
UAE	38.00 ± 0.89 [bc]	6.59 ± 0.99 [d]	18.80 ± 0.52 [b]	46.87 ± 0.17 [c]	3.61 ± 0.21 [a]
AE+US	38.84 ± 0.21 [c]	4.90 ± 0.05 [c]	23.07 ± 0.06 [c]	46.93 ±0.61 [c]	3.66 ± 0.15 [a]

US+AE, ultrasonic treatment followed by alkaline extraction; UAE, ultrasonic treatment combined with alkaline extraction; AE+US, alkaline extraction followed by ultrasonic treatment. Mean values ± standard deviations (n = 4) in a column labeled with different letters differ significantly ($p < 0.05$).

The results are consistent with the data of Vinayashree et al. [7], showing a pumpkin protein solubility of <10% at pH 4. Chavan et al. [52] observed that protein from beach pea had its lowest solubility at pH 4.5, and Dabbour et al. [16] reported sunflower protein solubility as low as 0.39% at pH 5, which is close to the isoelectric point of pH 4.5.

It was already shown that ultrasonic treatment improves protein solubility [14,27,53]. Compared to the control, the solubility of protein obtained by additionally applying ultrasound was significantly higher at each pH (see Table 3). This might be because proteins form aggregates in their native state, and cavitation is a physical factor that reduces hydrophobic interactions, necessary for the intermolecular association of protein molecules. It was also reported that ultrasound improves solubility by promoting the formation of monomers from insoluble or soluble protein aggregates, by affecting non-covalent interactions, and by releasing polar residues [23,53,54].

The AE+US treatment resulted in the highest protein solubility, presumably because of the impact of alkalinity on protein–water interactions prior to ultrasonic application as well as because of the changes in particle size. Ultrasonic treatment of soy protein isolates decreased particle size [14], and a protein solubility increase after ultrasonic treatment was also demonstrated for sunflower protein isolates [24].

3.6. Water Binding and Foaming Capacity

The effect of ultrasound on water-binding capacity is also shown in Table 3. When applying ultrasound during or after alkaline extraction, the WBC of the isolated protein was significantly lower. Considering conformational changes in protein molecules after an ultrasonic treatment and the subsequent formation of more soluble protein aggregates, it is likely that changes in the hydrophobic surface area are responsible for these differences. In a study on ultrasound-assisted sunflower protein isolation, an increase in protein solubility but a decrease in WBC was explained by the changes in surface hydrophobicity [24].

Foaming capacity (FC) refers to a protein's ability to unfold quickly and to dissolve, forming a cohesive layer that surrounds gas bubbles, whereas foaming stability (FS) refers to the ability to generate stable foams by forming a continuous intermolecular polymer network that envelopes air cells [5]. Mechanisms such as rearrangements, penetration, and molecular movement at the interface significantly contribute to FC and FS.

Ultrasonic treatment caused a significant increase in FC (Figure 3). Due to structural unfolding, FC largely depends on protein diffusivity at the gas–liquid interface, and an increase may be related to the fact that ultrasound enhances dispersibility during foam generation by mechanically homogenizing the protein particles. Ultrasound also induces partial structural changes in proteins, leading to more rapid protein adsorption at the gas–liquid interface and, consequently, to an increased foaming capacity [23]. In studies with sunflower [24] and soy protein isolate [55], an FC increase was attributed to particle size reduction caused by ultrasound. This might also be the case for the pumpkin protein extracted in our study. According to Xiong et al. [29], pea protein interaction was improved by ultrasonic treatment, and FC increased from 58% to 73.3%. In a study conducted on protein isolation from evening primrose by-products, ultrasound-assisted alkaline extraction also increased the foaming capacity [56].

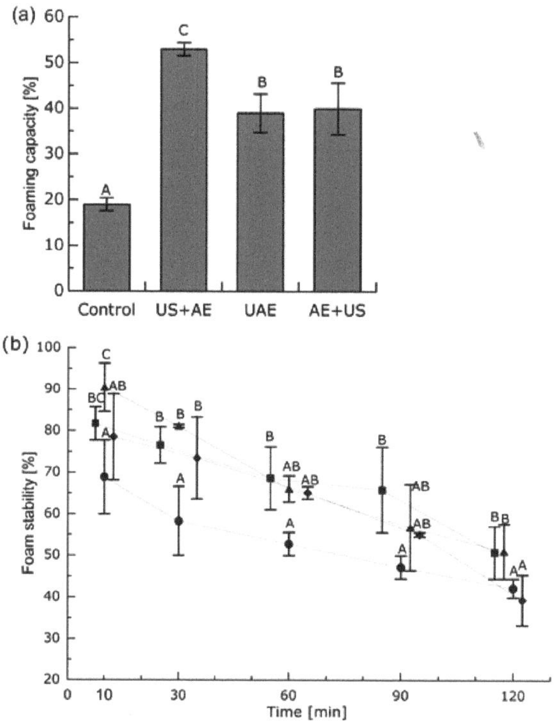

Figure 3. Foaming capacity (**a**) and foam stability (**b**) of protein isolates. Control (circles), alkaline extraction; US+AE (triangles), ultrasonic treatment followed by alkaline extraction; UAE (squares), ultrasonic treatment combined with alkaline extraction; AE+US (rhomboids), alkaline extraction followed by ultrasonic treatment. For better eye guidance, symbols are connected by dotted lines and are slightly shifted along the x-axis. Mean values ± standard deviations (n = 4) for the same period labeled with different letters differ significantly (p < 0.05).

The foaming stability of pumpkin protein isolate also increased when ultrasound was applied (see Figure 3). Li et al. [57] reported that ultrasonic treatment enhanced the FS of brewer's spent grain protein by 26.6% compared to the conventional extraction. It has been mentioned that ultrasound improves molecular flexibility, lowers surface tension, and facilitates the creation of strong elastic films around dispersed gas bubbles [58]. Ultrasonic support also increased the FS of sunflower protein isolates [24]. In the case of pea proteins, a smooth film at the air–water interface was responsible for the enhanced foam stability [29].

3.7. Stability of the Proteins

Table 4 summarizes the effects of the different extraction procedures on the thermal properties of the extracted protein. Protein denaturation temperature was the highest after control alkaline extraction, and significantly lower after the application of ultrasound, with the lowest T_{den} = 85.70 °C obtained in the US+AE treatment.

Table 4. Thermal properties of pumpkin seed protein isolates.

	Control	US+AE	UAE	AE+US
T_{den} (°C)	99.05 ± 4.15 ^{c}	85.70 ± 2.29 ^{a}	94.20 ± 2.84 ^{b}	92.44 ± 3.58 ^{b}
ΔH (J/g)	308.6 ± 9.6 ^{a}	371.0 ± 21.6 ^{b}	362.2 ± 10.5 ^{b}	360.8 ± 2.6 ^{b}

T_{den}, denaturation temperature; ΔH, enthalpy. Control, alkaline extraction; US+AE, ultrasonic treatment followed by alkaline extraction; UAE, ultrasonic treatment combined with alkaline extraction; AE+US, alkaline extraction followed by ultrasonic treatment. Mean values ± standard deviations (n = 4) in a row labeled with different letters differ significantly ($p < 0.05$).

These findings demonstrate that ultrasonication caused structural changes of the proteins that resulted in a change in their thermal behavior. The decrease in T_{den} after ultrasound application confirmed the loss of protein–protein bonds. A reduction in denaturation temperature because of conformational changes was also determined in the studies where sunflower protein was extracted using ultrasound [24], and similarly during the isolation of soy protein [59].

The energy needed to denature or unfold protein structure is also related to the enthalpy required for this process. ΔH was 308. 6 J/g for the control sample and increased by approx. 20% when ultrasound was applied during protein extraction. The reason for this might be protein aggregation after prolonged sonication. Sonication of whey protein concentrates for up to 5 min resulted in a decline in enthalpy, implying the destruction of protein bonds. However, after more than 5 min sonication, the enthalpy increased, also indicating the possibility of re-aggregations [60].

The extracted pumpkin seed protein had a main molecular mass of approx. 35 kDa. Ultrasonic treatment did not cause significant differences in the electrophoresis profiles compared to the control treatment, demonstrating that the application of ultrasound does not modify the main structure of the protein. Similar results were observed in SDS-PAGE analyses of pumpkin seed protein isolate [30], peanut protein isolate [21], pea protein isolate [29], and soybean protein isolate [59,61].

3.8. Color Properties

Table 5 illustrates the color properties of the isolates, which were significantly different ($p < 0.05$). Ultrasound affects color as a result of cavitation. It could have a favorable or unfavorable impact on the color pigments in foods. While it may accelerate a structural release of the pigments, it can also induce light absorption by causing changes in the pigment-containing structures [28]. The color difference of the protein isolates was most significant after applying the AE+US treatment (2.72 ± 0.14).

Table 5. Color properties of pumpkin seed protein isolates.

	Control	US+AE	UAE	AE+US
ΔE^*		2.02 ± 1.26	1.84 ± 0.34	2.72 ±0.14
L^*	71.20 ± 0.48 ^{b}	72.26 ± 1.76 ^{c}	69.64 ± 0.10 ^{a}	68.78 ± 0.17 ^{a}
h_{ab}	78.70 ± 0.10 ^{a}	80.33 ± 1.19 ^{b}	81.05 ± 0.33 ^{bc}	81.52 ± 0.08 ^{c}
C^*	21.66 ± 0.38 ^{b}	20.57 ± 1.04 ^{a}	22.08 ± 0.51 ^{b}	21.29 ± 0.97 ^{ab}

ΔE^*, color difference; L^*, lightness; h_{ab}, hue angle; C^*, chroma. Control, alkaline extraction; US+AE, ultrasonic treatment followed by alkaline extraction; UAE, ultrasonic treatment combined with alkaline extraction; AE+US, alkaline extraction followed by ultrasonic treatment. Mean values ± standard deviations (n = 6) in a row labeled with different letters differ significantly ($p < 0.05$).

There have been limited studies on changes in protein color caused by ultrasonication. A study conducted by Chittapalo and Noomhorm [62] on de-oiled rice bran demonstrated that sonicated rice bran protein concentrate was lighter. In case of protein extracted from album seed, ultrasonication also resulted in a higher lightness [28]. The hue angle expresses color quality, and $0^\circ < h_{ab} < 90^\circ$ indicates that protein isolate color is in the red–yellow quadrant of the $L^*a^*b^*$ color space. The increase in h_{ab} after ultrasound-assisted extraction indicates increased contributions of the yellow part of the spectrum. Chroma as a measure of color saturation was hardly affected by the impact of cavitation.

4. Conclusions

In this study, optimum alkaline extraction parameters (pH, solvent-to-solid ratio, and time) for protein extraction from pumpkin seed press cake were determined at pH 11, solvent-to-solid ratio of 29, and extraction time of 60 min. This research showed that the use of ultrasonication in combination with alkaline extraction enhances the protein yield. The protein recovery yield was 50.7 ± 1.5% with the US+AE and 55.2 ± 1.8% with the UAE, while AE+US provided the highest protein yield at 57.8 ± 2.0%.

The solubility of the protein obtained by applying ultrasound was considerably higher than the control and the highest when the AE+US treatment was applied (46.93 ± 0.61% at pH 9). Foaming capacity and stability significantly increased after ultrasonic treatment, and the highest foaming capacity was observed after using the US+AE treatment (53.0 ± 1.41%). Protein denaturation temperature decreased after ultrasonic treatment, as did the water binding capacity. In addition, ultrasound showed a significant impact on the size of the particles in the suspension.

Ultrasound is a promising and excellent alternative for assisting alkaline extraction of pumpkin press cake, and the protein obtained had appropriate techno-functional properties. When compared to the control, ultrasonic treatments increased solubility, foaming capacity, and stability while decreasing water binding capacity and particle size. The AE+US treatment was considered to be the most effective treatment among the three different ultrasonic treatments because it provided the highest increase in protein yield as well as high impacts on techno-functional properties. It can be assumed that protein yield and functionality can further be improved when ultrasound conditions are varied with respect to power input and treatment time.

Supplementary Materials: Information concerning the optimization procedure for the extraction of pumpkin seed protein can be downloaded at: https://www.mdpi.com/article/10.3390/foods11244029/s1, Protein Extraction Optimization.

Author Contributions: Conceptualization, D.S. and S.S.; methodology, D.S. and S.S.; investigation, D.S.; writing—original draft preparation, D.S.; writing—review and editing, S.S. and H.R.; supervision, H.R. All authors have read and agreed to the published version of the manuscript.

Funding: This research received no external funding.

Institutional Review Board Statement: Not applicable.

Informed Consent Statement: Not applicable.

Data Availability Statement: The data that support the conclusions of this study are available upon reasonable request from the corresponding author.

Acknowledgments: The authors are thankful to the Republic of Turkey—Ministry of National Education for funding Deniz Sert's PhD studies. We are also thankful to Estyria Naturprodukte GmbH (St. Ruprecht an der Raab, Austria) for providing pumpkin seed press cake for this work.

Conflicts of Interest: The authors declare no conflict of interest.

References

1. Ochoa-Rivas, A.; Nava-Valdez, Y.; Serna-Saldívar, S.O.; Chuck-Hernández, C. Microwave and ultrasound to enhance protein extraction from peanut flour under alkaline conditions: Effects in yield and functional properties of protein isolates. *Food Bioprocess Technol.* **2017**, *10*, 543–555. [CrossRef]
2. Henchion, M.; Hayes, M.; Mullen, A.M.; Fenelon, M.; Tiwari, B. Future protein supply and demand: Strategies and factors influencing a sustainable equilibrium. *Foods* **2017**, *6*, 53. [CrossRef] [PubMed]
3. Kumar, M.; Tomar, M.; Potkule, J.; Punia, S.; Dhakane, J.; Singh, S.; Kennedy, J.F. Functional characterization of plant-based protein to determine its quality for food applications. *Food Hydrocoll.* **2021**, *123*, 106986. [CrossRef]
4. Wang, J.S.; Wang, A.B.; Zang, X.P.; Tan, L.; Xu, B.Y.; Chen, H.H.; Ma, W.H. Physicochemical, functional and emulsion properties of edible protein from avocado (*Persea americana* Mill.) oil processing by-products. *Food Chem.* **2019**, *288*, 146–153. [CrossRef]
5. Jiang, Y.; Zhou, X.; Zheng, Y.; Wang, D.; Deng, Y.; Zhao, Y. Impact of ultrasonication/shear emulsifying/microwave-assisted enzymatic extraction on rheological, structural, and functional properties of *Akebia trifoliata* (Thunb.) Koidz. seed protein isolates. *Food Hydrocoll.* **2021**, *112*, 106355. [CrossRef]
6. Cisneros-Yupanqui, M.; Chalova, V.I.; Kalaydzhiev, H.R.; Mihaylova, D.; Krastanov, A.I.; Lante, A. Preliminary characterisation of wastes generated from the rapeseed and sunflower protein isolation process and their valorisation in delaying oil oxidation. *Food Bioprocess Technol.* **2021**, *14*, 1962–1971. [CrossRef]
7. Vinayashree, S.; Vasu, P. Biochemical, nutritional and functional properties of protein isolate and fractions from pumpkin (*Cucurbita moschata* var. Kashi Harit) seeds. *Food Chem.* **2021**, *340*, 128177. [CrossRef] [PubMed]
8. Food and Agricultural Organization of the United Nations Statistics (FAOSTAT). 2020. Available online: https://www.fao.org/faostat/en/#data/QCL/visualize (accessed on 14 May 2022).
9. European Commission, eAmbrosia, the EU Geographical Indications Register. 2022. Available online: https://ec.europa.eu/info/food-farming-fisheries/food-safety-and-quality/certification/quality-labels/geographical-indications-register/ (accessed on 15 November 2022).
10. Wang, H.; Chen, K.; Cheng, J.; Jiang, L.; Yu, D.; Dai, Y.; Wang, L. Ultrasound-assisted three phase partitioning for simultaneous extraction of oil, protein and polysaccharide from pumpkin seeds. *LWT-Food Sci. Technol.* **2021**, *151*, 112200. [CrossRef]
11. Quanhong, L.; Caili, F. Application of response surface methodology for extraction optimization of germinant pumpkin seeds protein. *Food Chem.* **2005**, *92*, 701–706. [CrossRef]
12. The American Heart Association (AHA). 2018. Available online: https://www.heart.org/ (accessed on 1 December 2022).
13. Bučko, S.; Katona, J.; Popović, L.; Vaštag, Ž.; Petrović, L. Functional properties of pumpkin (*Cucurbita pepo*) seed protein isolate and hydrolysate. *J. Serb. Chem. Soc.* **2016**, *81*, 35–46. [CrossRef]
14. Arzeni, C.; Martínez, K.; Zema, P.; Arias, A.; Pérez, O.E.; Pilosof, A.M.R. Comparative study of high intensity ultrasound effects on food proteins functionality. *J. Food Eng.* **2012**, *108*, 463–472. [CrossRef]
15. Bernardi, S.; Lupatini-Menegotto, A.L.; Kalschne, D.L.; Moraes Flores, É.L.; Bittencourt, P.R.S.; Colla, E.; Canan, C. Ultrasound: A suitable technology to improve the extraction and techno-functional properties of vegetable food proteins. *Plant Foods Hum. Nutr.* **2021**, *76*, 1–11. [CrossRef] [PubMed]
16. Dabbour, M.; He, R.; Ma, H.; Musa, A. Optimization of ultrasound assisted extraction of protein from sunflower meal and its physicochemical and functional properties. *J. Food Proc. Eng.* **2018**, *41*, 12799. [CrossRef]
17. Shirsath, S.R.; Sonawane, S.H.; Gogate, P.R. Intensification of extraction of natural products using ultrasonic irradiations—A review of current status. *Chem. Eng. Proc. Proc. Intensif.* **2012**, *53*, 10–23. [CrossRef]
18. Karki, B. Use of High-Power Ultrasound During Soy Protein Production and Study of Its Effect on Functional Properties of Soy Protein Isolate. Ph.D. Thesis, Iowa State University, Ames, IA, USA, 2009.
19. Silventoinen, P.; Sozer, N. Impact of ultrasound treatment and pH-shifting on physicochemical properties of protein-enriched barley fraction and barley protein isolate. *Foods* **2020**, *9*, 1055. [CrossRef]
20. Wang, F.; Zhang, Y.; Xu, L.; Ma, H. An efficient ultrasound-assisted extraction method of pea protein and its effect on protein functional properties and biological activities. *LWT-Food Sci. Technol.* **2020**, *127*, 109348. [CrossRef]
21. Zhang, Q.T.; Tu, Z.C.; Xiao, H.; Wang, H.; Huang, X.Q.; Liu, G.X.; Lin, D.R. Influence of ultrasonic treatment on the structure and emulsifying properties of peanut protein isolate. *Food Bioprod. Proc.* **2014**, *92*, 30–37. [CrossRef]
22. Li, Y.; Cheng, Y.; Zhang, Z.; Wang, Y.; Mintah, B.K.; Dabbour, M.; Jiang, H.; He, R.; Ma, H. Modification of rapeseed protein by ultrasound-assisted pH shift treatment: Ultrasonic mode and frequency screening, changes in protein solubility and structural characteristics. *Ultrasonics Sonochem.* **2020**, *69*, 105240. [CrossRef]
23. Dabbour, M.; He, R.; Mintah, B.; Xiang, J.; Ma, H. Changes in functionalities, conformational characteristics and antioxidative capacities of sunflower protein by controlled enzymolysis and ultrasonication action. *Ultrasonics Sonochem.* **2019**, *58*, 104625. [CrossRef]
24. Malik, M.A.; Sharma, H.K.; Saini, C.S. High intensity ultrasound treatment of protein isolate extracted from dephenolized sunflower meal: Effect on physicochemical and functional properties. *Ultrasonics Sonochem.* **2017**, *39*, 511–519. [CrossRef]
25. Li, K.; Ma, H.; Li, S.; Zhang, C.; Dai, C. Effect of ultrasound on alkali extraction protein from rice dreg flour. *J. Food Proc. Eng.* **2017**, *40*, 12377. [CrossRef]
26. Zhang, Y.; Wang, B.; Zhang, W.; Xu, W.; Hu, Z. Effects and mechanism of dilute acid soaking with ultrasound pretreatment on rice bran protein extraction. *J. Cereal Sci.* **2019**, *87*, 318–324. [CrossRef]

27. Zhu, Z.; Zhu, W.; Yi, J.; Liu, N.; Cao, Y.; Lu, J.; McClements, D.J. Effects of sonication on the physicochemical and functional properties of walnut protein isolate. *Food Res. Int.* **2018**, *106*, 853–861. [CrossRef]
28. Mir, N.A.; Riar, C.S.; Singh, S. Physicochemical, molecular and thermal properties of high-intensity ultrasound (HIUS) treated protein isolates from album (*Chenopodium album*) seed. *Food Hydrocoll.* **2019**, *96*, 433–441. [CrossRef]
29. Xiong, T.; Xiong, W.; Ge, M.; Xia, J.; Li, B.; Chen, Y. Effect of high intensity ultrasound on structure and foaming properties of pea protein isolate. *Food Res. Int.* **2018**, *109*, 260–267. [CrossRef] [PubMed]
30. Du, H.; Zhang, J.; Wang, S.; Manyande, A.; Wang, J. Effect of high-intensity ultrasonic treatment on the physicochemical, structural, rheological, behavioral, and foaming properties of pumpkin (*Cucurbita moschata* Duch.)-seed protein isolates. *LWT-Food Sci. Technol.* **2022**, *155*, 112952. [CrossRef]
31. Stone, A.K.; Karalash, A.; Tyler, R.T.; Warkentin, T.D.; Nickerson, M.T. Functional attributes of pea protein isolates prepared using different extraction methods and cultivars. *Food Res. Int.* **2015**, *76*, 31–38. [CrossRef]
32. Association of Official Analytical Chemists (AOAC). *Official Methods of Analysis*, 15th ed.; Association of Official Analytical Chemists: Arlington, VA, USA, 1990.
33. Gonzalez-Perez, S. Physico-Chemical and Functional Properties of Sunflower Proteins. Ph.D. Thesis, Wageningen University, Wageningen, The Netherlands, 2003.
34. Chen, J.Y.; Piva, M.; Labuza, T.P. Evaluation of water binding capacity (WBC) of food fiber sources. *J. Food Sci.* **1984**, *49*, 59–63. [CrossRef]
35. Moure, A.; Sineiro, J.; Domínguez, H. Extraction and functionality of membrane-concentrated protein from defatted Rosa rubiginosa seeds. *Food Chem.* **2001**, *74*, 327–339. [CrossRef]
36. Laemmli, U.K. Cleavage of structural proteins during the assembly of the head of bacteriophage T4. *Nature* **1970**, *227*, 680–685. [CrossRef]
37. Rohm, H.; Jaros, D. Colour of hard cheese. *Z. Lebensm.-Unters. Forsch.* **1996**, *203*, 241–244. [CrossRef]
38. Bučko, S.; Katona, J.; Popović, L.; Vaštag, Ž.; Petrović, L.; Vučinić–Vasić, M. Investigation on solubility, interfacial and emulsifying properties of pumpkin (*Cucurbita pepo*) seed protein isolate. *LWT-Food Sci. Technol.* **2015**, *64*, 609–615. [CrossRef]
39. Kotecka-Majchrzak, K.; Sumara, A.; Fornal, E.; Montowska, M. Oilseed proteins–Properties and application as a food ingredient. *Trends Food Sci. Technol.* **2020**, *106*, 160–170. [CrossRef]
40. Bárta, J.; Bártová, V.; Jarošová, M.; Švajner, J.; Smetana, P.; Kadlec, J.; Kozak, M. Oilseed cake flour composition, functional properties and antioxidant potential as effects of sieving and species differences. *Foods* **2021**, *10*, 2766. [CrossRef]
41. Sá, A.G.A.; da Silva, D.C.; Pacheco, M.T.B.; Moreno, Y.M.F.; Carciofi, B.A.M. Oilseed by-products as plant-based protein sources: Amino acid profile and digestibility. *Future Foods* **2021**, *3*, 100023. [CrossRef]
42. Lovatto, N.D.M.; Loureiro, B.B.; Bender, A.B.B.; Loureiro, C.B.; Goulart, F.R.; Speroni, C.S.; Silva, L.P.D. Phosphorylated protein concentrates pumpkin seed (*Cucurbita moschata*): Optimization by response surface methodology and nutritional characterization. *Ciência Rural* **2020**, *50*, 20190093. [CrossRef]
43. Tu, G.L.; Bui, T.H.N.; Tran, T.T.; Ton, N.M.N.; Le, V.V.M. Comparison of enzymatic and ultrasonic extraction of albumin from defatted pumpkin (*Cucurbita pepo*) seed powder. *Food Technol. Biotechnol.* **2015**, *53*, 479–487. [CrossRef]
44. Kim, M.Y.; Kim, E.J.; Kim, Y.N.; Choi, C.; Lee, B.H. Comparison of the chemical compositions and nutritive values of various pumpkin (*Cucurbitaceae*) species and parts. *Nutr. Res. Pract.* **2012**, *6*, 21–27. [CrossRef]
45. Gao, Z.; Shen, P.; Lan, Y.; Cui, L.; Ohm, J.B.; Chen, B.; Rao, J. Effect of alkaline extraction pH on structure properties, solubility, and beany flavor of yellow pea protein isolate. *Food Res. Int.* **2020**, *131*, 109045. [CrossRef]
46. Ivanova, P.; Chalova, V.; Koleva, L.; Pishtiyski, I.; Perifanova-Nemska, M. Optimization of protein extraction from sunflower meal produced in Bulgaria. *Bulgarian J. Agric. Sci.* **2012**, *18*, 153–160.
47. Ghodsvali, A.; Khodaparast, M.H.; Vosoughi, M.; Diosady, L.L. Preparation of canola protein materials using membrane technology and evaluation of meals functional properties. *Food Res. Int.* **2005**, *38*, 223–231. [CrossRef]
48. Hadidi, M.; Khaksar, F.B.; Pagan, J.; Ibarz, A. Application of Ultrasound-Ultrafiltration-Assisted alkaline isoelectric precipitation (UUAAIP) technique for producing alfalfa protein isolate for human consumption: Optimization, comparison, physicochemical, and functional properties. *Food Res. Int.* **2019**, *130*, 108907. [CrossRef] [PubMed]
49. Xu, Y.; Li, Y.; Bao, T.; Zheng, X.; Chen, W.; Wang, J. A recyclable protein resource derived from cauliflower by-products: Potential biological activities of protein hydrolysates. *Food Chem.* **2017**, *221*, 114–122. [CrossRef] [PubMed]
50. Yanjun, S.; Jianhang, C.; Shuwen, Z.; Hongjuan, L.; Jing, L.; Lu, L.; Jiaping, L. Effect of power ultrasound pretreatment on the physical and functional properties of reconstituted milk protein concentrate. *J. Food Eng.* **2014**, *124*, 11. [CrossRef]
51. Du, Y.; Jiang, Y.; Zhu, X.; Xiong, H.; Shi, S.; Hu, J.; Sun, W. Physicochemical and functional properties of the protein isolate and major fractions prepared from Akebia trifoliata var. australis seed. *Food Chem.* **2012**, *133*, 923–929. [CrossRef]
52. Chavan, U.D.; McKenzie, D.B.; Shahidi, F. Functional properties of protein isolates from beach pea (*Lathyrus maritimus* L.). *Food Chem.* **2001**, *74*, 177–187. [CrossRef]
53. Zhang, Z.; Regenstein, J.M.; Zhou, P.; Yang, Y. Effects of high intensity ultrasound modification on physicochemical property and water in myofibrillar protein gel. *Ultrasonics Sonochem.* **2017**, *56*, 960–967. [CrossRef]
54. Jiang, L.; Wang, J.; Li, Y.; Wang, Z.; Liang, J.; Wang, R. Zhang, M. Effects of ultrasound on the structure and physical properties of black bean protein isolates. *Food Res. Int.* **2014**, *62*, 595–601. [CrossRef]

55. Morales, R.; Martínez, K.D.; Ruiz-Henestrosa, V.M.P.; Pilosof, A.M. Modification of foaming properties of soy protein isolate by high ultrasound intensity: Particle size effect. *Ultrasonics Sonochem.* **2015**, *26*, 48–55. [CrossRef]
56. Hadidi, M.; Ibarz, A.; Pouramin, S. Optimization of extraction and deamidation of edible protein from evening primrose (*Oenothera biennis* L.) oil processing by-products and its effect on structural and techno-functional properties. *Food Chem.* **2021**, *334*, 127613. [CrossRef]
57. Li, W.; Yang, H.; Coldea, T.E.; Zhao, H. Modification of structural and functional characteristics of brewer's spent grain protein by ultrasound assisted extraction. *LWT-Food Sci. Technol.* **2021**, *139*, 110582. [CrossRef]
58. Sharma, G.M.; Su, M.; Joshi, A.U.; Roux, K.H.; Sathe, S.K. Functional properties of select edible oilseed proteins. *J. Agric. Food Chem.* **2010**, *58*, 5457–5464. [CrossRef] [PubMed]
59. Karki, B.; Lamsal, B.P.; Grewell, D.; Pometto, A.L., III; Van Leeuwen, J.; Khanal, S.K.; Jung, S. Functional properties of soy protein isolates produced from ultrasonicated defatted soy flakes. *J. Am. Oil Chem. Soc.* **2009**, *86*, 1021–1028. [CrossRef]
60. Chandrapala, J.; Zisu, B.; Palmer, M.; Kentish, S.; Ashokkumar, M. Effects of ultrasound on the thermal and structural characteristics of proteins in reconstituted whey protein concentrate. *Ultrasonics Sonochem.* **2011**, *18*, 951–957. [CrossRef] [PubMed]
61. Hu, H.; Wu, J.; Li-Chan, E.C.; Zhu, L.; Zhang, F.; Xu, X.; Pan, S. Effects of ultrasound on structural and physical properties of soy protein isolate (SPI) dispersions. *Food Hydrocollois* **2013**, *30*, 647–655. [CrossRef]
62. Chittapalo, T.; Noomhorm, A. Ultrasonic assisted alkali extraction of protein from defatted rice bran and properties of the protein concentrates. *Int. J. Food Sci. Technol.* **2009**, *44*, 1843–1849. [CrossRef]

foods

Article

Structure-Function Guided Extraction and Scale-Up of Pea Protein Isolate Production

Lucy Hansen, Fan Bu and Baraem P. Ismail *

Food Science and Nutrition Department, University of Minnesota, 1334 Eckles Ave., Saint Paul, MN 55108, USA
* Correspondence: bismailm@umn.edu; Tel.: +1-612-625-0147

Abstract: The lack of adequate guidance and control of the extraction conditions as well as the gap between bench- and industrial-scale production, contributes to the poor functionality of commercial pea protein isolate (cPPI). Therefore, pea protein extraction conditions were evaluated and scaled up to maximize protein purity and yield, while maintaining structural integrity, following mild alkaline solubilization with isoelectric precipitation and salt solubilization coupled with membrane filtration. Both extraction methods resulted in high protein yield (>64%) and purity (>87%). Structure-function characterization illustrated the preserved structural integrity of PPI samples and their superior solubility, gelation, and emulsification properties compared to cPPI. Results confirmed, for the first time, that double solubilization at mild pH (7.5) can replace single solubilization at high alkalinity and achieve a similar yield while preserving structural integrity. Additionally, this study demonstrated, the scalability of the benchtop salt extraction coupled with ultrafiltration/diafiltration. Scaling up the production eliminated some structural and functional differences between the salt-extracted PPI and pH-extracted PPI. Scaling-up under mild and controlled conditions resulted in partial denaturation and a low degree of polymerization, coupled with the superior functionality of the produced isolates compared to cPPI. Results of this work can be used as a benchmark to guide the industrial production of functional pea protein ingredients.

Keywords: pea protein isolate; salt extraction; pH extraction; ultrafiltration; benchtop versus pilot plant production; protein structure and functionality

Citation: Hansen, L.; Bu, F.; Ismail, B.P. Structure-Function Guided Extraction and Scale-Up of Pea Protein Isolate Production. *Foods* **2022**, *11*, 3773. https://doi.org/10.3390/foods11233773

Academic Editors: Norbert Raak, Ruifen Li and Laura Roman

Received: 16 October 2022
Accepted: 15 November 2022
Published: 23 November 2022

1. Introduction

Produced from yellow field peas (*Pisum sativum* L.), pea protein has similar profile and nutritional quality compared to soy protein [1,2]. Therefore, pea protein ingredients are being incorporated into several food and beverage applications, replacing soy and animal-based protein ingredients. However, the functional properties of pea protein namely solubility, gelation, and emulsification are inferior to that of soy protein [3], hindering its expanded use. The food industry, accordingly, is searching for effective measures to enhance the functionality of pea protein for successful incorporation in various applications.

While soy protein ingredients have gone through years of research and development to improve functionality through processing and breeding, pea protein ingredient development is yet to catch up. The protein in commercially available pea protein isolates (PPI) is generally completely denatured, extensively polymerized, and has a relatively high surface hydrophobicity [4]. These structural characteristics contribute to challenging functionality [4]. Preserving the protein structural integrity during the production of PPI may contribute to desired functional attributes and better inclusion prospects in different applications.

The functionality of pea protein is not only affected by the inherent protein profile, but also by the structural changes that take place under certain extraction and processing conditions during the protein isolation process. Commercial PPI is mostly produced by pH extraction, utilizing alkaline solubilization of the protein to separate it from starch and fiber, followed by precipitation at the protein's isoelectric point to remove soluble compounds and enhance the purity [5–8]. Differences in pH treatment, holding time, number of solubilizations, heat treatments, and drying conditions, contribute to differences in the protein profile and structure, ultimately impacting functionality. Several studies investigated pea protein extractions at pH levels above pH 8 [7,9–11] and the impact on protein yield; however, high alkalinity coupled with adverse heat treatment during production cause protein denaturation and subsequent polymerization, thus negatively impacting functionality.

Protein isolates can also be produced by salt extraction, utilizing a salt solubilization step followed by "salting out" to concentrate the protein. However, this extraction process results in high waste streams and salty protein isolates that require several washes and diafiltration [12]. A few studies investigated salt extraction with "salting out" step for the production of protein isolate [13–15]. However, such isolation method is currently not adapted in industry due to the high waste stream, high use of water, and low efficiency. Alternative concentration techniques such as membrane filtration may replace the "salting out" step, which could make this extraction method more feasible. There is limited research on the optimization of salt extraction conditions, including salt concentration and type of membrane filtration, for optimal pea protein purity and yield [10,16–19]. A few of these studies were based on benchtop trials following micellar precipitation and/or using dialysis membranes [11,17], while only one study was based on pilot plant trials using ultrafiltration/diafiltration membranes [18–20]. However, there are no reported comparative studies to determine scalability of a benchtop salt extraction process involving membrane purification. Additionally, there are limited reports on the structural and functional properties of pea protein as impacted by pH extraction versus salt extraction under industry feasible conditions [10,19].

Protein extraction conditions should be selected and controlled based on their efficiency in extracting protein, while maintaining structural integrity. Extraction efficiency can be determined by tracking protein purity and yield. Mild and feasible extraction conditions that produce PPI of high protein purity and yield and acceptable functionality need to be explored. Therefore, the objectives of this work were: (1) evaluate pea protein extraction conditions to maximize protein purity and yield following mild alkaline solubilization with isoelectric precipitation and salt solubilization coupled with membrane filtration; (2) characterize the impact of the two extraction methods on the protein structure and functionality; (3) and for the first time, translate controlled benchtop extraction conditions to pilot scale production of PPI, while evaluating the impact of scaling on the structure and functionality of the protein.

2. Materials and Methods

2.1. Materials

Yellow field pea flour (~20% protein content) was provided by AGT Foods (Regina, SK, Canada) and commercial pea protein isolate (cPPI, 81.2% protein, 3.86% ash), PURIS™ Pea Protein, was provided by Puris Foods (Minneapolis, MN, USA). Commercial whey protein isolate (cWPI, 94.6% protein, 4.10% ash), BiPro®, was provided by Agropur Ingredients (Eden Prairie, MN, USA). Defatted soy flour (~53% protein) and commercial soy protein isolate (cSPI, ~90.7% protein), ProFam® 974, were provided by Archer Daniels Midland (ADM) (Decatur, IL, USA). All aforementioned samples were stored at −20 °C prior to usage. SnakeSkin™ dialysis tubing (3.5 kDa cut off) and Sudan Red 7B were purchased from Thermo Fisher Scientific™ (Waltham, MA, USA). Criterion™ TGX™ 4–20% precast gels, Laemmli 4X loading buffer, Imperial™ Protein Stain, 10X Tris/Glycine/SDS running buffer and Precision Plus molecular weight marker were purchased from Bio-Rad Laboratories, Inc. (Hercules, CA, USA). Vivaflow® membrane ultrafiltration cross-flow cassettes (3 kDa MWCO) were purchased from Sartorius™ (Gottingen, Germany). All other chemical reagents and supplies were purchased from Thermo Fisher Scientific and Sigma-Aldrich.

2.2. Benchtop Pea Protein Extractions

2.2.1. PPI Production Following pH Extraction

Solubilization pH, precipitation pH, solubilization time, number of solubilizations, and use of dialysis were evaluated for high protein purity and yield. In triplicate, pea flour was suspended in double deionized water (DDW, 1:10 *w*/*v*) and adjusted to pH 7, 7.5 or 8 with 2N NaOH. The suspensions were stirred for 1 h or 2 h then centrifugated at 5000× *g* for 30 min, and the supernatant was collected. The residual pellet was either lyophilized directly or was redispersed in DDW for another solubilization cycle. The supernatant from the second solubilization was combined with the first supernatant. The pH of the supernatant (or combined supernatants) was adjusted to the isoelectric point (pH 4.5 or 5) and centrifuged at 5000× *g* for 10 min. The precipitate was then redispersed in DDW (1:4 *w*/*v*) and the pH was adjusted to 7 prior to lyophilization. Fractions obtained from every step (the pellet from the first centrifugation step, the supernatant from the second centrifugation step, and the final neutralized PPI fraction) were lyophilized, weighed, and analyzed for protein content following a Dumas method (AOAC 990.03) using a LECO® FP828 nitrogen analyzer (LECO, St. Joseph, MI, USA). Mass balance was tracked and used to determine the final protein yield. The effect of dialysis was also investigated, where the neutralized PPI fraction was either lyophilized as is or was dialyzed at 4 °C against DDW using a 3.5-kDa cutoff membrane. Ash content (AOAC method 942.05) was determined to evaluate the efficiency of dialysis. When not in use, the samples were stored at −20 °C.

2.2.2. PPI Production Following a Salt Solubilization Coupled with Membrane Filtration Method

Pea protein extraction pretrials using salt solubilization coupled with membrane filtration (salt extraction) were performed to select the salt concentration. Selected salt concentration was based on extraction yields. The lowest salt concentration (0.5 M) that produced over 80% protein yield was chosen. The conditions for membrane filtration and protein purification were then evaluated by comparing the effects of membrane ultrafiltration (UF), dialysis, or UF followed by dialysis. The use of membrane filtration and/or dialysis was investigated as an alternative way to "salting out" for the concentration and purification of proteins solubilized in dilute a salt solution. To produce salt extracted PPI, pea flour, in triplicate, was solubilized in 0.5 M NaCl (1:20 *w*/*v*) for 1 h at the natural pH (6.4) and at room temperature (23 °C). The suspension was then centrifuged at 5000× *g* for 30 min to separate insoluble starch and fiber. The supernatant was neutralized and subjected to tangential ultrafiltration (UF), dialysis, or UF coupled dialysis to assess the efficiency of salt reduction. For ultrafiltration, the benchtop Sartorius Vivaflow® 200 system was used with two Vivaflow® membrane (3 kDa MWCO) cassettes running in parallel to increase the speed of filtration. The system was set up according to manufacturer instructions, with the protein solution in a feed reservoir and the feed tube connected to a peristaltic pump (Masterflex Easy Load Pump Head- Size 15, Masterflex Economy Drive Peristaltic Pump 230 V, Sartorius) to pump the feed solution under pressure (2.5 bars) across the membranes, to concentrate the samples down to 50 mL, followed by diafilteration using 6 volumes of DDW (300 mL total) to continually decrease the concentration of salt and other low molecular weight (MW) compounds. After diafiltration, the solution was further concentrated to 25 mL. After filtration and/or dialysis, the protein solution was lyophilized and stored at −20 °C before subsequent analysis. The protein purity, protein yield, and ash content of PPI were determined following the abovementioned methods and used to determine the favorable extraction conditions.

2.3. Scaled-Up Pea Protein Extractions in the Pilot Plant

The selected pH and salt extraction methods were scaled up in the Joseph J. Warthesen Food Processing Center at the University of Minnesota to determine how the benchtop extractions translated to larger scale production. The ultimate goal of scaling up pea protein extraction under controlled conditions is to produce functional PPI that can be mass produced by the food industry for widespread use. Therefore, it was important to examine how well the selected benchtop extraction conditions translate to a larger scale. For both the pH- and salt-extractions, there were some unavoidable differences between the benchtop and scaled-up (SU) extractions. Differences were encountered when benchtop equipment did not have a direct analog in the pilot plant. First, the centrifuges used differed in their separation power. The benchtop Beckman floor centrifuge forms dry, compact pellets and a clear supernatant, while the horizontal decanter centrifuge (Westfalia Separator AG, 3.8 L/min, GEA Westfalia Separator Group Gmbh, Oelde, Germany) in the pilot plant cannot achieve the same level of separation. To help improve separation, a desludging disc centrifuge (Westfalia SB7, 3.8 L/min, GEA Westfalia Separator Group Gmbh, Oelde, Germany) was used in sequence to clarify the supernatant further. Additionally, large scale dialysis is infeasible, so ultrafiltration/diafiltration (UF/RO unit, 15–20 psi inlet, 10–15 psi outlet, PTI Advanced Filtration, PTI Technologies, St. Louis, MO, USA with tangential/cross flow and a spiral wound membrane, 3 kDa MWCO) was used instead. Total solids of the permeate was constantly measured using a CEM AVC-80 Microwave Moisture/Solids Balance Analyzer (CEM, Charlotte, NC, USA) to monitor salt removal. Furthermore, ingredients produced in the pilot plant must be food-grade, so the PPI was pasteurized after filtration. Following pasteurization, the PPI was homogenized to help improve ease of drying. Lyophilizing is not commonly done in industry because it is time consuming and costly, so the scaled-up PPI samples were dried using a SPX Flow Anhydro Spray Dryer (9.5% TS, 180 °C inlet, 90 °C outlet, 9 L/h) with a wheel type atomizer (24,500 rpm) (SPX Flow Inc., Charlotte, NC, USA). Lastly, extractions in the pilot plant spanned across two days of processing, so precautionary steps were added to the extraction methods to prevent microbial growth overnight.

2.3.1. Scaled-Up pH-Extraction

Pea flour was dispersed in deionized (DI) water (1:10, *w*/*v*; pH 7.5) and agitated in a jacketed tank (568 L) with automated stirrer for 1 h at room temperature (23 °C). The solution was then separated using a horizontal decanter centrifuge and clarified with a desludging disc centrifuge. The separated liquid was set aside in the cold room (6–8 °C). The precipitate was weighed, and the total solids (%TS) was measured. The pellet was resuspended at 1:10 *w*/*v* and the pH adjusted to 7.5, followed by agitation in the jacketed tank for another hour. The solution was then passed through the decanter centrifuge and desludging centrifuge again, and the separated liquid was collected and combined with the separated liquid from earlier in a clean jacketed tank. The combined solution was then adjusted to pH 4.5 and agitated for 10 min. The solution was then passed through the decanter centrifuge, and the proteinaceous precipitate was collected. The supernatant was pumped through the desludging centrifuge to ensure that no additional solids precipitated. The combined solids were transferred to a jacketed tank and reconstituted in DI water at 1:4 *w*/*v*. The solution was adjusted to pH 3 and left in the cold room (6–8 °C) overnight. Leaving the protein at its isoelectric point overnight would cause protein denaturation, leading to aggregation and irreversible polymerization, while neutralizing the supernatant could allow for microbial growth overnight. At the beginning of the second day, the solution was neutralized and agitated for 1 h at room temperature (23 °C). Ultrafiltration/diafiltration was performed in place of dialysis using a UF/RO unit described above. The protein solution was diafiltered with a set up similar to the benchtop diafiltration. DI water was added (40 L at a time) until the % TS of the exiting permeate stream was 0.00%. Following ultrafiltration/diafiltration, the retentate was pasteurized by running the solution through a high temperature short time (HTST; 73 °C for 15 s) processing system (MicroThermics® Electric Model 25HV Hybrid, 57–170 L-/h, MicroThermics® Inc., Raleigh, NC, USA), followed by two-stage homogenization (Gaulin 125 L, 2500 psi, 227 L/h, Manton-Gaulin Mfg. Co. Inc., Everett, MA, USA). The solution was then spray dried using a SPX Flow Anhydro Spray Dryer described above. Protein purity and ash content of the final SU-pH PPI were determined by Dumas and dry ashing, respectively, to assess how the pilot plant extraction compared to the benchtop extraction. When not in use, the sample was stored at −20 °C.

2.3.2. Scaled-Up Salt Extraction

Pea flour was dispersed in 0.5 M NaCl (1:20 *w*/*v*) at its natural pH (pH 6.4) and agitated in a jacketed tank (568 L) with automated stirrer for 1 h at room temperature (23 °C). The solution was then separated using a horizontal decanter centrifuge and clarified with a desludging centrifuge. The separated liquid was set aside in the cold room (6–8 °C). The precipitate was weighed, and the total solids (%TS) was measured. The pellet was resuspended in 0.5 M NaCl at 1:5 *w*/*v* and agitated in a jacketed tank for 30 min. The solution was then passed through the decanter centrifuge and desludging centrifuge again, and the separated liquid was collected and combined with the separated liquid from earlier in a jacketed tank. The solution was neutralized and left stirring in the cold room (6–8 °C) overnight. The salt content of the solution and the cool temperature were assumed to be sufficient to prevent microbial growth. The following day, the protein solution was ultrafiltered/diafiltered pasteurized, homogenized, and spray dried as described above. It is worth noting that a membrane MWCO of 3 kDa was selected to avoid loss of small molecular weight proteins and peptides in contrast to 10, 20, and 50 kDa MWCO that have been used previously [18–20]. Additionally, when scaling, it was essential to simulate the benchtop filtration setup. Protein purity and ash content of the final SU-salt PPI were determined by Dumas and dry ashing, respectively, to assess the scalability of salt extraction. When not in use, the sample was stored at −20 °C.

2.4. Structural Characterization

2.4.1. Protein Profiling by SDS-PAGE and SE-HPLC

The protein subunit distribution in the different PPI samples and references was visualized using SDS polyacrylamide gel electrophoresis (SDS-PAGE) as previously described by Laemmli [21] and modified by Bu [4]. Sample aliquots (5 μL; containing ~50 μg protein) and MW standard (10 μL) were loaded onto a 4–20% acrylamide gel under non-reducing and reducing conditions. The gel was electrophoresed, stained, de-stained, and imaged as reported previously by Boyle [22].

Size exclusion high performance liquid chromatography (SE-HPLC), as previously described by Bu [4], was used to evaluate the molecular weight distribution of the proteins in the different samples. Pea protein (0.1 g) was solubilized in 10 mL of 0.05 M sodium phosphate with 0.1 M sodium chloride buffer at pH 7 for two hours, after which the samples were filtered through a 0.45 μm filter. The same Shimadzu system, analytical column, MW calibration standards, chromatographic method, and identification process reported by Bu [4] were used to obtain the molecular weight distribution.

2.4.2. Protein Denaturation as Determined by Differential Scanning Calorimetry (DSC)

The denaturation temperature and enthalpy of the different protein samples were determined using DSC (Mettler Toledo, Columbus, OH, USA) equipped with the STARe software, version 11, as described by Boyle [22] and modified by Bu [4].

2.4.3. Measurement of Zeta Potential and Surface Hydrophobicity

A Zetasizer Nano Z instrument (Malvern Panalytical, Malvern, UK), equipped with Malvern's Zetasizer software (version 7.13), was used to determine the zeta potential of the protein samples as described previously Bu [4]. A fluorescence-based method previously reported by Boyle [22], without modification, was used to determine the surface hydrophobicity of the protein samples.

2.5. Functional Properties

The solubility of the protein samples (5% protein concentration, *w*/*v*) was determined at pH 3.4 and 7 with and without heating at 80 °C for 30 min following the procedure outlined by Wang [23], without modification. Solubility was expressed as the percentage of protein in the supernatant to the total protein in the initial solution as determined by the Dumas method.

Thermally induced gels were prepared by heating protein solutions (at 15% or 20% protein *w*/*v*) at 95 °C for 10 min or 20 min. The samples were then cooled to room temperature and the gel strength was determined using a TA-TX Plus texture Analyzer (Stable Micro Systems LTD, Surrey, UK) with a 100 mm diameter probe, at a test speed of 1 mm s^{-1} and distance of 0.5 mm from the plate. The force (N) required to rupture the gel was reported as gel strength.

Emulsification capacity (EC, at 1% or 2% protein in DDW, *w*/*v*), activity index (EAI), and stability (ES) of the different protein samples at pH 7 were determined following the methods outlined by Boyle [22] and modified by Hinnenkamp [24].

2.6. Statistical Analysis

All measurements were performed in triplicate, and analysis of variance (ANOVA) was performed using RStudio software version 1.1.463 for Mac (Rstudio, Inc., Boston, MA, USA). When determining significant difference between two means, a student's unpaired t-test was used ($p \leq 0.05$). For determining significant differences among three or more means, Tukey–Kramer Honest Significant Difference (HSD) multiple means comparison test was ($p \leq 0.05$).

3. Results and Discussion

3.1. Effect of Different pH Extraction Conditions on the Efficiency of PPI Production

Solubilization pH, isoelectric precipitation pH, solubilization duration, number of solubilizations, and use of dialysis were each evaluated, in turn. With all other conditions kept constant, the solubilization pH was set at either pH 7, 7.5, or 8. Compared to PPI samples extracted at pH 7.5 and 8, PPI extracted at pH 7 had the lowest protein yield with significantly higher protein content and % residual protein in the discarded pellet fraction (Table 1), demonstrating that pH 7 was least effective in extracting proteins from pea flour. Solubilizing the protein at pH 8, on the other hand, resulted in significantly higher PPI protein yield compared to PPI samples extracted at pH 7 or 7.5. While it is known that plant proteins have increased solubility at elevated pH levels [25], high pH can coextract other non-protein constituents such as polysaccharides [26,27], and can negatively impact the functionality of the protein due to inducing protein denaturation and subsequent aggregation [28]. Additionally alkaline pH promotes oxidation that may lead to browning, off-flavor, and further protein polymerization [29–33]. Therefore, pH 7.5 was selected as the solubilization pH, as it resulted in a comparable and acceptable PPI protein purity (>85% protein) and yield.

Next, the isoelectric precipitation pH was evaluated. Pea protein, similar to soy protein, has low solubility over a wide range of acidic pH 3.5–5.5 [34,35], with isoelectric precipitation commonly performed at pH 4.5 or 5 [35,36]. A comparative testing was performed to determine the direct impact, if any, of the precipitation pH on the protein yield. Though there were no significant differences in the final PPI protein purity or yield, the supernatant fraction that was discarded after protein precipitation at pH 5 had a significantly higher protein content and % protein lost than the supernatant discarded after protein precipitation at pH 4.5 (Table 1). Therefore, pH 4.5 was selected for protein precipitation.

With regard to solubilization duration, there was no significant differences in protein yield and purity when solubilized for one hour versus two (Table 1). Therefore, one hour of solubilization was chosen for industrial efficiency. Similarly, Hoang [25] did not observe significant increase in pea protein extraction yield with increasing solubilization duration from 15 to 45 min. Pea flour, however, was solubilized at pH 10, where the proteins are more readily soluble than at pH 7.5, the chosen pH for the present study. It is assumed that some protein was retained with the pellet post solubilization at pH 7.5. Therefore, additional round of solubilization with fresh solvent was tested next.

Table 1. Percent protein purity and distribution in the PPI, pellet, and supernatant fractions from pH and salt extractions under different conditions, as well as ash content (%) of each PPI sample.

pH Extractions

Extraction Treatment					pH-PPI			Discarded Pellet [1]		Discarded Supernatant [2]	
Solubilization pH	**Precipitation pH**	**Solubilization Duration (h)**	**Number of Solubilizations**	**Dialysis of PPI**	**Protein Purity (%)**	**Protein Yield [3] (%)**	**Ash (%)**	**Protein Purity (%)**	**Protein Residue [4] (%)**	**Protein Purity (%)**	**Protein Lost [5] (%)**
7	4.5	1	1	No	89.7 ^{a6}	56.3 ^{eB7}	5.11 ^{a}	8.08 ^{aA}	22.1 ^{aA}	29.5 ^{b}	18.7 ^{d}
8	4.5	1	1	No	86.5 ^{ab}	60.9 ^{cA}	5.02 ^{a}	6.76 ^{abB}	18.9 ^{aB}	29.5 ^{b}	19.7 ^{cd}
7.5	4.5	1	1	No	88.3 ^{ab}	58.1 ^{cdeB}	5.03 ^{a}	6.31 ^{abcB}	18.2 ^{aB}	28.7 ^{bc}	19.3 ^{cd}
7.5	5	1	1	No	89.4 ^{a}	57.5 ^{de}	4.24 $^{b\wedge 8}$	5.82 ^{bc}	17.1 ^{a}	31.5 $^{a\wedge}$	23.4 $^{a\wedge}$
7.5	4.5	2	1	No	85.0 ^{ab}	60.5 ^{cd}	5.07 ^{a}	5.66 ^{bc}	16.6 ^{ab}	27.7 ^{bc}	20.1 ^{c}
7.5	4.5	1	2	No	84.5 ^{b}	69.9 $^{a\dagger 9}$	5.19 ^{a}	3.00 ^{d†}	8.11 ^{c†}	27.4 ^{c†}	21.8 ^{b†}
7.5	4.5	1	2	Yes	87.6 ^{ab}	64.7 ^{b*10}	4.96 ^{a}	4.30 ^{cd}	11.2 ^{bc}	28.0 ^{bc}	21.6 ^{b}

Salt Extractions

Purification Treatment		Salt-PPI			Discarded Pellet	
Ultrafiltration	Dialysis of PPI	Protein Purity (%)	Protein Yield (%)	Ash (%)	Protein Purity (%)	Protein Residue (%)
Yes	No	67.9 ^{c}	76.1 ^{a}	11.4 ^{a}	7.97 ^{a}	25.2 ^{a}
No	Yes	86.9 ^{b}	69.7 ^{b}	7.19 ^{b}	7.98 ^{a}	25.3 ^{a}
Yes	Yes	92.8 $^{a\wedge *}$	72.0 ^{ab*}	1.56 $^{c\wedge *}$	7.70 ^{a}	24.2 ^{a}

[1] Pellet discarded after alkaline or salt solubilization; [2] Supernatant discarded after isoelectric precipitation; [3] Protein yield (%) represents the amount of protein extracted relative to the total amount of protein in the starting pea flour; [4] Protein residue (%) represents the amount of protein left in the discarded pellet relative to the total amount of protein in the starting pea flour; [5] Protein lost (%) represents the amount of protein lost to the discarded supernatant relative to the total amount of protein in the starting pea flour during pH extractions; [6] Means (n = 3) in each column with different lowercase letters indicate significant differences across extraction treatments; [7] Means with different capital letters indicate significant differences among different solubilization pHs; means with no capital letters indicate no significant differences, according to the Tukey–Kramer multiple means comparison test ($p < 0.05$); [8] A carrot (^) in pH extractions designates a significant difference among corresponding samples precipitated at pH 4.5 and 5, while a carrot (^) in salt extractions designates a significant difference among corresponding samples with and without ultrafiltration, as tested by the student's two-sample unpaired t-test ($p < 0.05$); [9] A cross of Lorraine (☨) designates a significant difference among corresponding samples solubilized once or twice, as tested by the student's two-sample unpaired t-test ($p < 0.05$); [10] An asterisk (*) designates a significant difference among a corresponding dialyzed and non-dialyzed sample, as tested by the student's two-sample unpaired t-test ($p < 0.05$).

Double solubilization, for 1 h each, was indeed effective in extracting additional protein from the first pellet, which ultimately led to ~12% increase in protein yield (58.1 vs. 69.9%), accompanied by a significant reduction (10%) in % residual protein in the discarded pellet fraction (Table 1). The reported PPI protein yield obtained from alkaline solubilization with isoelectric precipitation is around 60–70%, mainly achieved by solubilizing proteins at extreme alkaline conditions (pH 8–10) [17]. In contrast, the relatively similar or superior yield observed in this study when employing double solubilization was achieved using mild alkaline solubilization pH, hence preserving the protein structure. This is the first study to discover and illustrate that double solubilization at mild pH can replace single solubilization at high alkalinity and achieve similar yield, while potentially preserving structural integrity.

The effect of dialysis was investigated next to determine the impact of desalting and removal of small molecules on the purity and yield. PPI with high protein purity is desirable from a nutritional standpoint and may demonstrate superior functional properties due to the reduced content of other interfering components. Dialysis, however, caused a 5% decrease in the protein yield (Table 1), attributed to an additional transfer step and incomplete recovery of the protein from the dialysis tubing, and did not successfully decrease the ash content. While the difference in yield is statistically significant, the yield remains relatively high. Reducing the salt content using ultrafiltration/diafiltration is more effective compared to dialysis; therefore, it was decided to pursue membrane filtration during scale up trials to enhance upon the protein functionality even though the yield might be slightly compromised.

Based on these findings, the final extraction conditions were double solubilization of pea flour for one hour at pH 7.5, protein precipitation at pH 4.5, and the use of filtration,

These nondenaturing extraction conditions were chosen for the scaled-up production of pH extracted PPI.

3.2. Effect of Membrane Filtration on the Production Efficiency of Salt Extracted PPI

Salt extraction of plant proteins is less researched than alkaline extraction and is not currently used in industry for the production of PPI. Studies investigating salt extraction mostly recover the PPI by "salting out", which is not industry feasible due the use of excessive salt and water. Other studies report micellar precipitation (also called "hydrophobic out"), in which the proteinaceous supernatant from "salting in" gets diluted with cold water, leading to increased hydrophobic interactions causing protein precipitation and reduced protein solubility [10,12,17,37].

In this study, membrane filtration, in place of "salting out", was used to purify the protein after salt solubilization. For the first time, the utilization of membrane ultrafiltration (UF) compared to dialysis, separately, and in combination, were evaluated in terms of protein purity, yield, and ash content. The initial salt solubilization step was the same for all three purification treatments, thus both the protein purity and % protein residue of the discarded pellet were not impacted (Table 1). Ultrafiltration alone, using the benchtop membrane system, resulted in the lowest protein purity and highest ash content, demonstrating that it was not sufficient to completely remove all salt from the proteinaceous supernatant. Dialysis was significantly more effective in increasing protein purity and decreasing ash content, though the ash content was still relatively high, indicating that not all salt was removed (Table 1). The combination of UF and dialysis had comparative high yield and had significantly the highest PPI protein purity and the lowest ash content of the three purification treatments (Table 1), thus was selected as the purification treatment. This outcome could translate to successful use of UF coupled with diafiltration on a pilot or industrial scale.

To the best of our knowledge, very limited work has been done on the use of solubilization in dilute NaCl, followed by UF, to produce PPI. While ultrafiltration has been used following an alkaline or acidic solubilization step, or as further purification after isoelectric precipitation [11,16], it is not well characterized as a protein concentration step following salt solubilization. Tian [19] and Taherian [20] reported the use of ultrafiltration post salt solubilization of pea protein, however, 20 and 50 kDa MWCO membrane were used, respectively, which could potentially lead to loss of a considerable amount of relatively small molecular weight proteins.

Scalability of UF coupled with diafiltration could potentially be considered costly to operate. However, membrane filtration is commonly utilized in the dairy industry to produce whey protein concentrates and isolates of high quality and value. Additionally, in an economic evaluation study comparing alkaline extraction with isoelectric precipitation to membrane isolation for the production of soy protein isolate, there were no significant differences between in the cost of production [38]. Therefore, it is essential to investigate the potential differences between benchtop extractions and scaled-up extractions, under controlled conditions with small MWCO (3 kDa), to determine the feasibility of such extraction process and its impact on protein profile, structure, and functionality in comparison to the conventional pH extraction protocol.

3.3. Pilot Production of PPI Following the Selected Extraction/Purification Conditions

Pilot scale production of PPI was performed closely following the selected benchtop extraction/purification conditions, with slight modifications to accommodate industrial practices, including the use of diafiltration instead of dialysis, pasteurization for safety measures, and spray drying instead lyophilization. The SU production achieved similar PPI purity to the benchtop counterpart. SU-pH PPI had a protein purity of 88.7%, while SU-salt PPI had a protein purity of 92.4%. Ash content of the SU-salt PPI (1.66%) was comparable to that of the benchtop salt-PPI (1.56%), while ash content of the SU-pH PPI (2.94%) was significantly ($p < 0.05$) lower than the benchtop pH-PPI (4.96%). In their study on pilot scale production of salt extracted PPI, Tian [19] reported a lower PPI protein purity (81.1%), which could be attributed to higher retention of starch and fiber potentially due to less ideal separation mechanism, coupled with probable loss of protein when using high MWCO membrane (20 kDa). While there are a couple of reports on scaling up of salt extracted PPI, the literature still lacks a thorough and comparative investigation of the impact of different extraction methods, performed under mild conditions, and their scaling up, on the structural changes and consequent functionality compared to commercially available PPI.

3.4. Effect of Extraction Conditions and Scaling up on Protein Profile

Protein bands corresponding to lipoxygenase (~94 kDa), convicilin (~72 kDa), vicilin (~50 kDa), and legumin (~60 kDa) (Figure 1A) were apparent in all extracted PPIs, similar to previous reports [29,39,40]. Since convicilin and vicilin do not contain disulfide linkages, their corresponding bands were similar under reducing and nonreducing conditions (Figure 1A,B). On the other hand, legumin was cleaved into its acidic (~40 kDa) and basic (~20 kDa) subunits under reducing conditions (Figure 1B).

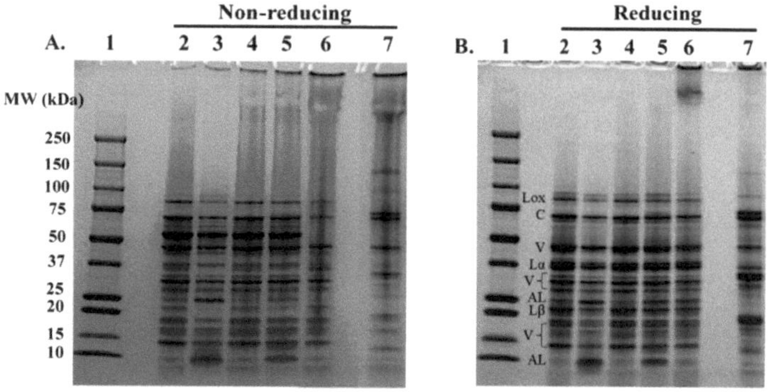

Figure 1. SDS-PAGE gel visualization of the protein profiles of protein isolate samples under non-reducing (**A**) and reducing (**B**) conditions. Lane 1: molecular weight (MW) marker; Lane 2: pH-PPI; Lane 3: salt-PPI; Lane 4: SU-pH PPI; Lane 5: SU-salt PPI; Lane 6: cPPI; Lane 7: cSPI. Lox: lipoxygenase; C: convicilin; V: vicilin; Lα: acidic subunit of legumin; Lβ: basic subunit of legumin; AL: albumin.

While benchtop extracted PPI and corresponding SU PPI samples had similar protein bands, there were a few notable differences. The most notable difference was between the benchtop salt-PPI and SU-salt PPI under nonreducing conditions. The upper part of the lane for SU-salt PPI had longitudinal smearing compared to that of the benchtop salt-PPI, indicating the presence of large protein aggregates (Figure 1A, lanes 3&5). Both SU-pH PPI and pH-PPI had smearing in the upper region of their respective lanes under nonreducing conditions (Figure 1B, lanes 2&4), indicating the presence of protein aggregates. However, the smearing was more intense for the SU-pH PPI, with a much darker band at the very top of the lane compared to the benchtop pH-PPI. The smearing and the intensity of the high molecular weight bands at the top of the lanes in the SU PPI samples were markedly less visible under reducing conditions (Figure 1B, lanes 4&5), indicating that these aggregates were formed primarily through disulfide linkages due to the thermal treatment during pilot plant production (pasteurization and spray drying).

When comparing salt-extracted PPI to pH-extracted PPI, unique bands around 25 and 9 kDa were present in salt PPI and SU-salt PPI and absent or less intense in the pH PPI and SU-pH PPI, under both reducing and nonreducing conditions (Figure 1A,B, compare lanes 3&5 to lanes 2&4). This observation indicated that these 25 and 9 kDa proteins have an isoelectric point further from pH 4.5, allowing them to remain soluble during the isoelectric precipitation step of the pH extraction, and hence are lost in the discarded supernatant. These proteins are likely albumins, which have an isoelectric point of around pH 5.5–6 [41–43]. Albumins have different structural properties than globulins, so their presence in the salt-extracted PPIs may cause some differences in functionality.

Both commercial references, cPPI and cSPI had intense smearing and high molecular weight bands compared to the benchtop and SU PPI samples. Additionally, the 60 kDa band corresponding to legumin and glycinin in cPPI and cSPI, respectively, was largely absent under nonreducing conditions (Figure 1A, lanes 6&7). However, under reducing conditions, the corresponding acidic and basic subunits around 40 and 20 kDa, respectively, were present in both lanes. Additionally, the smearing in the upper region of the lanes was markedly reduced. The reduction in smearing coupled with the appearance of the acidic and basic subunits indicated that the excessive polymerization, involving the legumin in cPPI and glycinin in cSPI, occurred mostly via disulfide linkages. These observations were similar to those reported previously [40,44], when comparing native PPI and isolated legumin fraction to excessively heated samples.

High molecular weight bands (>250 kDa) were still apparent under reducing conditions, in both cPPI and cSPI. The persistent presence of these large molecular weight bands indicated that the polymerization in these samples involved not only disulfide linkages, but also other covalent linkages [45–47]. Therefore, it was concluded that cPPI and cSPI were most likely subjected to high alkalinity and excessive heat treatment during processing, which could impair their functional properties. However, it was noted that under reducing conditions, the bands corresponding to glycinin subunits in cSPI had a relatively higher intensity than legumin counterparts in all PPI samples, indicating that the legumin to vicilin ratio is higher in SPI than in PPI, likely contributing to better functionality of cSPI compared to cPPI, as will be discussed in later sections.

The protein molecular weight distribution and the relative abundance of major protein fractions in the different isolates were determined using SE-HPLC (Figure 2 and Table 2). Compared to benchtop PPI samples, SU PPI samples had significantly higher relative abundance of soluble aggregates, regardless of extraction type. This observation is likely attributed to the additional processing that the SU samples were subjected to (pasteurization, homogenization and spray drying). Bu [4] reported enhanced functionality upon the formation of soluble aggregates. Therefore, the presence of soluble aggregates in SU PPI might contributed to better functionality compared to benchtop PPI. While cPPI also showed relatively high abundance of soluble aggregates, it had the least relative abundance of functional proteins across all PPI samples (Figure 2), confirming the SDS-PAGE observation (Figure 1A). Proteins in cPPI not only formed soluble aggregates, but they formed larger insoluble aggregates that were most likely filtered out when the sample was passed through the 0.45 μm filter prior to the SE-HPLC analysis, thus had no apparent chromatographic peak. In contrast, all benchtop and SU PPIs had high relative abundance of functional proteins. High level of insoluble aggregates along with the loss in functional proteins would likely impair the functionality of cPPI, whereas the presence of functional proteins and modest amounts of soluble aggregates in the pH and salt PPI samples extracted under selected conditions could contribute to better functionality.

3.5. Protein Denaturation as Impacted by Extraction Method and Scale

The impact of the extraction method (pH- vs. salt- extraction) and scale (benchtop vs. SU) on the protein denaturation temperature and enthalpy was assessed using DSC. The benchtop and SU PPI samples had two endothermic peaks corresponding to vicilin and legumin (Table 3). It is likely that the endothermic peak for convicilin overlapped with that of vicilin, as they are structurally similar. Isolated vicilin and convicilin fractions were reported to have the same denaturation temperature [31]. Moreover, convicilin represents a much smaller portion (4–8%) of the total pea proteins than vicilin (up to 52%), thus the majority of the enthalpy of denaturation for the first peak is attributed to vicilin [39].

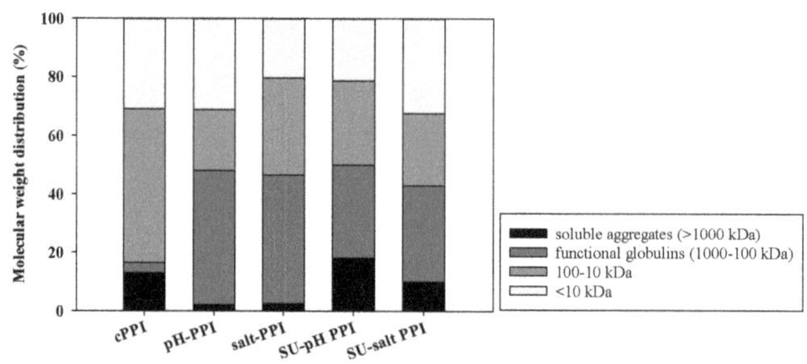

Figure 2. Percent relative abundance of different protein fractions in cPPI reference and PPI samples produced at bench scale and pilot scale. Samples were dissolved in pH 7 phosphate buffer and analyzed by SE-HPLC. Bars distribution represents means of $n = 3$.

Table 2. Relative abundance of soluble aggregates and functional proteins present in cPPI reference, and PPI samples produced at bench scale and pilot scale.

Protein Fractions [1]	Relative Abundance (%) [2]				
	cPPI	pH-PPI	Salt-PPI	SU-pH PPI	SU-Salt PPI
Soluble aggregates (Association of legumin, vicilin and other protein fractions)	13.06 [b3]	2.07 [d]	2.55 [d]	18.06 [a]	9.91 [c]
Legumin	*[4]	28.23 [a]	25.41 [a]	19.47 [b]	19.67 [b]
Convicilin	*	7.30 [a]	7.23 [a]	5.57 [b]	5.18 [b]
Vicilin	3.42 [e]	10.44 [b]	11.38 [a]	6.84 [c]	5.63 [d]

[1] Samples were dissolved in pH 7 phosphate buffer and analyzed by high-performance size exclusion chromatography (SE-HPLC); [2] Relative abundance (%) is the area of a specific peak divided by the total peak area for that sample; [3] Means ($n = 3$) in each row with different lowercase letters indicate significant differences according to the Tukey–Kramer multiple means comparison test ($p < 0.05$); [4] An asterisk (*) indicates that no peak was detected in this molecular weight range.

Table 3. Denaturation temperatures and enthalpy, surface hydrophobicity and surface charge of commercial protein references and the different PPI samples produced at bench scale and pilot scale.

Plant Protein Isolate	Denaturation Temperature and Enthalpy				Surface Hydrophobicity	Surface Charge
	Denaturation Temperature (Td, °C)	Enthalpy of Denaturation (ΔH, J g^{-1})	Denaturation Temperature (Td, °C)	Enthalpy of Denaturation (ΔH, J g^{-1})	RFI	mV
	Vicilin		Legumin			
cPPI	*[1]	*	*	*	12,718 [ab]	−31.1 [cd]
pH-PPI	83.3 [b2]	6.21 [a]	91.6 [ab]	0.81 [b]	9667 [cd]	−40.2 [a]
salt-PPI	88.5 [a]	6.82 [a]	93.5 [a]	1.54 [a]	7161 [d]	−30.4 [d]
SU-pH PPI	81.9 [c]	3.69 [b]	90.0 [b]	0.52 [b]	15,131 [a]	−34.5 [b]
SU-salt PPI	82.4 [c]	4.15 [b]	90.2 [b]	0.47 [b]	13,114 [ab]	−32.7 [bc]
	β-conglycinin		Glycinin			
cSPI	*	*	*	*	11,662 [bc]	−41.1 [a]

[1] An asterisk (*) represents no peak of denaturation observed; [2] Means ($n = 3$) in each column with different lowercase letters indicate significant differences among samples, according to the Tukey–Kramer multiple means comparison test ($p < 0.05$).

The legumin In the salt-PPI had slightly higher enthalpy of denaturation compared to that in the pH-PPI (Table 3). This observation compliments the SDS-PAGE observation, where more smearing in the upper part of the lane was noted for pH-PPI compared to salt-PPI (Figure 1A, lanes 2&3). Both observations indicated that salt-extraction was less denaturing than the pH-extraction. On the other hand, vicilin and legumin in both benchtop PPI samples had higher denaturation enthalpies than those in the SU PPI samples (Table 3). This observation confirmed that the processes during pilot plant production resulted in partial denaturation of the proteins, mostly attributed to the thermal treatment during pasteurization and spray drying.

While the SU PPI samples may have been partially denatured compared to the benchtop PPI samples, both were less denatured than the commercial references (cPPI and cSPI). Neither cPPI nor cSPI showed any endothermic peaks, indicating that the vicilin and legumin proteins were completely denatured. This finding supports the protein profiling observations (Figures 1 and 2, Table 2), which showed that both commercial references had excessive aggregation of denatured proteins, as was also reported by other researchers [4,32,48]. While the extraction and processing conditions of these commercial proteins are not known, it is presumed that the commercial references had undergone harsher extraction and thermal processing conditions compared to the selected SU conditions of this study, causing complete denaturation. Specifically, the difference in the denaturation state between the SU PPI samples and the references could be attributed to higher solubilization pH, overnight storage at the isoelectric point, thermal concentration step, and/or different spray drying conditions. The differences in the degree of denaturation will help explain differences in other structural properties and consequently functionality.

3.6. Protein Surface Properties as Impacted by the Extraction Method and Scale

Extraction method did not have a significant impact on surface hydrophobicity, while the extraction scale did (Table 3), confirming that thermal treatment had a major impact on the protein structure. This observation is consistent with the denaturation state of SU PPI samples compared to their benchtop counterparts. Denaturation disrupts electrostatic interactions, hydrogen bonding, and hydrophobic interactions that stabilize the native globular protein structure, causing unfolding and exposure of the hydrophobic core [49]. While cPPI in comparison to SU PPI samples was completely denatured (Table 3), there were no significant differences in surface hydrophobicity among these samples. Legumin protein was excessively polymerized in cPPI (Figure 1, lane 6). Upon complete denaturation, surface hydrophobicity will reach a maximum, and with subsequent and progressive polymerization induced by denaturation, it will be reduced [23]. Denatured proteins are attracted to each other via hydrophobic forces; once in close proximity, hydrophobic interactions occur and facilitate disulfide polymerization [50]. While both the SU PPIs and cPPI had protein aggregates, protein aggregation was more extensive in cPPI (Figure 1). Higher degree of denaturation and polymerization will have a detrimental impact on some functional properties, such as solubility, as will be discussed in later sections. Similarly, cSPI was completely denatured (Table 3), thus had high surface hydrophobicity. SPI, however, is unique in that it has a relatively high surface hydrophobicity along with fairly high surface charge [51]. The balance of surface charge and surface hydrophobicity affects how the protein interacts with its surrounding, thus, it impacts functional behavior such as solubility, gelation, and emulsification, as will be discussed in later sections.

Zeta potential is an indication of surface charge, which influences not only the protein solubility but also its intermolecular interactions. All tested proteins carried a net negative charge at pH 7, which is above their isoelectric point. The benchtop pH-PPI had the highest surface charge among all PPI samples, while the benchtop salt-PPI had the lowest surface charge (Table 3). This observation agrees with previous reports of legume proteins produced following pH extraction versus salt extraction [11]. The observed differences in surface charge may be explained by the protein composition of the pH-PPI compared to that of the salt-PPI. The pH-PPI was purified by isoelectric precipitation, so the overall pI of pH-PPI is at pH 4.5, where it carries no net charge. The salt-PPI, on the other hand, contained albumin proteins (Figure 1), which have an isoelectric point around pH 5.5–6 as explained earlier. Therefore, the overall pI of salt-PPI is likely higher than pH 4.5. Proteins carry more charge as they are farther from their isoelectric point, thus the pH-PPI had higher net negative charge than the salt-PPI.

The surface charge of SU-pH PPI was significantly lower than that of benchtop pH-PPI. This observation is most likely attributed to the partial denaturation incurred during

SU production. Upon denaturation of globular proteins, uncharged hydrophobic groups get exposed, causing a reduction in the measured zeta potential. On the other hand, while SU-salt PPI was also partially denatured, it had slightly higher surface charge than the benchtop salt-PPI. Although statistically significant, this difference could not be explained by the partial unfolding of the protein. Difference in ash content and the protein subunits that get involved in polymerization could be behind this observed difference. Polymerization, as noted by SDS-PAGE, involved partially the 25 kDa albumin subunit (Figure 1A, lane 3 vs. 5). Albumins, as mentioned before, would have less surface charge at pH 7 compared to globulins, thus polymerization involving albumin proteins may lead to the perceived higher net surface charge of SU-salt PPI compared to benchtop salt-PPI.

cPPI had the lowest surface charge compared to the other PPI samples, except for salt-PPI. cPPI was completely denatured (Table 3). The unraveling of the protein structure resulted in exposure of hydrophobic groups and consequently extensive protein aggregation (Figure 1). Although cSPI was similarly denatured and polymerized, it maintained higher net negative surface charge than most other proteins.

3.7. Protein Functionality as Impacted by the Extraction Method and Scale

Protein solubility of the different PPI samples was evaluated against not only commercial PPI and SPI, but also against a commercial WPI (cWPI). WPI has been considered the "gold standard" for use in high protein beverages, as it is highly soluble (at 5–10% protein) even post-thermal treatment, at both neutral and acidic pH [23]. WPI maintains high solubility over a wide range of pH since it has relatively low surface hydrophobicity (RFI of 4051) compared to PPI and SPI samples (Table 3). Accordingly, all PPI samples and reference cPPI and cSPI had significantly lower solubility than cWPI at pH 7, and more so at pH 3.4 (Table 4). Solubility was measured at both pH 7 and pH 3.4 to indicate potential use of pea protein in neutral and acidic high protein beverages as a replacement for WPI.

Table 4. Solubility, gel strength and emulsification properties of commercial protein references and PPI samples produced at bench scale and pilot scale.

Protein Isolate	% Protein Solubility (5% Protein)				Gel Strength (15 or 20% Protein) [1]	Emulsification Capacity (1 or 2% Protein) [2]	EAI	ES
	pH 7		*pH 3.4*					
	Not-Heated	Heated at 80 °C	Not-Heated	Heated at 80 °C	N	mL Oil/g of Protein	m^2/g Oil	Min
cWPI	99.7 [a3]	99.7 [a]	99.4 [a]	100.0 [a]	N/A [4]	N/A	N/A	N/A
cSPI	79.7 [d]	86.9 [b*5]	23.7 [f]	25.9 [d]	17.3 [a]	1085 [a]	131.8 [b]	32.5 [bc]
cPPI	22.2 [f]	39.0 [d*]	8.9 [g]	20.4 [d*]	N/A	260 [d]	88.7 [c]	45.4 [a]
pH-PPI	87.4 [b]	85.7 [b]	43.6 [e]	65.4 [c*]	6.1 [c]	441 [b]	149.1 [b]	36.9 [b]
salt-PPI	84.1 [c]	72.0 [c*]	88.7 [b]	90.1 [b]	14.6 [b]	452 [b]	197.4 [a]	26.0 [d]
SU-pH PPI	88.4 [b]	86.8 [b]	68.4 [c]	61.6 [c]	7.5 [c]	359 [c]	95.1 [c]	36.7 [bc]
SU-salt PPI	70.4 [e]	71.5 [c]	64.0 [d]	68.6 [c]	5.7 [c]	403 [bc]	72.4 [c]	30.8 [cd]

[1] SPI gel samples were prepared at 15% protein (*w/v*) and heated for 10 min at 95 °C, while PPI gels were prepared at 20% protein (*w/v*) and heated for 20 min at 95 °C; [2] SPI EC samples were prepared at 1% protein (*w/v*), while PPI EC samples were prepared at 2% protein (*w/v*); [3] Means ($n = 3$) in each column with different lowercase letters indicate significant differences among samples, according to the Tukey–Kramer multiple means comparison test ($p < 0.05$); [4] N/A: not available; [5] An asterisk (*) indicates a significant difference between a not-heated and heated sample ($p < 0.05$).

At pH 7, the pH-extracted PPI samples had significantly higher protein solubility than the salt-extracted PPI samples, which could partially be attributed to their relatively higher surface charge (Table 3). The observed difference could also be attributed to the selective solubilization of proteins at pH 7.5 during the pH-extraction method. During salt extraction, the solubilized proteins would not necessarily have selective solubility at a pH close to 7. On the other hand, partial denaturation of the pH-extracted PPI imparted by SU production did not majorly impact protein solubility at pH 7 under both not-heated and heated conditions. On the other hand, the solubility of SU-salt PPI at pH 7 was significantly lower than that of salt-PPI when not-heated yet was comparable when heated. This observation could be attributed to the combined effect of surface charge and denaturation. The thermal treatment caused partial denaturation of the major proteins in SU-salt PPI, and significantly increased the surface hydrophobicity compared to that of the benchtop salt-PPI (Table 3). However, heating the protein solutions at 80 °C for 30 min most likely resulted in partial denaturation of the proteins in the benchtop salt-PPI, but did not cause further changes to the proteins in SU-salt PPI.

At pH 3.4, salt-PPI, when not-heated and when heated, had significantly higher solubility than pH-PPI (Table 4). Since pH-extraction selected for proteins with an isoelectric point of 4.5, pH-PPI demonstrated comparatively lower solubility at pH 3.4 than salt-PPI. The lower net negative charge of salt-PPI at pH 7 compared to that of pH-PPI (Table 3) indicated that the pI of salt-PPI was higher than 4.5, potentially due to higher albumin content as explained earlier. Thus, the salt-PPI would carry higher net positive charge at pH 3.4 than pH-PPI. Upon partial protein denaturation in the SU samples, the potential difference in net positive charge between the salt extracted and pH extracted samples had no apparent impact on protein solubility. However, partial denaturation did result in significantly lower protein solubility of SU-salt PPI compared to salt-PPI, with no apparent influence of another factor. The ash contents of the SU-salt PPI and salt-PPI were not statistically different (1.66% ash for SU-salt PPI, 1.56% ash for benchtop salt-PPI), thus could not have contributed to differences in solubility. At pH 3.4, the net (positive) surface charge would be lower than the net (negative) surface charge at pH 7, resulting in potentially less repulsion among the partially denatured proteins, allowing for hydrophobic interactions and subsequent aggregation. On the other hand, The SU-pH PPI had significantly higher solubility at pH 3.4 when not-heated than the benchtop pH-PPI. This observation could be attributed to their different ash contents (4.96% ash for benchtop pH-PPI, 2.94% ash for SU-pH PPI; $p < 0.05$). At low levels, salt can help increase protein solubilization by "salting in" the proteins. However, after a certain point, salt content may be high enough to shield surface charge on the protein, thereby decreasing solubility [13,52]. This effect was not seen at pH 7, likely because the proteins carried higher charge further from their isoelectric point, so the salt content might not have had a major impact. However, at a pH closer to pH 4.5, the proteins carried less net charged, so the effects of shielding by salt ions were relatively more pronounced.

Regardless of extraction method or scale, PPI samples had comparable solubility to cSPI and superior solubility to cPPI at pH 7 and had superior solubility to both cSPI and cPPI at pH 3.4 (Table 4). Shand [32] similarly observed lower solubility of commercial PPI and SPI compared to native PPI and SPI. The authors attributed this observation to the denaturation and polymerization upon spray drying [32]. However, the SU PPI samples produced in the current study were also spray dried, indicating that the poor solubility of cPPI and cSPI were caused by harsh extraction/processing conditions other than spray drying. The high alkalinity (pH 9–11) used in different studies [7,9–11] and in industry to enhance extraction yield, is the likely cause of reduced solubility. These findings confirm that SU production of PPI under controlled conditions (e.g., lower pH coupled with double solubilization) will contribute to relatively high and acceptable solubility compared to current industrial processes.

While preserving the structural integrity of pea protein through controlled extraction conditions had a pronounced impact on solubility, its effect on gelation was modest. Gelation is impacted greatly by the inherent protein profile and characteristics. Due to relatively higher legumin (glycinin) to vicilin (β-conglycinin) ratio [1,32,53] in soy protein compared to pea protein, cSPI formed the strongest gel compared to all PPI samples (Table 4). In contrast to vicilin, legumin proteins contain cysteine residues that can form inter- and intramolecular linkages, contributing to a uniform structure and strength to a gel matrix [54].

Accordingly, PPI gels are believed to be primarily formed through hydrophobic interactions among vicilin proteins, with minor disulfide stabilization by the small amount of legumin present [32,53]. Due to this difference in the inherent protein profile, cSPI formed a relatively strong gel at 15% protein, while 20% protein concentration was needed for the PPI samples to form cohesive gels. Even at 20% protein concentration, the PPI gels formed were significantly weaker than the cSPI gel (at 15% protein). In fact, cPPI formed a very weak get that fell apart before a gel rupture force can be recorded. cPPI had low surface charge, was completely denatured, extensively aggregated, and irreversibly polymerized. It also had very low solubility at pH 7 compared to the other samples. Accordingly, heating cPPI caused further, yet random, aggregation of the proteins in the form of coagulum that was unable to entrap water. Taherian [20] similarly reported that commercial PPI did not form a gel at 20% protein, an observation they attributed to poor solubility due to denaturation and polymerization that occurred during protein extraction.

Both the benchtop and SU PPIs were superior to cPPI in their ability to form a gel, as they were less denatured during extraction and processing. The benchtop salt-PPI approached cSPI in terms of gel strength and had significantly the highest gel strength among the PPI samples (Table 4). Salt-PPI had slightly lower surface charge than the other PPI samples, and considerably lower surface hydrophobicity (Table 3), indicating a potentially more suitable hydrophilic to hydrophobic balance for a balanced protein-water and protein–protein interactions. Protein–protein interactions are required for a gel matrix to form. However, the presence of pre-formed large polymers/aggregates among denatured protein may have a negative effect on gel strength [55]. On the other hand, proteins that are less denatured and not polymerized will further denature during heating and then proceed to form soluble aggregate, leading to the formation of an ordered protein network [56,57]. The SU-pH and SU-salt PPI samples were partially denatured and aggregated (Table 3 and Figure 1), and the pH-PPI showed slight aggregation. While SU-pH PPI, SU-salt PPI, and pH-PPI still formed a gel, the gels likely formed through random associations compared to that of the salt-PPI, which was the least denatured and did not have any signs of aggregation (Figure 1 and Table 3).

cSPI also outperformed all PPI samples in most of the emulsification properties (Table 4), attributed to its moderate surface hydrophobicity and relatively high surface charge, resulting in a superior hydrophilic to lipophilic balance (HLB) compared to the PPI samples. All PPI samples did not form emulsion at 1% protein, thus had to be tested at 2% protein instead. The poor solubility, low surface charge, high surface hydrophobicity, and excessive polymerization inhibited cPPI from effectively interacting with either phase, resulting in the least EC among the samples. On the other hand, benchtop and SU PPIs had comparable EC values that were significantly higher than that of cPPI (Table 4).

Compared to cSPI, benchtop pH-SPI had similar EAI, and salt-PPI had significantly higher EAI. This observation indicated that salt-SPI, compare to all other samples, more readily migrated to the water/oil interface, and oriented their hydrophobic residues to the oil phase and hydrophilic residues to the water. Proteins that have flexible structures and good solubility have relatively high EAI [26,58,59]. Salt-PPI was the least denatured and polymerized among the samples, potentially contributing to ease of migration to the interface.

While SU PPI samples had significantly lower EAI than cSPI, they had comparable ES. ES is related to the thickness of the protein film at the interface and the charge on the surface of the oil droplets needed to cause repulsion among emulsified oil droplets, preventing coalescence. Results indicated that SU PPI samples were able to form a relatively thick film at the interface, leading to good ES, although they had lower EC and EAI than cSPI.

Of all protein isolates tested, cPPI had the highest ES but the lowest EAI (Table 4). As cPPI was highly denatured and aggregated, they are relatively less flexible to easily migrate to the water/oil interface. The poor solubility and high extent of denaturation and aggregation in cPPI would cause the protein to interact with each other rather than efficiently migrating to the water/oil interface, thus increasing the viscosity of the continuous phase, and contributing to the stability of the formed oil droplets [60].

4. Conclusions

Pea protein extraction was evaluated, with a focus on selecting conditions that are scalable and can preserve the structural integrity of the protein. Both pH- and salt- extractions under controlled conditions resulted in protein purity and yield comparable or superior to what has been reported previously for pea protein extractions. Particularly, extraction at close to neutral pH coupled with double solubilization, not only resulted in good protein yield and purity, but also resulted in preserved structure and superior functionality to cPPI. Results of this work, confirmed for the first time that double solubilization at mild pH can replace single solubilization at high alkalinity, and achieve similar yield, while preserving structural integrity. While there were some structural and functional differences in salt-PPI compared to pH-PPI, scaling up the production eliminated such differences. Additionally, this study confirmed, for the first time in pea, the scalability of the benchtop salt extraction coupled with membrane filtration. Scaling up under mild and controlled conditions resulted in partial denaturation and low degree of polymerization compared to cPPI. Both SU PPI samples had superior functionality compared to cPPI, produced under harsh pH-extraction conditions. These results provided valuable insights in scalability of specific benchtop parameters, which will guide industrial production of functional pea protein ingredient. In comparison to cSPI, both SU PPI had comparable solubility to cSPI at pH 7 and superior solubility at pH 3.4, making them suitable for acidic protein beverages; however, both SU PPI had inferior gelation and emulsification properties. In order to enhance the functionality of pea protein and expand their utilization, structural modification through physical, enzymatic, and chemical (such as Maillard glycation) approaches can be further explored. Given the inherent protein composition and distribution of the different pea protein fractions, selective breeding with targeted protein phenotyping could be another plausible approach to further enhance the prospects of pea protein.

Author Contributions: Conceptualization, B.P.I.; methodology, L.H. and F.B.; software, L.H. and F.B.; validation, B.P.I., L.H. and F.B.; formal analysis, L.H. and F.B.; data curation, L.H. and F.B.; writing—original draft preparation, L.H.; writing—review and editing, B.P.I.; supervision, B.P.I.; funding acquisition, B.P.I. All authors have read and agreed to the published version of the manuscript.

Funding: This work was supported by funds from the United States Department of Agriculture/ Agriculture Research Station—Pulse Crop Health Initiative (Agreement number 58-3060-8-024).

Data Availability Statement: Data available upon reasonable request from the corresponding author.

Acknowledgments: The authors would like to acknowledge Maneka Malalgoda for her help with molecular weighed distribution analysis by size exclusion chromatography.

Conflicts of Interest: The authors declare no conflict of interest.

Abbreviations

cPPI: commercial pea protein isolate; cSPI, commercial soy protein isolate; cWPI, commercial whey protein isolate; DSC, differential scanning calorimetry; EAI, emulsification activity index; EC, emulsification capacity; ES, emulsification stability; pH-PPI, pH extracted pea protein isolate; salt-PPI, salt extracted pea protein isolate; SDS-PAGE, sodium dodecyl sulfate-polyacrylamide gel electrophoresis; SE-HPLC, size-exclusion high-performance liquid chromatography; SU, scaled-up; SU-pH PPI, scaled-up pH extracted pea protein isolate; SU-salt PPI, scaled-up salt extracted pea protein isolate. UF, ultrafiltration; WPI, whey protein isolate.

References

1. Tulbek, M.C.; Lam, R.S.H.; Wang, Y.; Asavajaru, P.; Lam, A. Chapter 9—Pea: A Sustainable Vegetable Protein Crop. In *Sustainable Protein Sources*; Nadathur, S.R., Wanasundara, J.P.D., Scanlin, L., Eds.; Academic Press: San Diego, CA, USA, 2017; pp. 145–164.
2. Chang, C.; Tu, S.; Ghosh, S.; Nickerson, M. Effect of pH on the inter-relationships between the physicochemical, interfacial and emulsifying properties for pea, soy, lentil and canola protein isolates. *Food Res. Int.* **2015**, *77*, 360–367. [CrossRef]
3. Zhao, H.; Shen, C.; Wu, Z.; Zhang, Z.; Xu, C. Comparison of wheat, soybean, rice, and pea protein properties for effective applications in food products. *J. Food Biochem.* **2020**, *44*, e13157. [CrossRef]
4. Bu, F.; Nayak, G.; Bruggeman, P.; Annor, G.; Ismail, B.P. Impact of plasma reactive species on the structure and functionality of pea protein isolate. *Food Chem.* **2022**, *371*, 131135. [CrossRef]
5. Adebiyi, A.P.; Aluko, R.E. Functional properties of protein fractions obtained from commercial yellow field pea (*Pisum sativum* L.) seed protein isolate. *Food Chem.* **2011**, *128*, 902–908. [CrossRef]
6. Arntfield, S.D.; Maskus, H.D. 9-Peas and other legume proteins. In *Handbook of Food Proteins*; Phillips, G.O., Williams, P.A., Eds.; Woodhead Publishing: Sawston, UK, 2011; pp. 233–266.
7. Cui, L.; Bandillo, N.; Wang, Y.; Ohm, J.-B.; Chen, B.; Rao, J. Functionality and structure of yellow pea protein isolate as affected by cultivars and extraction pH. *Food Hydrocolloid* **2020**, *108*, 106008. [CrossRef]
8. Pelgrom, P.J. *Dry Fractionation for Sustainable Production of Plant Protein Concentrates*; Wageningen University and Research: Wageningen, The Netherlands, 2015.
9. Gao, Z.; Shen, P.; Lan, Y.; Cui, L.; Ohm, J.-B.; Chen, B.; Rao, J. Effect of alkaline extraction pH on structure properties, solubility, and beany flavor of yellow pea protein isolate. *Food Res. Int.* **2020**, *131*, 109045. [CrossRef]
10. Tanger, C.; Engel, J.; Kulozik, U. Influence of extraction conditions on the conformational alteration of pea protein extracted from pea flour. *Food Hydrocolloid* **2020**, *107*, 105949. [CrossRef]
11. Karaca, A.C.; Low, N.; Nickerson, M. Emulsifying properties of chickpea, faba bean, lentil and pea proteins produced by isoelectric precipitation and salt extraction. *Food Res. Int.* **2011**, *44*, 2742–2750. [CrossRef]
12. Murray, E.; Myers, C.; Barker, L.; Maurice, T. Functional attributes of proteins—A noncovalent approach to processing and utilizing plant proteins. *Util. Protein Resour.* **1981**. Available online: https://agris.fao.org/agris-search/search.do?recordID=US201301965539 (accessed on 14 November 2022).
13. Duong-Ly, K.C.; Gabelli, S.B. Salting out of proteins using ammonium sulfate precipitation. In *Methods in Enzymology*; Elsevier: Amsterdam, The Netherlands, 2014; Volume 541, pp. 85–94.
14. Xu, H.-N.; Liu, Y.; Zhang, L. Salting-out and salting-in: Competitive effects of salt on the aggregation behavior of soy protein particles and their emulsifying properties. *Soft Matter* **2015**, *11*, 5926–5932. [CrossRef] [PubMed]
15. Deak, N.; Murphy, P.; Johnson, L. Effects of NaCl concentration on salting-in and dilution during salting-out on soy protein fractionation. *J. Food Sci.* **2006**, *71*, C247–C254. [CrossRef]
16. Hayati Zeidanloo, M.; Ahmadzadeh Ghavidel, R.; Ghiafeh Davoodi, M.; Arianfar, A. Functional properties of Grass pea protein concentrates prepared using various precipitation methods. *J. Food Sci. Technol.* **2019**, *56*, 4799–4808. [CrossRef] [PubMed]
17. Stone, A.K.; Karalash, A.; Tyler, R.T.; Warkentin, T.D.; Nickerson, M.T. Functional attributes of pea protein isolates prepared using different extraction methods and cultivars. *Food Res. Int.* **2015**, *76*, 31–38. [CrossRef]
18. Gao, L.L.; Nguyen, K.D.; Utioh, A.C. Pilot scale recovery of proteins from a pea whey discharge by ultrafiltration. *LWT-Food Sci. Technol.* **2001**, *34*, 149–158. [CrossRef]
19. Tian, S.; Kyle, W.S.; Small, D.M. Pilot scale isolation of proteins from field peas (*Pisum sativum* L.) for use as food ingredients. *Int. J. Food Sci. Technol.* **1999**, *34*, 33–39. [CrossRef]
20. Taherian, A.R.; Mondor, M.; Labranche, J.; Drolet, H.; Ippersiel, D.; Lamarche, F. Comparative study of functional properties of commercial and membrane processed yellow pea protein isolates. *Food Res. Int.* **2011**, *44*, 2505–2514. [CrossRef]
21. Laemmli, U.K. Cleavage of structural proteins during the assembly of the head of bacteriophage T4. *Nature* **1970**, *227*, 680–685. [CrossRef] [PubMed]
22. Boyle, C.; Hansen, L.; Hinnenkamp, C.; Ismail, B.P. Emerging camelina protein: Extraction, modification, and structural/functional characterization. *J. Am. Oil Chem. Soc.* **2018**, *95*, 1049–1062. [CrossRef]
23. Wang, Q.; Ismail, B. Effect of Maillard-induced glycosylation on the nutritional quality, solubility, thermal stability and molecular configuration of whey protein. *Int. Dairy J.* **2012**, *25*, 112–122. [CrossRef]
24. Hinnenkamp, C.; Ismail, B.P. Enhancing emulsion stability: The synergistic effect of combining Procream and partially hydrolyzed whey protein. *Int. Dairy J.* **2021**, *119*, 105059. [CrossRef]
25. Hoang, H.D. Evaluation of Pea Protein and Modified Pea Protein as Egg Replacers. Ph.D. Thesis, North Dakota State University, Fargo, ND, USA, 2012.
26. Feyzi, S.; Milani, E.; Golimovahhed, Q.A. Grass Pea (*Lathyrus sativus* L.) Protein Isolate: The Effect of Extraction Optimization and Drying Methods on the Structure and Functional Properties. *Food Hydrocolloid* **2018**, *74*, 187–196. [CrossRef]
27. Reinkensmeier, A.; Bußler, S.; Schlüter, O.; Rohn, S.; Rawel, H.M. Characterization of individual proteins in pea protein isolates and air classified samples. *Food Res. Int.* **2015**, *76*, 160–167. [CrossRef]
28. Lee, H.; Htoon, A.; Uthayakumaran, S.; Paterson, J. Chemical and functional quality of protein isolated from alkaline extraction of Australian lentil cultivars: Matilda and Digger. *Food Chem.* **2007**, *102*, 1199–1207. [CrossRef]

29. Barac, M.; Cabrilo, S.; Stanojevic, S.; Pesic, M.; Pavlicevic, M.; Zlatkovic, B.; Jankovic, M. Functional properties of protein hydrolysates from pea (*Pisum sativum*, L.) seeds. *Int. J. Food Sci. Technol.* **2012**, *47*, 1457–1467. [CrossRef]
30. Boye, J.; Aksay, S.; Roufik, S.; Ribéreau, S.; Mondor, M.; Farnworth, E.; Rajamohamed, S. Comparison of the functional properties of pea, chickpea and lentil protein concentrates processed using ultrafiltration and isoelectric precipitation techniques. *Food Res. Int.* **2010**, *43*, 537–546. [CrossRef]
31. O'Kane, F.E.; Happe, R.P.; Vereijken, J.M.; Gruppen, H.; Van Boekel, M.A.J.S. Characterization of Pea Vicilin. 1. Denoting Convicilin as the α-Subunit of thePisumVicilin Family. *J. Agric. Food Chem.* **2004**, *52*, 3141–3148. [CrossRef]
32. Shand, P.J.; Ya, H.; Pietrasik, Z.; Wanasundara, P.K.J.P.D. Physicochemical and textural properties of heat-induced pea protein isolate gels. *Food Chem.* **2007**, *102*, 1119–1130. [CrossRef]
33. Sumner, A.; Nielsen, M.; Youngs, C. Production and evaluation of pea protein isolate. *J. Food Sci.* **1981**, *46*, 364–366. [CrossRef]
34. Gueguen, J.; Cerletti, P. Proteins of some legume seeds: Soybean, pea, fababean and lupin. In *New and Developing Sources of Food Proteins*; Springer: Berlin, Germany, 1994; pp. 145–193.
35. Taherian, A.R.; Mondor, M.; Lamarche, F. Enhancing nutritional values and functional properties of yellow pea protein via membrane processing. *Peas Cultiv. Var. Nutr. Uses* **2012**, 1–48.
36. Lu, Z.X.; He, J.F.; Zhang, Y.C.; Bing, D.J. Composition, physicochemical properties of pea protein and its application in functional foods. *Crit. Rev. Food Sci. Nutr.* **2019**, *60*, 1–13. [CrossRef]
37. Owusu-Ansah, Y.J.; McCurdy, S.M. Pea proteins: A review of chemistry, technology of production, and utilization. *Food Rev. Int.* **1991**, *7*, 103–134. [CrossRef]
38. Hensley, D.; LAWHON, J.T. Economic evaluation of soy isolate production by a membrane isolation process. *Food Technol.* **1979**, *33*, 46–50.
39. Tzitzikas, E.N.; Vincken, J.-P.; De Groot, J.; Gruppen, H.; Visser, R.G.F. Genetic Variation in Pea Seed Globulin Composition. *J. Agric. Food Chem.* **2006**, *54*, 425–433. [CrossRef] [PubMed]
40. Mession, J.-L.; Chihi, M.L.; Sok, N.; Saurel, R. Effect of globular pea proteins fractionation on their heat-induced aggregation and acid cold-set gelation. *Food Hydrocolloid* **2015**, *46*, 233–243. [CrossRef]
41. Croy, R.R.; Hoque, M.S.; Gatehouse, J.A.; Boulter, D. The major albumin proteins from pea (*Pisum sativum* L.) Purification and some properties. *Biochem. J.* **1984**, *218*, 795–803. [CrossRef]
42. Swanson, B.G. Pea and lentil protein extraction and functionality. *J. Am. Oil Chem. Soc.* **1990**, *67*, 276–280. [CrossRef]
43. Makri, E.; Papalamprou, E.; Doxastakis, G. Study of functional properties of seed storage proteins from indigenous European legume crops (lupin, pea, broad bean) in admixture with polysaccharides. *Food Hydrocolloid* **2005**, *19*, 583–594. [CrossRef]
44. Mession, J.-L.; Sok, N.; Assifaoui, A.; Saurel, R.m. Thermal Denaturation of Pea Globulins (*Pisum sativum* L.) Molecular Interactions Leading to Heat-Induced Protein Aggregation. *J. Agric. Food Chem.* **2013**, *61*, 1196–1204. [CrossRef]
45. Phillips, G.O.; Williams, P.A. *Handbook of Food Proteins*; Elsevier: Amsterdam, The Netherlands, 2011.
46. Damodaran, S.; Parkin, K.L. Amino acids, peptides, and proteins. In *Fennema's Food Chemistry*; CRC Press: Boca Raton, FL, USA, 2017; pp. 235–356.
47. Friedman, M. Chemistry, biochemistry, nutrition, and microbiology of lysinoalanine, lanthionine, and histidinoalanine in food and other proteins. *J. Agric. Food Chem.* **1999**, *47*, 1295–1319. [CrossRef]
48. Lee, K.; Ryu, H.; Rhee, K. Protein solubility characteristics of commercial soy protein products. *J. Am. Oil Chem. Soc.* **2003**, *80*, 85–90. [CrossRef]
49. Foegeding, E.A.; Davis, J.P. Food protein functionality: A comprehensive approach. *Food Hydrocolloid* **2011**, *25*, 1853–1864. [CrossRef]
50. Rickert, D.; Johnson, L.; Murphy, P. Functional Properties of Improved Glycinin and β-nglycinin Fractions. *J. Food Sci.* **2004**, *69*, FCT303–FCT311. [CrossRef]
51. Lampart-Szczapa, E. Legume and oilseed proteins. *Chem. Funct. Prop. Food Proteins* **2001**, *2*, 407–436.
52. Zhou, H.X. Interactions of macromolecules with salt ions: An electrostatic theory for the Hofmeister effect. *Proteins Struct. Funct. Bioinform.* **2005**, *61*, 69–78. [CrossRef] [PubMed]
53. Sun, X.D.; Arntfield, S.D. Gelation properties of salt-extracted pea protein isolate induced by heat treatment: Effect of heating and cooling rate. *Food Chem.* **2011**, *124*, 1011–1016. [CrossRef]
54. Wolf, W.J. Soybean proteins. Their functional, chemical, and physical properties. *J. Agric. Food Chem.* **1970**, *18*, 969–976. [CrossRef]
55. Sun, X.D.; Arntfield, S.D. Gelation properties of salt-extracted pea protein induced by heat treatment. *Food Res. Int.* **2010**, *43*, 509–515. [CrossRef]
56. Tombs, M. Gelation of globular proteins. *Faraday Discuss. Chem. Soc.* **1974**, *57*, 158–164. [CrossRef]
57. Hermansson, A.-M. Aggregation and denaturation involved in gel formation. In *Functionality and Protein Structure*; ACS Publications: Washington, DC, USA, 1979.
58. Nakai, S. Structure-function relationships of food proteins: With an emphasis on the importance of protein hydrophobicity. *J. Agric. Food Chem.* **1983**, *31*, 676–683. [CrossRef]
59. Barac, M.; Cabrilo, S.; Pesic, M.; Stanojevic, S.; Zilic, S.; Macej, O.; Ristic, N. Profile and Functional Properties of Seed Proteins from Six Pea (*Pisum sativum*) Genotypes. *Int. J. Mol. Sci.* **2010**, *11*, 4973–4990. [CrossRef]
60. Nour, A.H. Emulsion types, stability mechanisms and rheology: A review. *Int. J. Innov. Res. Sci. Stud. (IJIRSS)* **2018**, *1*, 14–21.

foods

Review

Protein-Based Fat Replacers: A Focus on Fabrication Methods and Fat-Mimic Mechanisms

Niloufar Nourmohammadi [1], Luke Austin [2] and Da Chen [1,*]

[1] Department of Animals, Veterinary and Food Sciences, University of Idaho, Moscow, ID 83844, USA
[2] Department of Biological Sciences, University of Idaho, Moscow, ID 83844, USA
* Correspondence: dchen@uidaho.edu

Abstract: The increasing occurrence of obesity and other non-communicable diseases has shifted the human diet towards reduced calorie intake. This drives the market to develop low-fat/non-fat food products with limited deterioration of textural properties. Thus, developing high-quality fat replacers which can replicate the role of fat in the food matrix is essential. Among all the established types of fat replacers, protein-based ones have shown a higher compatibility with a wide range of foods with limited contribution to the total calories, including protein isolate/concentrate, microparticles, and microgels. The approach to fabricating fat replacers varies with their types, such as thermal–mechanical treatment, anti-solvent precipitation, enzymatic hydrolysis, complexation, and emulsification. Their detailed process is summarized in the present review with a focus on the latest findings. The fat-mimic mechanisms of fat replacers have received little attention compared to the fabricating methods; attempts are also made to explain the underlying principles of fat replacers from the physicochemical prospect. Finally, a future direction on the development of desirable fat replacers in a more sustainable way was also pointed out.

Keywords: protein; fat replacer; fat mimic; microparticulation; microgels; thermomechanical treatment

Citation: Nourmohammadi, N.; Austin, L.; Chen, D. Protein-Based Fat Replacers: A Focus on Fabrication Methods and Fat-Mimic Mechanisms. *Foods* **2023**, *12*, 957. https://doi.org/10.3390/foods12050957

Academic Editors: Norbert Raak, Ruifen Li and Laura Roman

Received: 1 January 2023
Revised: 9 February 2023
Accepted: 22 February 2023
Published: 23 February 2023

1. Introduction

With the prevalence of obesity and obesity-associated chronic diseases, customers become increasingly aware of the calorie intake of their diet. According to the National Health and Nutrition Examination Survey (2017–2020), ~42% of U.S. adults aged $\geq$20 have obesity [1]. The Dietary Guideline for Americans, 2020–2025, encourages the public to consume low- or non-fat foods for healthier diets [2]. Reducing calorie intake by decreasing the amount of fat in food, especially saturated ones, has been considered one of the strategies to reduce the occurrence of obesity [3]. Nevertheless, the commitment to consuming low-fat foods or maintaining a low-fat diet remains challenging because of their deteriorated texture and sensorial properties compared to those of full-fat ones. Hence, the development of fat replacers that imitate not only the functional role of fat but also its sensory features is essential to improve the quality attributes of low-fat foods. Depending on the properties and manufacturing approaches, fat replacers fall into two categories: (1) fat substitute: typically, biomolecules or their degraded products with little or no calories, functioning similarly to fat. They fall into three main groups based on their sources, which are carbohydrate-based, which can hold water and impart a creamy texture close to fat (such as starches and gums), protein-based (such as egg white, milk, and whey), and fat-based fat replacers, which are too large to be digested with little contribution to calories (such as Caprenin as cocoa butter fat substitute and Olestra) [4]. As a well-known fat substitute example, Olestra is a product composed of sucrose, and hexa-, hepta-, and octa-esters of saturated and unsaturated fatty acids [5], but its application has been limited due to the risks of causing gastrointestinal side effects, such as abdominal cramping [6]; (2) fat mimetic (FM): ingredients that partially mimic the organoleptic properties of animal fat, includes mainly food hydrocolloids (gums, cellulose microfibrils, pectins), proteins, protein aggregates, protein–polysaccharides composites, and emulsion gels.

Protein-based fat replacers have received increasing attention. They boost the protein nutrition of food products with a low-calorie contribution. According to the dietary reference, adequate intakes of at least 0.8 g/kg body weight of high-quality protein for sedentary adults, the optimum amount of 1.2–1.8 g/kg body weight for adults with moderate activity, and 1.8–2.2 g g/kg body weight for adults with hypertrophy and strength training per day would be an ideal goal for health enhancement. When it comes to the health-related impact of a high-protein diet, protein intakes above the recommended amount within a certain range could limit the appearance of sarcopenia, and reduce the loss of muscle mass [7]. Due to their highly reactive features towards pH, temperature, ions, and enzymes, protein-based fat replacers can be tuned with distinct physiochemical properties for expanded food applications, such as in yogurt, cream cheese, salad dressings, and frozen desserts. Common sources of proteins to develop fat replacers include egg white protein, whey protein, gelatin, soy, pea, and zein [8]. Animal proteins are considered as higher quality because of the well-balanced amino acid profiles and high digestibility and bioavailability. For plant proteins, they commonly lack cysteine and methionine; however, this nutritional deficiency could be overcome by mixing different types of plant proteins [9]. In addition, due to the presence of anti-nutritional factors, such as trypsin inhibitors, phenolic compounds, phytates, cyanogenic compounds, lectins, and saponins, the digestibility of plant proteins can be reduced unless properly processed [10–12]. Furthermore, the incorporation of protein-based fat replacers levels up the protein content in foods, which enhances protein nutrition. The protein-based fat replacements have advantages over carbohydrate-based ones in terms of flavor interactions [13] and the amount of fat that could be replaced [14]. However, protein-based fat replacers may not be suitable to be incorporated into over-processed food products [15,16] because of protein denaturation and interaction with other components (e.g., Maillard reaction), resulting in a loss of functionality and fat-like mouthful feelings [17].

The types of fat replacers, and their characterization and food applications have been reviewed systematically [12,18,19]. However, the approaches to develop protein-based fat replacers and the mechanism of their fat-mimic effects, which varies significantly with the source of proteins, their molecular weight, solubility, and surface chemistry, remain rarely summarized. This information is essential to design fat replacers with desirable functionality using the most appropriate approach. We attempt to cover the latest findings on those within the last five years. Future perspectives on the production of protein-based fat replacers with improved sustainability are also provided.

2. Types of Protein-Based Fat Replacers

2.1. Protein Concentrates and Isolates

Protein concentrates or isolates can be used directly as fat replacers (Figure 1). They may be slightly denatured during the manufacturing process [20]. The former contains ~30–80% protein, whereas the latter reaches 90–92% [21,22]. Extensive research has been conducted on developing low-fat dairy products including yogurt [23], ice cream [24], and cheese [25,26] using protein concentrates/isolates, especially whey protein due to its high compatibility with other dairy ingredients and a matching flavor profile [27,28]. A higher level (up to 6.8%) addition of whey protein could improve syneresis, yield stress, storage modulus (G′), viscosity, and creaminess of the low-fat yogurt with reduced serum separation compared to the control [29,30]. In low-fat cheese, the partial (3–8%) replacement of fat with whey proteins has been found to improve its hardness because of the formation of more compact structures [31].

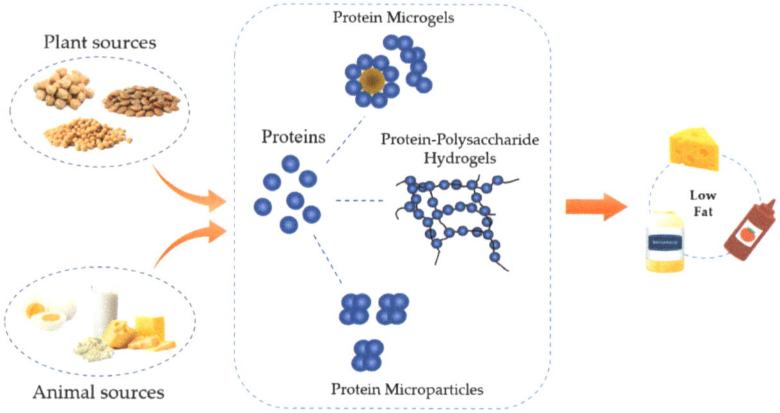

Figure 1. A schematic showing the types of protein-based fat replacers.

2.2. Microparticulated Proteins

Protein microparticulation is a process of aggregation. The size of the particles commonly ranges within 0.1–10 μm, but larger ones >10 μm have also been reported [32]. Protein particles with a size larger than 5 μm can be detected by oral mucosa [33]; thus, a smaller size is preferable to provide the smoothening mouthful feelings. The most widely used microparticulated proteins is from whey proteins, which was patented in 1988 and later commercialized with the brand name Simplesse® [34]. Some other proteins, such as soy protein, bovine serum albumin [35], egg white protein [36], gelatin [37], zein [38], wheat protein [39], pea protein, and potato protein [40], have also been used to produce fat replacers, as summarized in Table 1. Due to their spherical shape and size, they were claimed to mimic fat droplets and create a smooth and creamy mouthfeel through a "ball-bearing" mechanism, which will be discussed later.

Table 1. A summary of different types of protein-based fat replacers, their fabrication methods, and applications.

Type	Protein Source	Fabrication Method	Particle Size (μm)	Application	References
Protein concentrate/isolate	Whey concentrates	Ultrafiltration at 40–45 °C, membrane cut-off 10 kDa	-	Reduced fat cheese	[26]
	Soy protein fractions	Protein solubilization at pH 8 and 9 for 1 h, separate oil-rich cream by centrifuge, protein precipitation in pH 4.5 and 5 by adding 1 M HCl hold for 1 h	10–250	Meat analog	[41,42]
Protein microparticles	Microparticulated soy protein/egg white protein	Heated 95 °C for 5–15 min with continuous stirring	2–20	Drinkable or semi-solidprotein-rich foods	[36]
	Microparticulated whey, potato, and pea proteins	Heating to 10 °C above denature temperature of proteins, held for 10 min, and cooling in a concentric cylinder with 100–150 s^{-1} shear rate	20–250	-	[43]
	Microparticulated whey proteins	Extrusion at 90 °C with a screw speed of 200–1000 rpm	2–7	Reduced-fat yogurt	[44]

Table 1. *Cont.*

Type	Protein Source	Fabrication Method	Particle Size (μm)	Application	References
	Potato protein	Extrusion at 80 °C and 800 rpm screw speed, pH 6.9	9–110	Fat-reduced dessert	[40]
	Pea protein	Extrusion cooking at 100 °C with 600 rpm screw speed	10–75	Milk dessert	[45]
	Egg white protein	Heated at 75 °C for 13 min, followed by high-shear homogenization at 10,000 rpm for 60 s	9.4	Salad dressing	[46]
	Microparticulated whey protein	Heated at 85 °C for 15 min and sonicated at 20 kHz for 1 min	0.01–2	Reduced-fat cheese emulsion	[47]
	Soy protein hydrolysate	Alcalase 2.4 L hydrolysate followed by heating at 90 °C for 20 min and homogenized at 8000 rpm for 6 min	7.1–9.3	Ice cream	[24]
	Pea protein hydrolysate	Hydrolysis by Alcalase 2.4 L for 3 min, followed by heating at 85 °C for 10 min	-	-	[48]
	Zein/carboxymethyl dextrin	Zein was dissolved in 80% ethanol and then added to a dextrin solution, followed by the removal of ethanol	0.1–0.6	Sausage	[49,50]
Protein–polysaccharide complex	Oat β-glucan/marine collagen peptide	Oat β-glucan was dissolved in buffer solution at pH 3 (glycine–HCl buffer) at 12% concentration and mixed with marine collagen peptide solution at different ratios. The samples were stirred at 25 °C for 4 h, then subjected to high pressure at 400–500 MPa for 30 min	-	Sausage	[51]
	Pea protein/pectin	Pea protein powder was added to solutions containing different ratios of pectin to reach 15% (*w/v*) of protein concentrate. The dispersion was stirred at 1500 rpm for 1 h at room temperature followed by adjusting the pH to 6.5 using 2 N NaOH	-	-	[52]
Protein microgels	Canola protein microgels	Heated at 90 °C under stirring for 1 h, stored at 4 °C to form a gel followed by homogenization and complexation with polysaccharides	1–100	Pickering emulsion stabilization as a potential fat replacer	[53]

Table 1. *Cont.*

Type	Protein Source	Fabrication Method	Particle Size (μm)	Application	References
	Whey protein emulsion gel microparticles	Heated at 90 °C for 20 min, followed by the addition of oil, homogenization, and addition of glucono-delta-lactone	0.3–300	Low-fat yogurt	[54]
	Whey protein/alginate microgels	Whey protein was heated at 80 °C for 30 min followed by alginate addition	-	-	[55]
	Soy protein microgels	Salt-induced coacervation followed by heating at 80 °C for 30 min under stirring	1–3	-	[56]

Note: The "-" means "it is not reported".

2.3. Protein–Polysaccharides Hydrogel

Either proteins or polysaccharides have the capacity to form hydrogels. Protein-based hydrogels are mainly particle type, whereas those from polysaccharides are "thread or linear" type. When mixing the two, complexation could occur [57], which tailors protein functionality in foods by altering its surface chemistry and aggregating behavior [52,58,59] (Figure 1). Many studies have confirmed that the addition of polysaccharides prevented the coalescence and interaction between microparticulated proteins, either by shielding charged groups or by decreasing the collision rate between molecules through an increase in the viscosity of the system. The presence of polysaccharides can also bind a large amount of water to provide a creaminess sensation through the oral process [60]. Thus, a protein-based fat mimetic in combination with polysaccharides is usually formulated to replace fat to develop low-fat products [27,37,57]. The polysaccharides used in complexation with protein are gum arabic [61], pectin [52,62], alginate [63], and xanthan [64]. Nowadays, there is a growing trend to develop plant protein–polysaccharide hydrogel from peas [52], lentils [65], and soybeans [66] to increase their sustainability.

2.4. Microgel Particles

Protein-based microgel particles are novel types of fat replacers, with a size at nano- to micro-meter level [32]. They can be categorized into different types: fragments of protein hydrogels, protein assembles, emulsified gel droplets [67], and protein–polysaccharides coacervates [68] (Figure 1). The particles have the dual properties of particles and polymers, which endows ideal lubricating effects. Proteins can be used as the sole component of the microgel particles, and their elasticity depends on the types of proteins used and the interactions among protein molecules. In general, plant protein-based microgel particles have smaller elastic moduli compared to those of animal proteins due to the formation of less covalent bonds [69]. By changing environmental conditions, such as pH and ionic strength, the surface charge of protein molecules alters, which further affects their electrostatic interactions, resulting in distinct elasticity. For example, a 15-fold increment of elasticity of whey protein microgel particles was found at pH 3 and pH 5.5 compared to those at neutral pH [70]. Besides protein itself, other components could also be incorporated into microgel particles, such as polysaccharides and/or plant oils. Either of them contributes to the increased deformability of the particles. When plant oil was used, the system turns into an emulsion type, with Pickering emulsion being the typical example [53]. The emulsion-type protein-based microgels commonly have a more regular shape and better lubricating capacity than those of protein-only particles due to the formation of a lubrication film from the protein nanoparticles and base oils [71]. However, they are vulnerable to environmental change and long-term stability remains a challenge. Synthetic polymers or organic solvents might be used to decrease the surface tension or increase the stability of emulsion-type microgel particles, but they are not suitable for food applications.

3. Production of Protein-Based Fat Replacers

The way to fabricate fat replacers associates closely with their physicochemical properties and functionalities, such as particle size and shape, viscosity, gelling, emulsifying, foaming, and water-holding capacity. This further affects the color, texture, and other sensory characteristics of the final products. For proteins with high water solubility, such as whey proteins, thermal treatment to trigger protein aggregation under the controlled shear is commonly used. For those with limited water solubility, microparticulation could be achieved by breaking down large aggregates into smaller ones or improving their solubility/dispersibility first followed by precipitation. In the following sections, the common approaches used to produce fat replacers are discussed with a focus on protein microparticles and microgels.

3.1. Protein Concentrate or Isolate

The production of a protein concentrate or isolate has been well summarized [72–75] and will not be detailed here. Animal proteins, especially dairy proteins, are concentrated by ultrafiltration [26] or diafiltration followed by optional evaporation of the retentate prior to spray drying (Figure 2). The temperature of spray drying associates closely with the denaturation of the proteins, and their aggregation and particle morphology. The mean diameter of whey protein powders has been reported to increase from ~15 μm to ~20 μm when the outlet temperature was increased from 60 °C to 100 °C [76]. The morphology of whey protein concentrate particles was changed from spherical at a lower inlet air temperature to a deflated shape at high temperatures [77]. How the morphology and size of particles in a milk protein isolate or concentrate affect their fat-mimic capacity remains poorly studied. It has to be noted that the size and morphology of the particles in powders may change after they are incorporated into the food matrix due to the presence of water, salt, or other food components.

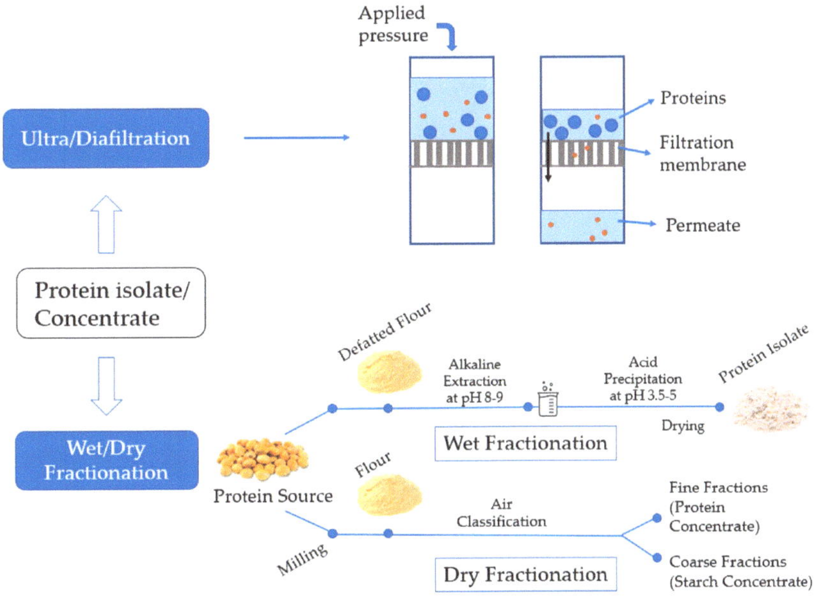

Figure 2. A schematic showing the production of protein isolates and concentrates.

For plant proteins, wet and dry fractionation has been widely used (Figure 2). Wet methods include mainly alkaline-isoelectric point precipitation, or water, salt, or acid extraction. By using acid or alkaline, the protein recovery efficiency is high due to the increased solubility of proteins at the pH values far away from its isoelectric point [41,42]. Alkaline-isoelectric precipitation is more commonly used than other methods in food industries to extract proteins followed by spray drying to produce a protein isolate [78], but it uses a large amount of water and generates excessive salt in wastewater, which potentially threatens the freshwater ecosystem. In addition, high pH used during extraction may cause denaturation, racemization, and lysinoalanine formation of plant proteins, resulting in deteriorated quality [78,79]. Dry fractionation is more environmentally friendly and specific to produce a protein concentrate (40–50% protein content). It takes advantage of the size and charge of proteins that differed from those of starch and dietary fiber and achieves the separation by sieving or electrostatic interactions forces [72]. Since no high pH and extensive heat are involved during isolation, the proteins tend to have better functionality than those from the alkaline method.

3.2. *Microparticulated Proteins*

3.2.1. Thermal–Mechanical Treatments

Thermomechanical treatment is the most common approach to develop microparticulated proteins as fat replacers. In general, a heating source is needed to unfold proteins by heating above their denaturation temperature (Figure 3). Unfolded proteins are then aggregated via covalent (S-S bonds) and non-covalent interactions (mainly hydrophobic interactions). Animal proteins are commonly heated at 75–95 °C for 20–40 min for microparticulation to occur [80–82]. More extensive heating (80–95 °C, ~30 min) is required for plant proteins due to their higher thermal stability [43,54]. The duration of heating shortens with the increase of temperature. For instance, pea protein particles were formed within a minute at 135 °C [45]. The morphology and size of the protein particle changes with the heating conditions. At lower temperatures, particles commonly have a smaller size with higher compactness [45,82–84]. Besides temperature, the shear force also applies to the system during heating unless at a low protein concentration (≤5%), otherwise, gelation could occur. The size of the particles negatively correlates with the shear force. High shear force results in more rigorous breaking of the aggregates and produces particles with smaller sizes [43], as demonstrated in whey proteins [82].

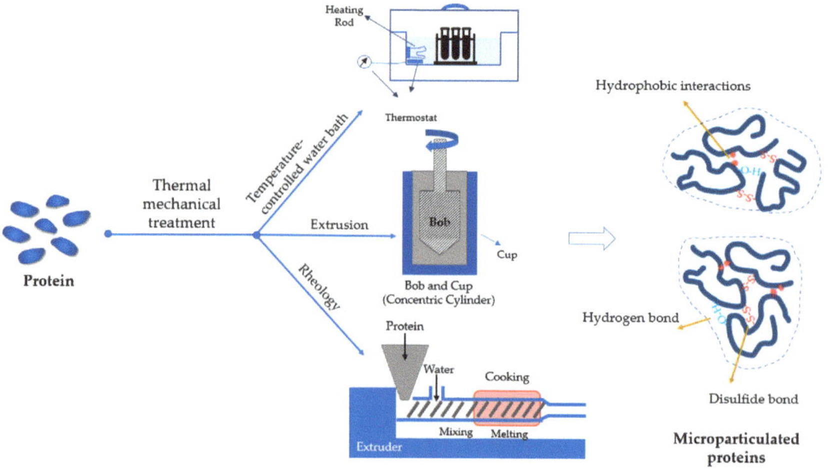

Figure 3. A schematic illustrating the production of microparticulated proteins using thermal–mechanical treatments.

Multiple strategies have been adopted to provide heat and shear simultaneously (Figure 3). The simplest setup is to stir the sample on a temperature-controlled water bath. The strength of the shear can be fulfilled through adjusting the stirring speed, but the heating and cooling rate may be beyond control unless a sophisticated heating/cooling system is used. The other shortcoming of the method is it fails to provide ideal mixing of the concentrated protein suspension because of the high viscosity. These can be overcome by using a concentric cylinder accessory (bob-cup) equipped on a rheometer. The shear rate of the bob can be adjusted to the required value while the temperature is controlled by the cup. Such a method has been used to design microparticulated structures from whey, potato, and pea proteins [43] (Table 1). Within a certain range, the increase of the shear rate reduces the size of the formed particles as it disrupts the accumulated aggregation of proteins. The concentric cylinder is also able to monitor the microparticulation progress and explore the effects of shear force and temperature on the viscoelasticity, but the large-scale production of microparticles using a concentric cylinder remains a challenge as the volume of the cup is limited (e.g., 25 mL or less). The extrusion process, another thermomechanical treatment, is widely applied in food industries with the features of large-scale and continuous production. Proteins are unfolded and aggregated under the control of heating and screw rotation. Compared to the water bath and concentric cylinder, the temperature inside the extruder can achieve above 100 °C due to the high-sealed environment [45]. This facilitates protein microparticulation within a short duration. In addition, the protein content in the fed materials can be adjusted to higher values ($\geq$20%), which significantly increases the yield of protein microparticles [85]. Whey protein particles produced through extrusion have been found to enhance the creaminess of low-fat, plain, stirred yogurt while maintaining its consumer acceptance with a less than 0.5% addition [44,85]. Similarly, the incorporation of pea protein microparticles derived from the same method into fat-reduced milk desserts resulted in approximately the same creamy perception as the full-fat milk dessert [45]. The conventional extrusion method consumes a large amount of energy to unfold proteins for subsequent aggregation. A hybrid technology, which takes advantage of the expanding properties of supercritical carbon dioxide (SC-CO_2) to induce aggregation at lower temperatures (<100 °C), is more energy efficient [86]. In supercritical fluid extrusion, proteins tend to expose the hydrophobic regions and aggregates driven by the surrounding non-polar environment, resulting in distinct properties of microparticles, such as the protein–protein interactions, surface hydrophobicity, and rheological properties [87].

High-pressure homogenization combined with heat treatment is another technique capable of applying a strong shear force and a sudden pressure modulation to tune protein aggregation (Table 1). Liu et al. used such a method (10,000 rpm for 60 s, 75 °C for 13 min) to produce microparticulated egg white proteins as a fat replacer in salad dressings, which are comparable to the commercial salad dressings' features [46]. Ultrasound-assisted heating could also induce protein microparticulation. The ultrasonication could be conducted simultaneously with the heating or after. A recent study found large whey protein aggregates (60–600 μm) were formed by heating, whose size was reduced to 0.01–2 μm upon sonication [47]. The treatment also resulted in higher surface hydrophobicity of the particles, which is possibly due to the exposure of more interior regions of protein aggregates.

3.2.2. Enzymatic Hydrolysis

Protein aggregation and microparticulation could also be triggered by enzymatic treatment with or without additional heating. Transglutaminase catalyzes acyl transfer between ε-amino groups of lysine residues and the γ-carboxyamide groups to form covalent cross-links, but the reaction is slow and requires hours to complete, which hinders large-scale production [88]. The breaking down of proteins to expose hydrophobic regions or reactive amino acid residues also induces aggregation [48], which can be fulfilled through limited enzymatic hydrolysis. Compared to heat-induced aggregation, proteolysis is a greener approach, which consumes less energy. Depending on the types of proteases and the proteins, the conditions to promote aggregation vary. For instance, small aggregates were formed in *Bacillus licheniforms* (BLP) hydrolyzed whey proteins when 70% of the proteins were intact [89]. Using the same enzyme, whey proteins were found to form soft and turbid aggregate gels at 50 °C for 1 h [90]. For pea proteins, when the degree of hydrolysis was controlled to around 6–7% by a short time (2–3 min) hydrolysis with alcalase at 50 °C, the obtained hydrolysates were aggregated rapidly in response to heat. Analysis of the aggregates found that non-covalent interactions were the dominant forces that drive aggregation [48]. This suggests the aggregates might be deformed easily to provide lubricating effects. Zang et al. studied the influence of limited enzymatic hydrolysis of rice bran proteins by trypsin, at the ratio of 1:100 (Enzyme:Substrate)(*v/w*), on the emulsifying properties. They found a significant enhancement of the emulsifiying properties of hydrolysates with a 3% degree of hydrolysis. This was due to the release and exposure of soluble peptides from insoluble aggregates resulting in the exposure of more ionizable amino groups [91].

3.2.3. Anti-Solvent Precipitation

Anti-solvent precipitation has been used to produce fine particles at the micro- and nano-scale level. The size and morphology of the particles could be well controlled by adjusting the protein concentration and polarity of the solvent [92]. The method is mainly applied to fabricate prolamin-based composites, such as zein and wheat gluten. Zein is the dominating protein found in maize, which can be extracted from dry-milled corn (DMC) or distillers-dried grains by using aqueous ethanol, acetic acid, or alkaline [93,94]. The hydrophobic nature of zein and its incomplete amino acid profile limit the food applications. Nevertheless, the inherent hydrophobicity and the heat-softening capacity of zein make it an ideal candidate for fat analog [95]. To achieve this, zein or its aggregates are suggested to micronize to a level similar to those of oil droplets by anti-solvent precipitation. Firstly, hydrophobic proteins are solubilized in an organic solvent, such as aqueous ethanol (70 to 90% *v/v*) or acetic acid/ethanol solution (e.g., at 55/45 ratio, *v/v*, pH 3.0) [96] under agitation [97,98]. Then, the concentration of the organic solvents is lowered down by adding water, resulting in increased polarity of the environment [99]. This triggers the aggregation of zein or wheat gluten mediated by hydrophobic interactions and precipitates out from the system when gravity overcomes electrostatic repulsions. Cui et al. have used zein nanoparticles prepared from anti-solvent precipitation as a stabilizer in low-fat emulsions. They solubilized zein in an 85% (*v/v*) aqueous ethanol at room temperature followed by the addition of phytic acid (PA) solutions to improve its stability. By stirring the mixture continuously for 30 min at 600 rpm, zein protein nanoparticles with a mean size of ~160 nm were formed [100]. The concentration of the organic solvent and the drying temperature directly links to the size and stiffness of the formed nanoparticles. Bisharat et al. (2018) found that the size of zein particles was increased from an average of 200–500 nm to 1 µm as the ethanol concentration was decreased from 90% to 70%. When the drying temperature was decreased from 55 °C to 40 °C and then to room temperature, the Young's modulus of the particles was dropped continuously, corresponding to a higher flexibility and less resistance to stretching [101,102]. Zein could also form particles containing polysaccharides using anti-solvent precipitation. The particles have been shown to stabilize oil droplets as a potential fat replacer in sausages [49,50].

3.2.4. Protein–Polysaccharides Hydrogel

When proteins and polysaccharides co-exist, depending on the environmental conditions and the concentration of each component, they behave distinctly. Thermodynamic incompatibility occurs mainly at high protein and polysaccharide concentrations due to their differed chemical nature and affinity towards solvent. Nevertheless, phase separation may not occur because of the high viscosity of the system. When gelation occurs fast, for instance, heating under a high temperature, the extent of phase separation or inhomogeneity of the gel would become less [60]. This results in higher consistency of the gel texture. Even though heating is not the sole condition to form a protein–polysaccharides hydrogel, it is the most common approach. Depending on the types of proteins and polysaccharides, they can both gel under heat, but not always. Most of the proteins are heat-sensitive and tend to unfold and aggregate to form a gel under heat, whereas polysaccharides could remained unchanged [103]. The proportion of the polysaccharides affects the rheological and fat-mimic capacity of the mixed gel and can be tuned accordingly. For instance, when mixing soy protein isolate and cellulose nanofibrils (CF), with the increasing of CF, the creaminess was increased [104]. If strong electrostatic interactions occur during the initial heating of proteins and polysaccharides, they can be gelled without extra heating. Since the complexation occurs rapidly, the protein or polysaccharides solution needs to be prepared separately before mixing. A high-pressure homogenization might be required to facilitate a homogenous distribution of polysaccharides and proteins [51]. Using this approach, Fan et al. (2020) developed oat β-glucan/marine collagen peptides mixed gels to replace the fat in sausage products [51]. Adjusting the environmental pH or ionic strength enables us to modify the surface charge of proteins, which further tunes the interactions between proteins and polysaccharides [103], and the textural properties of the gel. As non-covalent interaction are weak forces, the formed gels commonly possess high deformability against force and heat as desirable fat mimics.

3.3. Microgel Particles

For protein-only microgel particles, a hydrogel needs to be prepared followed by size reduction via homogenization or microfluidization to form micro- or nano-particles. Gelation could be conducted by using either the hot or cold method. Heat gelation occurs normally at 80–95 °C at a relatively high protein concentration (≥10%) for 20–30 min to induce protein aggregation and cross-linking (Figure 4). Cold gelation also requires heat to denature the protein first, followed by aggregating using calcium [105], salt [56], polysaccharides [55], or acid [106] at room temperature (Table 1). By changing the concentration of the reactants, the microgel particles display distinct physicochemical properties. For instance, when the concentration of $CaCl_2$ was increased from 0.02 to 0.1 M, the size of whey protein microgel particles was decreased with the increment of their viscoelasticity [105]. When using salt, the concentration should be high enough to screen the surface charge of proteins to promote coacervation [56], but not cause the "salt in" effects. Similar to salt, the charged polysaccharides could also mediate the aggregation of proteins by reducing their surface charge. The hydrogen bonding between proteins and polysaccharides may also contribute and possibly play more significant roles than the electrostatic interaction on large-scale aggregation [107].

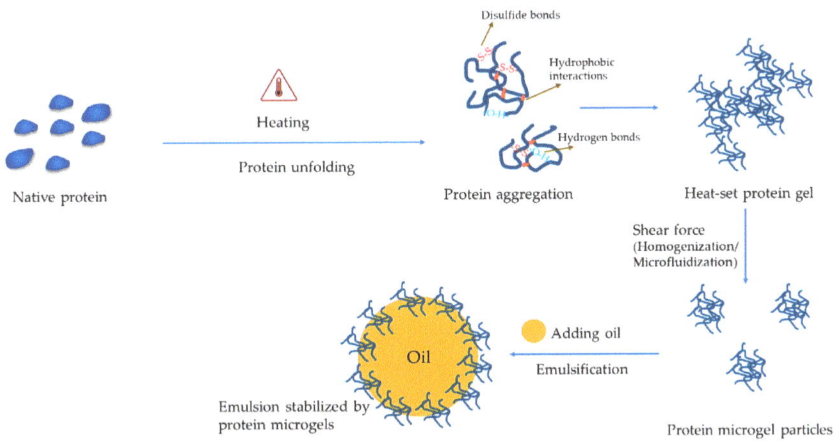

Figure 4. A schematic showing the formation of protein-based microgels.

Besides protein-only or protein-polysaccharides microgel particles, there is increasing interest to fabricate emulsion-based ones (Figure 4). Proteins could be gelled [53] or denatured via heating (e.g., 90 °C for 20 min) [54] followed by emulsification with plant oil using homogenization. Heating enables the exposure of the buried hydrophobic regions of proteins for increased non-polarity favoring the stabilization of oil droplets. When the concentration of denatured proteins is too low to form gels, additional gelling steps are required, such as acidification [54] and complexation with polysaccharides [108]. In some other cases, proteins are mixed with oil in their native states to form emulsions with or without the assistance of an emulsifier, then heated or acidified to trigger the gelation of proteins [109,110]. Thermal gelation has been shown to deliver aggregates or heterogeneous microgels, whereas gelation by acidification resulted in spherical and more homogenous particles with a better fat-mimic capacity [110]. To facilitate industrial applications, the microgel particles can be harvested by centrifugation or membrane filtration followed by spray drying to form powders.

4. The Fat-Mimicking Mechanisms

An ideal fat replacer should possess lubrication, flow properties, and heat melting capacity [111] analog to fat. Unfortunately, protein-based fat replacers can hardly mimic all of them. Providing lubrication for the reduced friction of food products is the main function of a fat replacer. One of the theories to explain the changes in textures is the "ball-bearing" effect. It refers to the protein microparticles or microgels employing a rolling mechanism similar to "ball bearings" (Figure 5). When force is applied, the balls rotate and/or slide to decrease frictions of the surrounding matrix [112]. The surrounding matrix could be the neighboring protein networks or the boundary regime between proteins and non-protein components [113–115]. The shape and size of the protein microparticles allow them to entrain in the narrow space between the tongue and palate in the mouth and provide creaminess and the perception of smoothness. The spherical shape and size of microparticulated fat replacers ranging from 0.2 to 9 μm are the key factors in providing the ball-bearing effect. This mechanism has been claimed to be dominant for microparticulated whey proteins in liquid and semi-liquid model foods [113]. Besides size and shape, the interior compactness of the protein particles is also highly desirable for the ball-bearing effect to occur. Sarkar et al. (2017) found negligible changes in the microstructure and size of whey protein microgel particles after the tribology test, implying their resistance towards friction force [115].

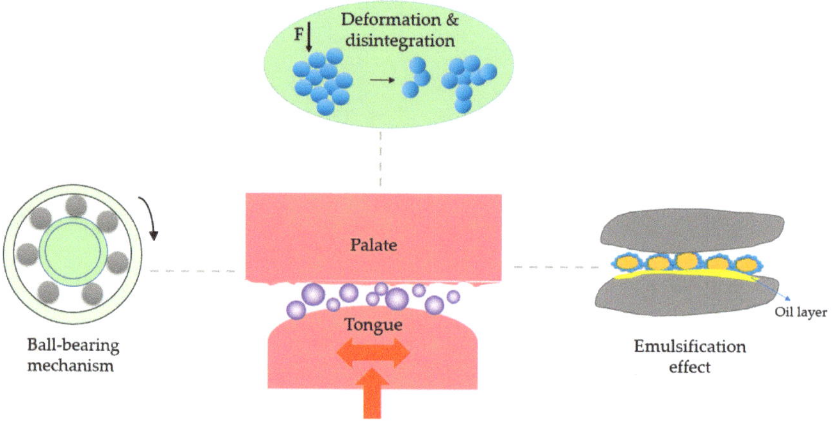

Figure 5. A schematic showing the mechanisms of fat-mimic effects.

For the particles with irregular shapes and larger sizes, such as aggregates from plant protein or protein hydrolysates, their deformation or even disintegration would contribute to a soft and/or smooth texture (Figure 5). Mechanical deformation of food through chewing and a biting motion in the oral cavity helps facilitate organoleptic perception, such as flavor release or the textural attributes of food. These movements, alongside the forces that the tongue makes by bouncing and excreting against the palate, could deform or even break the protein aggregates through disrupting non-covalent interactions [116]. Klost et al. (2020) have changed the proportion of soluble and insoluble pea protein aggregates in the fermented gels to mimic the texture of yogurt. They found the gels with higher insoluble aggregates showed higher flexibility and were easier to deform, showing smaller elastic moduli. After reducing the size of the aggregates by applying a larger intercycle strain, the system was converted from predominantly elastic to plastic behavior [58]. In a recent study, Chen and Campanella (2022) observed a substantial reduction of viscosity of pea protein hydrolysate gels under shear. Such shear-thinning behavior was partially due to the breaking down of the formed aggregates in addition to the realignment of the aggregates toward the shear flow direction [48].

Another assumption for the fat-mimic mechanism is the emulsification effects where the fat replacers may form an emulsifier layer or oil layer on the respective surfaces for lubrication, depending on the stability of emulsion gel microparticles (Figure 5). For those with high stability [117], the oil droplets are entrapped by interacting with the inner surface of protein particles during mastication. The outer surface of protein particles contacts with other macromolecules (e.g., starch, proteins) in the food matrix. Since the oil droplet inside the emulsion gel microparticles has a low glass transition and high flexibility, the outer layer that made up of protein particles can act as a filler. When experiencing force such as chewing, the particles slide to each other and deform to reduce the friction and increase the creaminess. For emulsion-type particles with limited stability, during mastication, they can be broken down into small fragments at a small deformation with the release of entrapped oil. The released oil lubricates the foods and provides a smooth mouth perception [118]. For a certain type of fat replacers, a set of fat-mimic mechanisms may occur which requires a comprehensive assessment [32,119].

5. Conclusions and Future Trends

In summary, designing protein-based fat replacers is essential to improve the quality attributes of low-fat and fat-free food products; although, they are more costly compared to those of carbohydrate-based ones. Microparticulated proteins, as the dominant fat replacers, are produced mainly by thermal–mechanical treatment. Other fabrication methods, such as anti-solvent precipitation and enzymatic hydrolysis, have also been used with less heat requirement. Microgel particles as a novel type of fat replacement receives increasing attention. With the incorporation of plant oil, it could provide a better lubrication effect and creaminess. The mechanisms of protein-based fat replacers remain unambiguous. The ball-bearing effect, protein aggregates/particle deformation and disassembly, and their emulsification contribute significantly to the lubrication and flow properties of fat-replacer incorporated food products for desirable texture and mouthful feeling.

Although much progress has been made in the production, modification, and/or application of fat replacers, the development of novel protein-based fat replacers in a greener way is still in its infancy. The thermal–mechanical treatment requires a large amount of energy input to trigger protein unfolding and aggregation, which, in turn, levels up the cost of the final products and is not environmentally friendly. The modification of the protein structure using processing or bioprocessing to promote their aggregating capacity under heat requires further investigation. Most of the protein-based fat replacers do not have the heat melting property except the prolamin-based ones. The incorporation of zein particles to those of others with high compatibility and ideal fat-mimic effects is worthwhile to explore. There is a growing trend of using plant proteins as fat replacers and each of them has its own structural, physiochemical, and mechanical properties. Yet, the maximum amount of fat that can be replaced with plant-based protein is ambiguous. The evaluation of different protein sources, as well as combining plant-based with other animal-based proteins, such as whey, or other non-protein ingredients, such as carbohydrates [38], to test their fat-mimicking potential, needs further attention. In addition, tribology combined with rheological textural analysis is dominant to assess the lubrication of fat replacer-incorporated low-fat products [119]. The difference between instrumental analysis and human mouth sensing implies a combination of the two is more accurate to reflect the true fat-mimic capacity of protein-based fat replacers and needs focused attention.

Author Contributions: N.N. and D.C. designed and wrote the manuscript. L.A. edited the manuscript. All authors have read and agreed to the published version of the manuscript.

Funding: This research received no external funding.

Data availability Statement: Not applicable.

Acknowledgments: The authors would like to acknowledge the support from the Department of Animal, Veterinary and Food Sciences in the University of Idaho.

Conflicts of Interest: The authors declare no conflict of interest.

References

1. Bryan, S.; Afful, J.; Carroll, M.; Te-Ching, C.; Orlando, D.; Fink, S.; Fryar, C.N.H.S.R. *NHSR 158. National Health and Nutrition Examination Survey 2017–March 2020 Pre-Pandemic Data Files*; National Center for Health Statistics (US): Hyattsville, MD, USA, 2021. [CrossRef]
2. U.S. Department of Agriculture and U.S. Department of Health and Human Services. *Dietary Guidelines for Americans, 2020–2025*, 9th ed.; Washington, DC, USA, 2020.
3. Temkov, M.; Mureșan, V. Tailoring the Structure of Lipids, Oleogels and Fat Replacers by Different Approaches for Solving the Trans-Fat Issue—A Review. *Foods* **2021**, *10*, 1376. [CrossRef] [PubMed]
4. Jacqueline, M.B. Chapter 6—Lipids Basics: Fats and Oils in Foods and Health: Healthy Lipid Choices, Roles and Applications in Nutrition. In *Culinary Nutrition*; Academic Press: Cambridge, MA, USA, 2013; pp. 231–277.
5. Jandacek, R.J.; Letton, J.C. Procter and Gamble Co. Compositions Containing Novel Solid, Nondigestible, Fat-Like Compounds. U.S. Patent No. 4,797,300, 10 January 1989.
6. Hunt, R.; Zorich, N.L.; Thomson, A.B. Overview of Olestra: A New Fat Substitute. *Can. J. Gastroenterol. Hepatol.* **1998**, *12*, 193–197. [CrossRef] [PubMed]
7. Phillips, S.M.; Chevalier, S.; Leidy, H.J. Protein "requirements" beyond the RDA: Implications for optimizing health. *Appl. Physiol. Nutr. Metab.* **2016**, *41*, 565–572. [CrossRef] [PubMed]
8. Marcus, J.B. *Culinary Nutrition: The Science and Practice of Healthy Cooking*; Academic Press: Cambridge, MA, USA, 2013; ISBN 978-0-12-391883-3.
9. Bohrer, B.M. Review: Nutrient density and nutritional value of meat products and non-meat foods high in protein. *Trends Food Sci. Technol.* **2017**, *65*, 103–112. [CrossRef]
10. Rodrigues, I.M.; Coelho, J.F.J.; Carvalho, M.G.V.S. Isolation and valorisation of vegetable proteins from oilseed plants: Methods, limitations and potential. *J. Food Eng.* **2012**, *109*, 337–346. [CrossRef]
11. Samtiya, M.; Aluko, R.E.; Dhewa, T. Plant food anti-nutritional factors and their reduction strategies: An overview. *Food Prod. Process. Nutr.* **2020**, *2*, 6. [CrossRef]
12. Vélez-Erazo, E.M.; Okuro, P.K.; Gallegos-Soto, A.; da Cunha, R.L.; Hubinger, M.D. Protein-based strategies for fat replacement: Approaching different protein colloidal types, structured systems and food applications. *Food Res. Int.* **2022**, *156*, 111346. [CrossRef]
13. Schirle-Keller, J.P.; Reineccius, G.A.; Hatchwell, L.C. Flavor Interactions with Fat Replacers: Effect of Oil Level. *J. Food Sci.* **1994**, *59*, 813–815. [CrossRef]
14. Yashini, M.; Sunil, C.K.; Sahana, S.; Hemanth, S.D.; Chidanand, D.V.; Ashish, R. Protein-based Fat Replacers—A Review of Recent Advances. *Food Rev. Int.* **2021**, *37*, 197–223. [CrossRef]
15. Pokorny, J. Substrate influence on the frying process. *Grasas Aceites* **1998**, *49*, 265–270. [CrossRef]
16. Aggarwal, D.; Sabikhi, L.; Sathish Kumar, M.H. Formulation of reduced-calorie biscuits using artificial sweeteners and fat replacer with dairy–multigrain approach. *NFS J.* **2016**, *2*, 1–7. [CrossRef]
17. O'Connor, T.P.; O'Brien, N.M. Fat Replacers. In *Reference Module in Food Science*; Elsevier: Amsterdam, The Netherlands, 2016; ISBN 978-0-08-100596-5.
18. Lucca, P.A.; Tepper, B.J. Fat replacers and the functionality of fat in foods. *Trends Food Sci. Technol.* **1994**, *5*, 12–19. [CrossRef]
19. De Souza Paglarini, C.; Vidal, V.A.S.; Martini, S.; Cunha, R.L.; Pollonio, M.A.R. Protein-based hydrogelled emulsions and their application as fat replacers in meat products: A review. *Crit. Rev. Food Sci. Nutr.* **2022**, *62*, 640–655. [CrossRef] [PubMed]
20. Cui, L.; Kimmel, J.; Zhou, L.; Chen, B.; Rao, J. Improving the functionality of pea protein isolate through co-spray drying with emulsifying salt or disaccharide. *Food Hydrocoll.* **2021**, *113*, 106534. [CrossRef]

21. Nehete, J.; Narkhede, M.; Bhambar, R.; Gawali, S. Natural proteins: Sources, isolation, characterization and applications. *Pharmacogn. Rev.* **2013**, *7*, 107. [CrossRef]
22. Goulding, D.A.; Fox, P.F.; O'Mahony, J.A. Milk proteins: An overview. In *Milk Proteins*, 3rd ed.; Boland, M., Singh, H., Eds.; Academic Press: Cambridge, MA, USA, 2020; pp. 21–98. ISBN 978-0-12-815251-5.
23. Fang, T.; Shen, X.; Hou, J.; Guo, M. Effects of polymerized whey protein prepared directly from cheese whey as fat replacer on physiochemical, texture, microstructure and sensory properties of low-fat set yogurt. *LWT* **2019**, *115*, 108268. [CrossRef]
24. Liu, R.; Wang, L.; Liu, Y.; Wu, T.; Zhang, M. Fabricating soy protein hydrolysate/xanthan gum as fat replacer in ice cream by combined enzymatic and heat-shearing treatment. *Food Hydrocoll.* **2018**, *81*, 39–47. [CrossRef]
25. Urgu, M.; Türk, A.; Ünlütürk, S.; Kaymak-Ertekin, F.; Koca, N. Milk Fat Substitution by Microparticulated Protein in Reduced-fat Cheese Emulsion: The Effects on Stability, Microstructure, Rheological and Sensory Properties. *Food Sci. Anim. Resour.* **2019**, *39*, 23–34. [CrossRef]
26. Borges, A.R.; Pires, A.F.; Marnotes, N.G.; Gomes, D.G.; Henriques, M.F.; Pereira, C.D. Dairy by-products concentrated by ultrafiltration used as ingredients in the production of reduced fat washed curd cheese. *Foods* **2020**, *9*, 1020. [CrossRef]
27. Protte, K.; Weiss, J.; Hinrichs, J.; Knaapila, A. Thermally stabilised whey protein-pectin complexes modulate the thermodynamic incompatibility in hydrocolloid matrixes: A feasibility-study on sensory and rheological characteristics in dairy desserts. *LWT* **2019**, *105*, 336–343. [CrossRef]
28. Lesme, H.; Rannou, C.; Famelart, M.-H.; Bouhallab, S.; Prost, C. Yogurts enriched with milk proteins: Texture properties, aroma release and sensory perception. *Trends Food Sci. Technol.* **2020**, *98*, 140–149. [CrossRef]
29. Costa, M.P.; Frasao, B.S.; Rodrigues, B.L.; Silva, A.C.; Conte-Junior, C.A. Effect of different fat replacers on the physicochemical and instrumental analysis of low-fat cupuassu goat milk yogurts. *J. Dairy Res.* **2016**, *83*, 493–496. [CrossRef] [PubMed]
30. Torres, I.C.; Amigo, J.M.; Knudsen, J.C.; Tolkach, A.; Mikkelsen, B.Ø.; Ipsen, R. Rheology and microstructure of low-fat yoghurt produced with whey protein microparticles as fat replacer. *Int. Dairy J.* **2018**, *81*, 62–71. [CrossRef]
31. Sołowiej, B.G.; Nastaj, M.; Szafrańska, J.O.; Terpiłowski, K.; Małecki, J.; Mleko, S. The effect of fat replacement by whey protein microcoagulates on the physicochemical properties and microstructure of acid casein model processed cheese. *Int. Dairy J.* **2022**, *131*, 105385. [CrossRef]
32. Kew, B.; Holmes, M.; Stieger, M.; Sarkar, A. Review on fat replacement using protein-based microparticulated powders or microgels: A textural perspective. *Trends Food Sci. Technol.* **2020**, *106*, 457–468. [CrossRef]
33. Ziegler, G.R. Process for Producing Microparticulated Protein and the Product Thereof. U.S. Patent US5147677A, 15 September 1992.
34. Singer, N.S.; Yamamoto, S.; Latella, J. Protein Product Base. U.S. Patent US4734287A, 29 March 1988.
35. Singer, N.S.; Latella, J.; Shoji, Y. Fat Emulating Protein Products and Process. U.S. Patent US4961953A, 9 October 1990.
36. Zhang, T.; Guo, J.; Chen, J.-F.; Wang, J.-M.; Wan, Z.-L.; Yang, X.-Q. Heat stability and rheological properties of concentrated soy protein/egg white protein composite microparticle dispersions. *Food Hydrocoll.* **2020**, *100*, 105449. [CrossRef]
37. Niu, Y.; Fang, H.; Huo, T.; Sun, X.; Gong, Q.; Yu, L. A novel fat replacer composed by gelatin and soluble dietary fibers from black bean coats with its application in meatballs. *LWT* **2020**, *122*, 109000. [CrossRef]
38. Song, J.; Sun, C.; Zhang, J.; Xiong, Z.; Fang, Y. Fabrication, Characterization, and Formation Mechanism of Zein-Gum Arabic Nanocomposites in Aqueous Ethanol Solution with a High Ethanol Content. *J. Agric. Food Chem.* **2020**, *68*, 13138–13145. [CrossRef]
39. Zhang, Z.-Y.; Zhou, H.-M.; Bai, Y.-P. Use of glycosylated wheat protein in emulsions and its application as a fat replacer in microwave cakes. *J. Cereal Sci.* **2021**, *100*, 103256. [CrossRef]
40. Tanger, C.; Utz, F.; Spaccasassi, A.; Kreissl, J.; Dombrowski, J.; Dawid, C.; Kulozik, U. Influence of Pea and Potato Protein Microparticles on Texture and Sensory Properties in a Fat-Reduced Model Milk Dessert. *ACS Food Sci. Technol.* **2022**, *2*, 169–179. [CrossRef]
41. Geerts, M.E.J.; Dekkers, B.L.; van der Padt, A.; van der Goot, A.J. Aqueous fractionation processes of soy protein for fibrous structure formation. *Innov. Food Sci. Emerg. Technol.* **2018**, *45*, 313–319. [CrossRef]
42. Peng, Y.; Kersten, N.; Kyriakopoulou, K.; van der Goot, A.J. Functional properties of mildly fractionated soy protein as influenced by the processing pH. *J. Food Eng.* **2020**, *275*, 109875. [CrossRef]
43. Tanger, C.; Quintana Ramos, P.; Kulozik, U. Comparative Assessment of Thermal Aggregation of Whey, Potato, and Pea Protein under Shear Stress for Microparticulation. *ACS Food Sci. Technol.* **2021**, *1*, 975–985. [CrossRef]
44. Hossain, M.K.; Keidel, J.; Hensel, O.; Diakité, M. The impact of extruded microparticulated whey proteins in reduced-fat, plain-type stirred yogurt: Characterization of physicochemical and sensory properties. *LWT* **2020**, *134*, 109976. [CrossRef]
45. Tanger, C.; Schmidt, F.; Utz, F.; Kreissl, J.; Dawid, C.; Kulozik, U. Pea protein microparticulation using extrusion cooking: Influence of extrusion parameters and drying on microparticle characteristics and sensory by application in a model milk dessert. *Innov. Food Sci. Emerg. Technol.* **2021**, *74*, 102851. [CrossRef]
46. Liu, R.; Tian, Z.; Song, Y.; Wu, T.; Sui, W.; Zhang, M. Optimization of the Production of Microparticulated Egg White Proteins as Fat Mimetic in Salad Dressings Using Uniform Design. *Food Sci. Technol. Res.* **2018**, *24*, 817–827. [CrossRef]
47. Gamlath, C.J.; Leong, T.S.H.; Ashokkumar, M.; Martin, G.J.O. Incorporating whey protein aggregates produced with heat and ultrasound treatment into rennet gels and model non-fat cheese systems. *Food Hydrocoll.* **2020**, *109*, 106103. [CrossRef]
48. Chen, D.; Campanella, O.H. Limited enzymatic hydrolysis induced pea protein gelation at low protein concentration with less heat requirement. *Food Hydrocoll.* **2022**, *128*, 107547. [CrossRef]

49. Meng, R.; Wu, Z.; Xie, Q.-T.; Zhang, B.; Li, X.-L.; Liu, W.-J.; Tao, H.; Li, P.-J. Zein/carboxymethyl dextrin nanoparticles stabilized pickering emulsions as delivery vehicles: Effect of interfacial composition on lipid oxidation and in vitro digestion. *Food Hydrocoll.* **2020**, *108*, 106020. [CrossRef]
50. Li, X.-L.; Meng, R.; Xu, B.-C.; Zhang, B.; Cui, B.; Wu, Z.-Z. Function emulsion gels prepared with carrageenan and zein/carboxymethyl dextrin stabilized emulsion as a new fat replacer in sausages. *Food Chem.* **2022**, *389*, 133005. [CrossRef]
51. Fan, R.; Zhou, D.; Cao, X. Evaluation of oat β-glucan-marine collagen peptide mixed gel and its application as the fat replacer in the sausage products. *PLoS ONE* **2020**, *15*, e0233447. [CrossRef] [PubMed]
52. Zhang, D.; Chen, D.; Patel, B.; Campanella, O.H. Pectin as a natural agent for reinforcement of pea protein gel. *Carbohydr. Polym.* **2022**, *298*, 120038. [CrossRef] [PubMed]
53. Rezaee, M.; Ait Aider-Kaci, F.; Aider, M. Effects of Hydrocolloid Agar, Gelatin, Pectin, and Xanthan on Physicochemical and Rheological Properties of Pickering Emulsions Stabilized by Canola Protein Microgel as a Potential Animal Fat Replacer. *ACS Food Sci. Technol.* **2022**, *2*, 1681–1690. [CrossRef]
54. Li, H.; Zhang, L.; Jia, Y.; Yuan, Y.; Li, H.; Cui, W.; Yu, J. Application of whey protein emulsion gel microparticles as fat replacers in low-fat yogurt: Applicability of vegetable oil as the oil phase. *J. Dairy Sci.* **2022**, *105*, 9404–9416. [CrossRef] [PubMed]
55. Leon, A.M.; Aguilera, J.M.; Park, D.J. Mechanical, rheological and structural properties of fiber-containing microgels based on whey protein and alginate. *Carbohydr. Polym.* **2019**, *207*, 571–579. [CrossRef] [PubMed]
56. Chen, N.; Nicolai, T.; Chassenieux, C.; Wang, Y. pH and ionic strength responsive core-shell protein microgels fabricated via simple coacervation of soy globulins. *Food Hydrocoll.* **2020**, *105*, 105853. [CrossRef]
57. Sun, C.; Liu, R.; Liang, B.; Wu, T.; Sui, W.; Zhang, M. Microparticulated whey protein-pectin complex: A texture-controllable gel for low-fat mayonnaise. *Food Res. Int.* **2018**, *108*, 151–160. [CrossRef]
58. Klost, M.; Brzeski, C.; Drusch, S. Effect of protein aggregation on rheological properties of pea protein gels. *Food Hydrocoll.* **2020**, *108*, 106036. [CrossRef]
59. Li, A.; Gong, T.; Yang, X.; Guo, Y. Interpenetrating network gels with tunable physical properties: Glucono-δ-lactone induced gelation of mixed Alg/gellan sol systems. *Int. J. Biol. Macromol.* **2020**, *151*, 257–267. [CrossRef]
60. Le, X.T.; Rioux, L.-E.; Turgeon, S.L. Formation and functional properties of protein–polysaccharide electrostatic hydrogels in comparison to protein or polysaccharide hydrogels. *Adv. Colloid Interface Sci.* **2017**, *239*, 127–135. [CrossRef]
61. Yang, D.; Yao, X.; Wang, L.; Xu, K.; Li, D.; Liu, N.; Midgley, A.; Liu, D.; Katsuyoshi, N. Physicochemical stability of Pickering emulsion stabilized with spherical and fibrous iron ions loaded whey protein isolate/gum Arabic complexes. *Food Hydrocoll.* **2023**, *138*, 108471. [CrossRef]
62. Du, Q.; Zhou, L.; Lyu, F.; Liu, J.; Ding, Y. The complex of whey protein and pectin: Interactions, functional properties and applications in food colloidal systems—A review. *Colloids Surf. B Biointerfaces* **2022**, *210*, 112253. [CrossRef] [PubMed]
63. Wu, S.; Wang, L.; Zhao, Y.; Chen, B.; Qiu, D.; Sun, P.; Shao, P.; Feng, S. Fabrication of high strength cold-set sodium alginate/whey protein nanofiber double network hydrogels and their interaction with curcumin. *Food Res. Int.* **2023**, *165*, 112490. [CrossRef]
64. Pilevaran, M.; Tavakolipour, H.; Naji-Tabasi, S.; Elhamirad, A.H. Investigation of functional, textural, and thermal properties of soluble complex of whey protein–xanthan gum hydrogel. *J. Food Process Eng.* **2021**, *44*, e13751. [CrossRef]
65. Wang, Y.; Ghosh, S.; Nickerson, M.T. Effect of pH on the formation of electrostatic complexes between lentil protein isolate and a range of anionic polysaccharides, and their resulting emulsifying properties. *Food Chem.* **2019**, *298*, 125023. [CrossRef] [PubMed]
66. Dong, D.; Cui, B. Fabrication, characterization and emulsifying properties of potato starch/soy protein complexes in acidic conditions. *Food Hydrocoll.* **2021**, *115*, 106600. [CrossRef]
67. Torres, O.; Murray, B.; Sarkar, A. Design of novel emulsion microgel particles of tuneable size. *Food Hydrocoll.* **2017**, *71*, 47–59. [CrossRef]
68. Chu, Y.; Jo, Y.; Chen, L. Size-controllable core/shell whey protein microgels with narrow particle size distribution fabricated by a facile method. *Food Hydrocoll.* **2022**, *124*, 107316. [CrossRef]
69. Chen, D.; Kuzmenko, I.; Ilavsky, J.; Pinho, L.; Campanella, O. Structural evolution during gelation of pea and whey proteins envisaged by time-resolved ultra-small-angle X-ray scattering (USAXS). *Food Hydrocoll.* **2022**, *126*, 107449. [CrossRef]
70. Bahri, A.; Chevalier-Lucia, D.; Marchesseau, S.; Schmitt, C.; Gergely, C.; Martin, M. Effect of pH change on size and nanomechanical behavior of whey protein microgels. *J. Colloid Interface Sci.* **2019**, *555*, 558–568. [CrossRef]
71. Bao, Y.; Liu, K.; Zheng, Q.; Yao, L.; Xu, Y. A Review of Preparation and Tribological Applications of Pickering Emulsion. *J. Tribol.* **2021**, *144*, 011902. [CrossRef]
72. Schutyser, M.A.I.; van der Goot, A.J. The potential of dry fractionation processes for sustainable plant protein production. *Trends Food Sci. Technol.* **2011**, *22*, 154–164. [CrossRef]
73. Meena, G.S.; Singh, A.K.; Panjagari, N.R.; Arora, S. Milk protein concentrates: Opportunities and challenges. *J. Food Sci. Technol.* **2017**, *54*, 3010–3024. [CrossRef] [PubMed]
74. Kelly, P. Manufacture of Whey Protein Products: Concentrates, Isolate, Whey Protein Fractions and Microparticulated. In *Whey Proteins*; Deeth, H.C., Bansal, N., Eds.; Academic Press: Cambridge, MA, USA, 2019; pp. 97–122. ISBN 978-0-12-812124-5.
75. Minj, S.; Anand, S. Whey Proteins and Its Derivatives: Bioactivity, Functionality, and Current Applications. *Dairy* **2020**, *1*, 233. [CrossRef]
76. Anandharamakrishnan, C.; Rielly, C.D.; Stapley, A.G.F. Effects of Process Variables on the Denaturation of Whey Proteins during Spray Drying. *Dry. Technol.* **2007**, *25*, 799–807. [CrossRef]

77. Fang, Y.; Rogers, S.; Selomulya, C.; Chen, X.D. Functionality of milk protein concentrate: Effect of spray drying temperature. *Biochem. Eng. J.* **2012**, *62*, 101–105. [CrossRef]
78. Kumar, M.; Tomar, M.; Potkule, J.; Verma, R.; Punia, S.; Mahapatra, A.; Belwal, T.; Dahuja, A.; Joshi, S.; Berwal, M.K.; et al. Advances in the plant protein extraction: Mechanism and recommendations. *Food Hydrocoll.* **2021**, *115*, 106595. [CrossRef]
79. Sari, Y.W.; Mulder, W.J.; Sanders, J.P.M.; Bruins, M.E. Towards plant protein refinery: Review on protein extraction using alkali and potential enzymatic assistance. *Biotechnol. J.* **2015**, *10*, 1138–1157. [CrossRef]
80. Zhang, T.; McCarthy, J.; Wang, G.; Liu, Y.; Guo, M. Physiochemical properties, microstructure, and probiotic survivability of nonfat goats' milk yogurt using heat-treated whey protein concentrate as fat replacer. *J. Food Sci.* **2015**, *80*, M788–M794. [CrossRef]
81. Ni, Y.; Wen, L.; Wang, L.; Dang, Y.; Zhou, P.; Liang, L. Effect of temperature, calcium and protein concentration on aggregation of whey protein isolate: Formation of gel-like micro-particles. *Int. Dairy J.* **2015**, *51*, 8–15. [CrossRef]
82. Homer, S.; Lundin, L.; Dunstan, D.E. Heat-induced whey protein microparticulation under continuous shear in acidic conditions. *Food Hydrocoll.* **2021**, *121*, 107044. [CrossRef]
83. Wolz, M.; Kastenhuber, S.; Kulozik, U. High moisture extrusion for microparticulation of whey proteins –Influence of process parameters. *J. Food Eng.* **2016**, *185*, 56–61. [CrossRef]
84. Wolz, M.; Kulozik, U. System parameters in a high moisture extrusion process for microparticulation of whey proteins. *J. Food Eng.* **2017**, *209*, 12–17. [CrossRef]
85. Hossain, M.K.; Krah, A.; Hensel, O.; Diakite, M. Microparticulation of Whey Protein Concentrates using the Hot Extrusion Process: The Influence of Protein Concentrations and other Parameters. *J. Food Process Technol.* **2019**, *10*, 1–6. [CrossRef]
86. Rizvi, S.S.H.; Mulvaney, S. Extrusion Processing with Supercritical Fluids. U.S. Patent US5120559A, 9 June 1992.
87. Arora, B.; Singha, P.; Rizvi, S.S.H. Supercritical fluid extrusion: Die design and physicochemical properties of milk protein extrudates. *Innov. Food Sci. Emerg. Technol.* **2021**, *68*, 102637. [CrossRef]
88. Chen, C.-C.; Chen, L.-Y.; Chan, D.-S.; Chen, B.-Y.; Tseng, H.-W.; Hsieh, J.-F. Influence of Microbial Transglutaminase on Physicochemical and Cross-Linking Characteristics of Individual Caseins. *Molecules* **2020**, *25*, 3992. [CrossRef]
89. Otte, J.; Ju, Z.Y.; Skriver, A.; Qvist, K.B. Effects of Limited Proteolysis on the Microstructure of Heat-Induced Whey Protein Gels at Varying pH. *J. Dairy Sci.* **1996**, *79*, 782–790. [CrossRef]
90. Tarhan, O.; Spotti, M.J.; Schaffter, S.; Corvalan, C.M.; Campanella, O.H. Rheological and structural characterization of whey protein gelation induced by enzymatic hydrolysis. *Food Hydrocoll.* **2016**, *61*, 211–220. [CrossRef]
91. Zang, X.; Yue, C.; Wang, Y.; Shao, M.; Yu, G. Effect of limited enzymatic hydrolysis on the structure and emulsifying properties of rice bran protein. *J. Cereal Sci.* **2019**, *85*, 168–174. [CrossRef]
92. Joye, I.J.; McClements, D.J. Production of nanoparticles by anti-solvent precipitation for use in food systems. *Trends Food Sci. Technol.* **2013**, *34*, 109–123. [CrossRef]
93. Gupta, J.; Vadlani, P.V.; Lau, C.-S.; Madl, R.L.; Shi, Y.C. Innovative zein extraction from distillers' grains with solubles: Process development and product characterization studies. *Environ. Prog. Sustain. Energy* **2019**, *38*, 13093. [CrossRef]
94. Anderson, T.J.; Ilankovan, P.; Lamsal, B.P. Two fraction extraction of α-zein from DDGS and its characterization. *Ind. Crops Prod.* **2012**, *37*, 466–472. [CrossRef]
95. Gu, J.; Xin, Z.; Meng, X.; Sun, S.; Qiao, Q.; Deng, H. A "reduced-pressure distillation" method to prepare zein-based fat analogue for application in mayonnaise formulation. *J. Food Eng.* **2016**, *182*, 1–8. [CrossRef]
96. Liu, X.; Guo, J.; Wan, Z.-L.; Liu, Y.-Y.; Ruan, Q.-J.; Yang, X.-Q. Wheat gluten-stabilized high internal phase emulsions as mayonnaise replacers. *Food Hydrocoll.* **2018**, *77*, 168–175. [CrossRef]
97. Mattice, K.D.; Marangoni, A.G. Evaluating the use of zein in structuring plant-based products. *Curr. Res. Food Sci.* **2020**, *3*, 59–66. [CrossRef] [PubMed]
98. De Marco, I. Zein Microparticles and Nanoparticles as Drug Delivery Systems. *Polymers* **2022**, *14*, 2172. [CrossRef]
99. Li, M.; Yu, M. Development of a nanoparticle delivery system based on zein/polysaccharide complexes. *J. Food Sci.* **2020**, *85*, 4108–4117. [CrossRef]
100. Cui, S.; McClements, D.J.; Shi, J.; Xu, X.; Ning, F.; Liu, C.; Zhou, L.; Sun, Q.; Dai, L. Fabrication and characterization of low-fat Pickering emulsion gels stabilized by zein/phytic acid complex nanoparticles. *Food Chem.* **2023**, *402*, 134179. [CrossRef]
101. Pascoli, M.; de Lima, R.; Fraceto, L.F. Zein Nanoparticles and Strategies to Improve Colloidal Stability: A Mini-Review. *Front. Chem.* **2018**, *6*, 6. [CrossRef]
102. Bisharat, L.; Berardi, A.; Perinelli, D.R.; Bonacucina, G.; Casettari, L.; Cespi, M.; AlKhatib, H.S.; Palmieri, G.F. Aggregation of zein in aqueous ethanol dispersions: Effect on cast film properties. *Int. J. Biol. Macromol.* **2018**, *106*, 360–368. [CrossRef]
103. Makshakova, O.N.; Zuev, Y.F. Interaction-induced structural transformations in polysaccharide and protein-polysaccharide gels as functional basis for novel soft-matter: A case of carrageenans. *Gels* **2022**, *8*, 287. [CrossRef] [PubMed]
104. Sun, L.; Chen, W.; Liu, Y.; Li, J.; Yu, H. Soy protein isolate/cellulose nanofiber complex gels as fat substitutes: Rheological and textural properties and extent of cream imitation. *Cellulose* **2015**, *22*, 2619–2627. [CrossRef]
105. Lee, J.; Duggan, E. Whey protein microgels for stabilisation of foams. *Int. Dairy J.* **2022**, *132*, 105399. [CrossRef]
106. Kharlamova, A.; Chassenieux, C.; Nicolai, T. Acid-induced gelation of whey protein aggregates: Kinetics, gel structure and rheological properties. *Food Hydrocoll.* **2018**, *81*, 263–272. [CrossRef]
107. Turgeon, S.L.; Schmitt, C.; Sanchez, C. Protein–polysaccharide complexes and coacervates. *Curr. Opin. Colloid Interface Sci.* **2007**, *12*, 166–178. [CrossRef]

108. Li, B.; Gu, W.; Bourouis, I.; Sun, M.; Huang, Y.; Chen, C.; Liu, X.; Pang, Z. Lubrication behaviors of core-shell structured particles formed by whey proteins and xanthan gum. *Food Hydrocoll.* **2022**, *127*, 107512. [CrossRef]
109. Wen, P.; Zhu, Y.; Luo, J.; Wang, P.; Liu, B.; Du, Y.; Jiao, Y.; Hu, Y.; Chen, C.; Ren, F.; et al. Effect of anthocyanin-absorbed whey protein microgels on physicochemical and textural properties of reduced-fat Cheddar cheese. *J. Dairy Sci.* **2021**, *104*, 228–242. [CrossRef]
110. Lacroix, A.; Hayert, M.; Bosc, V.; Menut, P. Batch versus microfluidic emulsification processes to produce whey protein microgel beads from thermal or acidic gelation. *J. Food Eng.* **2022**, *312*, 110738. [CrossRef]
111. Roller, S.; Jones, S.A. *Handbook of Fat Replacers*; CRC Press: Boca Raton, FL, USA, 1996; ISBN 978-1-4200-4897-1.
112. Wardle, F. *Ultra-precision Bearings*; Elsevier: Amsterdam, The Netherlands, 2015; ISBN 978-0-85709-218-2.
113. Liu, K.; Tian, Y.; Stieger, M.; van der Linden, E.; van de Velde, F. Evidence for ball-bearing mechanism of microparticulated whey protein as fat replacer in liquid and semi-solid multi-component model foods. *Food Hydrocoll.* **2016**, *52*, 403–414. [CrossRef]
114. Chojnicka, A.; de Jong, S.; de Kruif, C.G.; Visschers, R.W. Lubrication Properties of Protein Aggregate Dispersions in a Soft Contact. *J. Agric. Food Chem.* **2008**, *56*, 1274–1282. [CrossRef]
115. Sarkar, A.; Kanti, F.; Gulotta, A.; Murray, B.S.; Zhang, S. Aqueous Lubrication, Structure and Rheological Properties of Whey Protein Microgel Particles. *Langmuir* **2017**, *33*, 14699–14708. [CrossRef] [PubMed]
116. Sethupathy, P.; Moses, J.A.; Anandharamakrishnan, C. Food Oral Processing and Tribology: Instrumental Approaches and Emerging Applications. *Food Rev. Int.* **2021**, *37*, 538–571. [CrossRef]
117. Baune, M.-C.; Schroeder, S.; Witte, F.; Heinz, V.; Bindrich, U.; Weiss, J.; Terjung, N. Analysis of protein-network formation of different vegetable proteins during emulsification to produce solid fat substitutes. *J. Food Meas. Charact.* **2021**, *15*, 2399–2416. [CrossRef]
118. Luo, N.; Ye, A.; Wolber, F.M.; Singh, H. In-mouth breakdown behaviour and sensory perception of emulsion gels containing active or inactive filler particles loaded with capsaicinoids. *Food Hydrocoll.* **2020**, *108*, 106076. [CrossRef]
119. Sarkar, A.; Krop, E.M. Marrying oral tribology to sensory perception: A systematic review. *Curr. Opin. Food Sci.* **2019**, *27*, 64–73. [CrossRef]

MDPI AG
Grosspeteranlage 5
4052 Basel
Switzerland
Tel.: +41 61 683 77 34

Foods Editorial Office
E-mail: foods@mdpi.com
www.mdpi.com/journal/foods

www.ingramcontent.com/pod-product-compliance
Lightning Source LLC
Chambersburg PA
CBHW041332210920
51850CB00013B/3063